Geometry for Teachers

ROBERT A. NOWLAN

ROBERT M. WASHBURN

Southern Connecticut State College

GEOMETRY FOR TEACHERS

Harper & Row, Publishers New York Evanston San Francisco London

Sponsoring Editor: George J. Telecki
Project Editor: Ralph Cato
Designer: Gayle Jaeger
Production Supervisor: Will C. Jomarrón

Geometry for Teachers

Library of Congress Cataloging in Publication Data

Nowlan, Robert A
Geometry for teachers.
1. Geometry. I. Washburn, Robert M., joint author.
II. Title.
QA445.N68 516'.2 74-7098
ISBN 0-06-044866-0

To our families

CONTENTS

PREFACE TO THE INSTRUCTOR

THE INTUITIVE APPROACH

Geometry for Teachers is an intentionally intuitive approach to the discipline, based on the conviction that students must develop geometric intuition before any true appreciation of a rigorous development can be attained. Our intended audience is prospective teachers of mathematics, both in elementary and middle schools. Our goal is to foster among these undergraduates a more mature understanding of geometric concepts and their interrelationships. We wish to both heighten their appreciation of mathematics in general and improve their understanding of geometric content.

Geometry for Teachers is based on material we used at Southern Connecticut State College over a four-year period. The approach embodied in this text was developed because we felt that schoolteachers' understanding of geometry was usually based on misconceptions arrived at when they were high school sophomores. The motivation for this text also reflects the fact that our discipline is seeing a reaffirmation of the need to stress intuition and analysis in the study of all mathematics. An examination of some of the newer elementary school mathematics series shows that geometry is being allotted an integral and important place in the curriculum.

Usually any book intended for prospective schoolteachers begins either with a study of sets or logic. Mercifully (for the student, at any rate), *we have decided to do neither.* We find that the majority of students entering college these days have already been exposed, at various points in their education, to concepts of set theory quite sufficient for our purposes. Perhaps the reader will find this omission at the beginning of the study a gesture of good faith.

As for logic, its inclusion at the beginning of a mathematical study may seem to be mathematically correct, *but we question its pedagogical correctness.* Just as we wish to build understanding and appreciation for proof throughout this book, we wish to polish the logical development *as we proceed.* This seems much more appropriate to natural development. Indeed, we find that our own students are not necessarily illogical, merely disorganized.

More specific special features of *Geometry for Teachers* are listed below.

1. Material is developed in a spiral fashion—key geometrical concepts are encountered in various places throughout the text. Each new encounter tends to extend and broaden the students' understanding of the concept by progressing from the concrete to the abstract.
2. Each chapter begins with a prologue, the purpose of which is not only to be informative and motivational, but also to remind students of precisely what point they have reached in the spiral, that is, what they have already learned about a particular concept and the direction they will head in their further development of the concept.
3. All chapters include one or more experiments. These are included to provide students with new and relevant concrete experiences with acquired concepts by identifying problems and questions that need to be solved. An atmosphere is created in which the readers are encouraged to discover solutions to these problems. Some of the experiments have predictable outcomes; others are open-ended, allowing different students to draw different conclusions. Some of the experiments serve as models for experiments that teachers can use in a mathematics laboratory. All have been selected with an eye toward setting a good example for our prospective teachers.
4. Congruence and similarity are treated through a transformation approach. It appears that transformation geometry is an idea whose time has finally arrived. The approach is not only more "modern," but it also seems more mathematically and pedagogically satisfactory.
5. Coordinate geometry is developed in three chapters, each separated by chapters that are largely synthetic in content and development. Our decision to depart from the usual practice of developing coordinate geometry in its entirety in a single section of the book is based on our firm conviction that a truly modern approach to geometry should be a blend of synthetic *and* analytic ideas. This approach also seems more consistent with the concept of spiral development.
6. The body of each chapter is mostly informative—developing answers to questions that have been raised in the prologue, establishing the need for good definitions, and providing procedures for problem solving. The chapter summary, of course, provides students with a quick review. The exercises,

found at the end of each section, not only pose questions that test students' understanding of the important ideas, but are designed to provide motivation for and transition into the next section.

7. Throughout the book, inductive reasoning is stressed. We find that students are more successful with deductive reasoning if *they* have played some role in the formulation of the relationships to be proven. The difficulty of the material is meant to increase slowly; the earlier chapters are fairly elementary, whereas later chapters include considerably more sophisticated ideas.
8. A key element of the spiral development is that *formal* deductive proof is not required until Chapter 15. Students are encouraged from the first page onward to provide explanations, justifications, and convincing arguments to support certain conclusions. As students progress through the text, their ability to *communicate* such arguments is steadily cultivated until it culminates with the treatment of formal proof in Chapter 15.
9. Throughout the manuscript, we have included certain asides entitled "In our opinion" and "Comment." The former include comments of a pedagogical and philosophical nature that we have found useful to share with our students. Frequently, these opinions have inspired interesting and, we believe, productive discussions. Their purpose is to help us reach our goal of affecting students' attitudes toward mathematics itself. The latter type of aside consists of short observations that should aid students in understanding and appreciating the topic under discussion.

Although this text was designed for use in a two-semester course, it is possible to leave out certain sections and teach a logically sound single semester or two-quarter course while retaining the features listed above. This can be done by omitting substantial portions that relate to coordinate and solid geometry. More specifically, a single semester course can be obtained by following this sequence of sections: 1.1–2.4, 2.6–4.3, 4.7–5.1, 5.3, 6.1–6.2, 6.4–8.5, 9.1–9.4, 10.2–10.3, 13.1–13.6, 15.1–15.8, 15.9.

We wish to express special gratitude to our typist, Mrs. Maryann Franco; to our colleagues Dorothy Schrader, Helen Bass, Leo Kuczynski, Bert Latil, Michael Meck, and Philip Smith for their encouragement and criticism of the manuscript; to our students at Southern Connecticut State College, particularly Cecilia Lowe, Mary Dereshiwsky, and Patricia Consoli; and to the editors and staff of Harper & Row for their many hours of assistance. We extend particular thanks to our families, whose faith in our ability to write this book and unfailing encouragement has allowed us to complete this project.

Robert A. Nowlan
Robert M. Washburn

PREFACE TO THE STUDENT

AN OPPORTUNITY TO THINK DEEPLY ABOUT SIMPLE, FAMILIAR THINGS

This book was written so that you might teach yourself geometry in the most rewarding and, we hope, enjoyable way. There are various levels of abstraction at which geometry can be appreciated and we expect that you have already experienced several in your previous mathematical studies. Our purpose is not to completely repeat any of these previous approaches to geometry. We do hope to assist you in developing an intelligent understanding of geometric concepts, a more sophisticated appreciation for the organization of geometry, and a feeling for the *priorities* of the geometric ideas that you may eventually teach.

For many students, the study of geometry is made difficult by teaching strategies that foster inadequate understanding. Part of the fault is traceable to the fact that, although the "how" and even the "why" of geometry are stressed, too little emphasis is placed on helping students develop a *proper perspective.* Often questions are answered before students have even thought to raise them in the first place! Logical developments are given without sufficient *intuitive* background. Proof is stressed when there is no understanding or appreciation of exactly what constitutes a "proof."

We do not pretend to have found definitive solutions to these thorny problems. *Geometry for Teachers* has, however, been written *with them in mind.* Our intention is to give you an opportunity to think deeply about simple, familiar things, looking at geometry from a *natural* rather than a completely rigorous point of view. We do *not* wish to merely inform you, nor to make you proficient at imitation, but rather to help develop your critical and analytical senses.

Whether we succeed or not will depend mostly on you. If you are sympathetic to our goals, the material and its presentation may prove useful to you not only as a student of mathematics but eventually as a teacher of mathematics.

Throughout, it will be assumed that you have at least an

intuitive familiarity with many of the concepts. Efforts will be made to *refine* this intuitive understanding by identifying subtleties, developing relationships, and determining relative significance of topics. Real-world analogs will be used to both motivate and illustrate the geometric ideas. Many terms and concepts will be used informally long before they are completely and rigorously defined.

For instance, everyone knows what a triangle or a circle is, or at least has an understanding sufficient for most purposes. We make use of this understanding to make conjectures and draw conclusions about related concepts. These terms are defined and described in more detail when the intuitive understanding later becomes insufficient for our purposes.

In a similar manner, we make use of the very familiar Pythagorean Theorem in advance of its proof because this is *convenient*. You may recall that the theorem states:

For any right triangle, the square of the length of the hypotenuse is equal to the sum of the squares of the lengths of the legs of the triangle.

You can teach yourself to utilize the results of this theorem long before you need to completely understand its meaning within a postulational system. Our approach will be to point out the various times that we can benefit from this most valuable theorem, and to "prove" the theorem only when it suits our purposes to do so later in the book. Although the concept of proof is not formally treated until Chapter 15, you are constantly encouraged to analyze, verify, justify, and explain why certain mathematical ideas are valid. The nature of validity is constantly discussed as you make conjectures and examine arguments made to support these conjectures.

Our purpose in avoiding a formal treatment of proof until late in the book derives from our conviction that it is not possible to teach the techniques, nature, and appreciation for proof by merely including a chapter on proof. Furthermore, we believe that a premature discussion of proof causes students to be *misdirected* into believing that the form and style of a proof are more important than the convincing argument that they are to communicate. Thus, the first 14 chapters lay the foundations for the chapter on proof, attempting to develop your ability *to create and communicate convincing arguments.* In numerous experiments and exercises, you are requested to make conjectures based on examination of data and observations. Further, you are asked to supply arguments that convince you of the validity of your conjectures.

Your own part in this course should be characterized by a desire to gain a more complete and mature understanding of mathematics. This will make you a better, more effective teacher. We encourage you to question, to experiment, and to ponder. Teacher, teach thyself.

R.A.N.
R.M.W.

Geometry
for
Teachers

POINTS IN SPACE

1

1.1 PROLOGUE

The concepts of Euclidean geometry have their origin and many of their most important applications in three-dimensional physical space. Our approach will be to examine certain problems of the so-called "real world" that can be described, generalized, and solved in an idealized world. This process allows us, among other things, to give an approximate description of the world in which we live.

We shall try to emphasize the idea that, although we shall be studying points, lines, and planes, these terms are subject to many different interpretations. Furthermore, we shall see that the relationships between points, lines, and planes are independent of any interpretation we choose to give. To illustrate this idea we will examine and compare some simple models that have been extracted from the real world. Finally, we shall modify one of these models so that it will more nearly seem to describe our intuitive concept of the real world.

in our opinion . . .

For the young child, the distinction between the physical world and the geometric world need not be explicitly stressed. Initially, a child needs to develop his ability to distinguish between physical objects, to gain a vocabulary, and to experience the notions of space, shape, and size. He must learn to draw upon his preschool experiences to sharpen his understanding of concepts such as big, small, long, short, tall, and so forth.

As the student matures in his mathematical understanding, he must learn to distinguish between physical objects and geometric ideas. Just as the student is expected to abstract the idea of number from his early experience of the collection of things, his experiences with the real physical world will help him to abstract geometrical ideas of shape, size, and space.

Example 1

Consider a set of people who all belong to the same club. This particular club has several committees. The bylaws of the club, given below, dictate the makeup of the committees.

1. The club has at least one committee.
2. Every committee contains exactly three members.
3. Any two members of the club belong to exactly one committee together.
4. Not all members of the club belong to the same committee.
5. Any two committees have exactly one member in common.

From the information given in the bylaws of the club, it is possible to determine how many people belong to the club, how many committees the club has, and how many committees each member of the club belongs to.

An analysis of the problem might go something like this. By Rule 1, we know there is at least one committee. By Rule 2, we know the committee has exactly three members. Let us call these members A, B, and C, and let $[A, B, C]$ denote the committee to which they belong. Rule 4 indicates that there must be at least one additional member of the club not belonging to the committee $[A, B, C]$. Let us call this member D.

Now, by Rule 3, there must be a committee with members A and D. This committee must have exactly one more member. This additional member cannot be either B or C, or Rule 3 would be contradicted; there must be a new member, say E, giving a committee $[A, D, E]$. By similar reasoning, we see that there must be a committee with members B and D. The third member of this committee cannot be A, C, or E. (Why not?) Thus, there is a sixth member of the club, F, and a committee $[B, D, F]$. In exactly the same way, we discover the existence of a member G and a committee $[C, D, G]$.

So far, we have

Committee 1 = $[A, B, C]$
Committee 2 = $[A, D, E]$
Committee 3 = $[B, D, F]$
Committee 4 = $[C, D, G]$

We also need committees with members A and F, A and G, B and E, B and G, C and E, C and F (check to see that these are the only remaining committees needed). Note that our needs are met by the formation of the following committees:

Committee 5 = $[A, F, G]$
Committee 6 = $[B, E, G]$
Committee 7 = $[C, E, F]$

A careful check will show that if the membership of the club is seven, arranged on committees as above, the rules of the club are satisfied. We note that each member of this club belongs to exactly three committees. It is possible to show that the club consists of *exactly* seven members and *exactly* seven committees, but we will not pursue this investigation.

Exercise 1.1.1

Suppose a college campus is being designed. The village zoning ordinances impose the following restrictions on the construction of buildings and sidewalks connecting these buildings.

1. The campus must have at least one sidewalk.
2. Every sidewalk connects exactly three buildings.
3. Any two buildings are connected by exactly one sidewalk.
4. Not all buildings can be connected by the same sidewalk.
5. Any two sidewalks have exactly one building in common.

Determine the minimum number of buildings and sidewalks that must be constructed

In considering Example 1 and Exercise 1.1.1, perhaps you noted that many things are irrelevant to their solution. For instance, it is unnecessary to know the nature of the club, the actual names of the members, the type of college, or where it is located. In fact, whether we are dealing with a club and committees of the club, or college buildings and sidewalks connecting them, is inconsequential. Both problems are similar to each other; their solutions are identical. The important elements are the rules that govern the relationship between the terms we are investigating. For example, if you replace the words "club," "committee," and "member" of Example 1 by "college campus," "sidewalks," and "buildings," respectively, you obtain the system described in Exercise 1.1.1.

Thus, the models described here have much in common. In order to characterize the system of Exercise 1.1.1 we really only need to reiterate the solution of Example 1. The solution does not depend on the interpretation given. It really does not make any difference what names we give to the elements of our system. The terms space, lines, and points could just as well be used instead of club, committees, and members or campus, sidewalks, and buildings. If we make this substitution we obtain the following system:

1. The space has at least one line.
2. Every line contains exactly three points.
3. Any two points of the space belong to exactly one line together.

4. Not all points of the space belong to the same line.
5. Any two lines have exactly one point in common.

Rule 3 might sound better if written: There is exactly one line through any two points. In Fig. 1.1.1 you can see a diagram that may assist you in visualizing the solutions of all of these systems. The letters *A, B, C, . . .* can represent members, buildings, or points. Find at least two other interpretations for space, line, and point. Rewrite the rules by making the obvious replacements. Examine the resulting model in the manner similar to that described in Example 1.

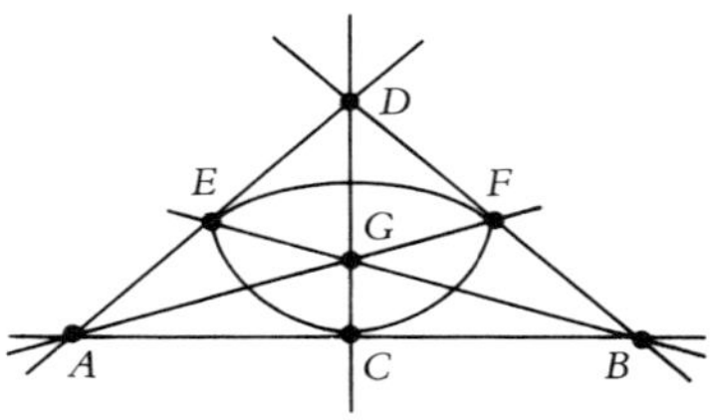

Figure 1.1.1

Experiment 1

We can construct another geometry by retaining the first four rules, but altering the fifth one.

1. There exists at least one line.
2. Every line contains exactly three points.
3. There is exactly one line through any two points.
4. Not all points are on the same line.
5. If l is a line and P a point not on line l, there is exactly one line containing P, which has no points in common with l.

Try and discover the least number of points in such a space, the number of lines, and the number of lines through any point. When you have done this, reconsider the system with the interpretations discussed in Example 1 and Exercise 1.1.1.

We hope that these interpretations have helped to demonstrate that an exact meaning of point, line, and space is unnecessary in determining relationships between these ideas. It is not necessary that we have a definition of space, line, or point to characterize the system.

1.2 BASIC CONCEPTS

We now proceed to the investigation of an infinite geometry. Our fundamental notions will be space, point, line, and plane. As in the case of the other models, these ideas will not be defined, but we can consider an interpretation for them.

A *point* can be thought of as a *location* and will be usually represented by a dot. The dot is a picture of a point; it has location but it has other features as well, such as color, width, length, and even height. *None* of these is to be considered as a property of a point.

When we speak of the idea of a *line*, we represent the idea with a picture such as in Fig. 1.2.1. The idea we are supposed to gather is something of *infinite length* in either sense of direction. (This is what the arrowheads are supposed to indicate.) The idea of line we have in mind is "straight," a description that can be imagined but not defined. Note as with the picture of a point, the picture of a line has features that are not characteristic of a line, such as color, width, and height. This is because the picture is drawn with ink which possesses these features.

Figure 1.2.1

Figure 1.2.2 is meant to suggest the idea of a *plane*. We are trying to relate the notion of a *flat surface* extending indefinitely in all directions. (This is the purpose of the arrowheads, but often we will picture only a part of a plane in our discussion.) As in the case of line, we are not able to actually picture the infinite extent of the idea of a plane, but it is important to keep in mind that it possesses this feature. Here, as before, our picture of a plane possesses features not to be associated with the idea of a plane.

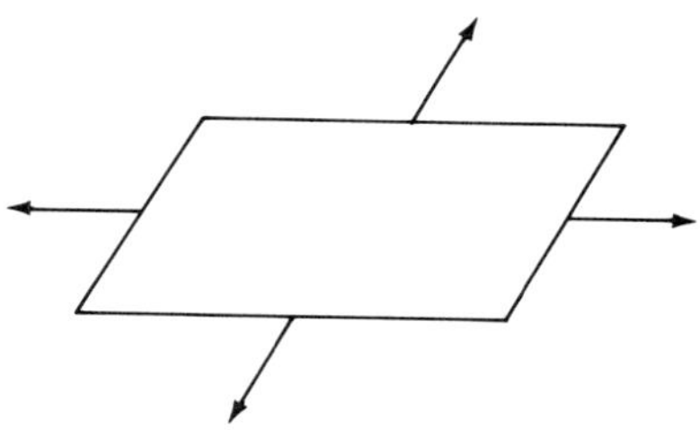

Figure 1.2.2

When attempting to suggest *space*, we will often use something like Fig. 1.2.3. Note the figure used to illustrate space is

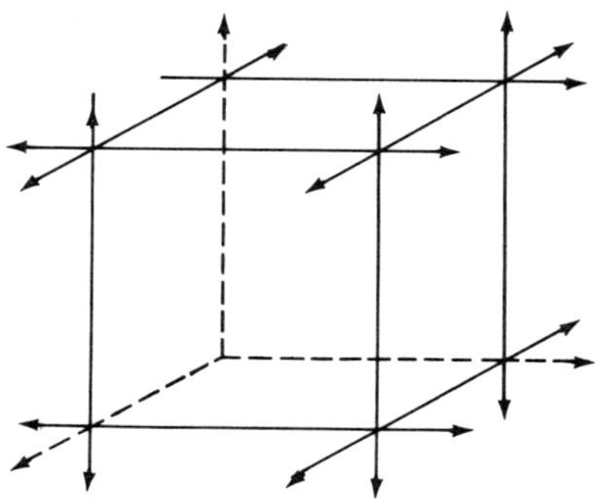

Figure 1.2.3

a "solid region" which extends indefinitely in all directions (once more, the purpose of the arrowheads).

There are two other terms that we shall be frequently using. Each system we have examined so far has been characterized by a set of rules. We shall refer to these rules as *postulates*. From the postulates we are able to deduce additional properties of these systems. These properties will be designated as *theorems*.

comment . . .

The elementary school student should discover physical objects that share features with point, line, plane, and space. He should also describe those features of the physical objects which are not peculiar to the geometric ideas.

Exercise 1.2.1

List some theorems of the system described in Example 1, Exercise 1.1.1, and Experiment 1.

1.3 INCIDENCE RELATIONS

Let us take the set of postulates of Experiment 1 and retain those postulates which seem to describe our physical world and alter those which do not. The first postulate still seems reasonable and so we will keep it.

1.3.1 Postulate

There is at least one line.

The second postulate, that a line contains exactly three points, does not seem to make much sense; in fact, it appears as if there should be an infinite set of points. Thus, we will have the following as our second postulate.

1.3.2 Postulate

Every line contains an infinite number of points.

In Fig. 1.3.1, a line is labeled l and one of the points, P. That P is a member of the set of points which is the line l can be stated in each of the following equivalent ways.

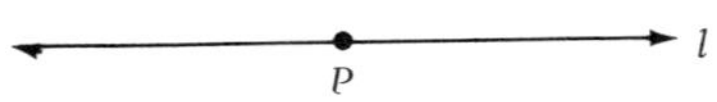

Figure 1.3.1

1. $P \varepsilon l$.
2. P lies on line l.
3. P is contained in line l.
4. The line l passes through the point P.

5. The point P is incident with the line l.
6. l contains P.

Since the third postulate seems to be quite reasonable, we shall adopt it, changing the wording slightly as was suggested earlier.

1.3.3 Postulate

There is exactly one line through any two points.

Actually, this statement is meant to reflect our understanding and expectations of the properties of a line. Recall that we are trying to think of a line as being "straight." One interpretation of the notion of a straight line is that for any two distinct points P and Q on the line, the shortest distance between P and Q is the path of a straight line. In Fig. 1.3.2 we have two points P and Q and various paths connecting P and Q.

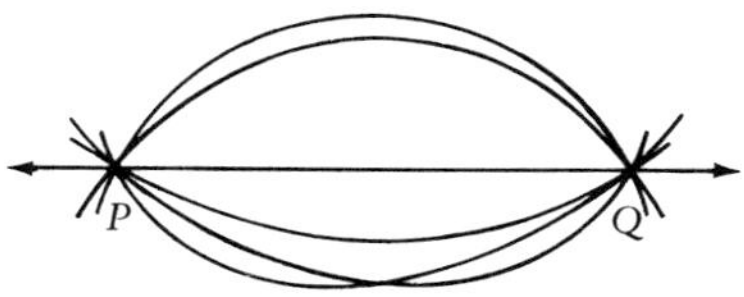

Figure 1.3.2

Since we are looking at a physical object, we use our intuition to conclude that the various "curved segments" which contain both P and Q would be paths whose lengths would be greater than the path of the "straight segment." Intuitively, we claim that between any two points, there is a unique path with shortest length which lies on a straight line containing P and Q.

Since we are agreeing that for any two distinct points in space, there is one and only one line containing both points, we can identify a line by naming any two distinct points of the line. As an example, suppose the line l (Fig. 1.3.3) contains points P, Q, and R. One symbol for the line is $\overleftrightarrow{PQ}$, read "line P, Q." The order in which we name the two points is not important, so that $\overleftrightarrow{QP}$ is another symbol which names the line l. Other names for the line are $\overleftrightarrow{PR}$, $\overleftrightarrow{RP}$, $\overleftrightarrow{QR}$, and $\overleftrightarrow{RQ}$. Of course, since there are an infinite number of points on the line l (why?) we could have an infinite number of symbols that would name the line.

Figure 1.3.3

The truth of the fourth postulate, that not all points of space are contained in the same line, depends on our interpretation of "space." As we shall soon see, we shall at times be wanting to restrict our "space" to one of three possible interpretations: a one-dimensional space (for example, a single line), a two-dimensional space (for example, a single plane), or a three-dimensional space (for example, the universe). Since in the two- and three-dimensional interpretations the postulate is true, we shall adopt it, keeping in mind that in the one-dimensional case it is not true. Thus, we have the following.

1.3.4 Postulate

Not all points of space belong to the same line.

Our fifth postulate considers the number of points two lines in space can have in common. If you consider Figs. 1.3.4, 1.3.5, and 1.3.6, you can readily see that they might have one point in common.

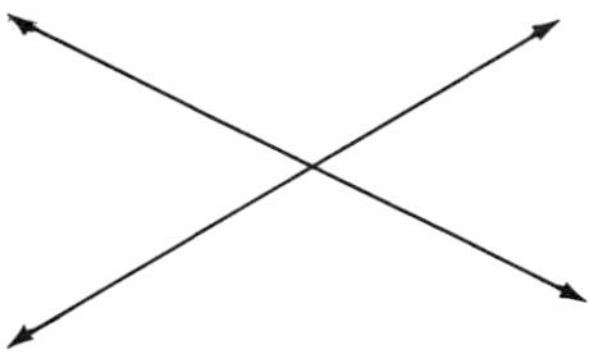

Figure 1.3.4

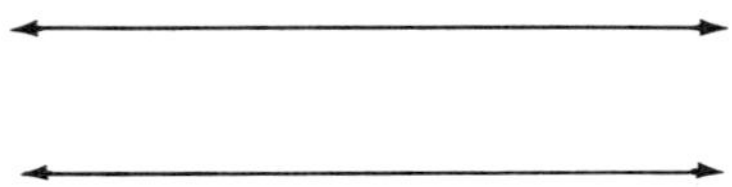

Figure 1.3.5

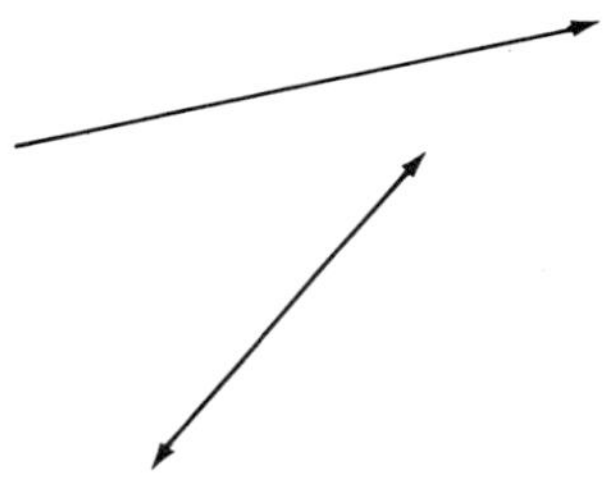

Figure 1.3.6

It is obvious what Figs. 1.3.4 and 1.3.5 are trying to depict. But Fig. 1.3.6 is subject to interpretation. If you consider the two lines to both be in the same plane (that is, both on the page), then the lines will have one point in common. On the other hand, if you can visualize one of the lines as being on the paper, but the second line as "piercing" the paper, then the lines have no points in common. This suggests our fifth postulate:

1.3.5 Postulate

Two distinct lines have either one or no points in common.

in our opinion . . .

Asking students to draw conclusions from pictures such as these can be setting a dangerous precedent. Our comments about the figures are based solely on our personal interpretations of the diagrams. We cannot say with conviction that lines never have a point in common if extended indefinitely. Rather, we are imagining these figures as examples of pairs of lines in space which have one or no points in common. Pictures are used to illustrate facts, not as a basis for conclusions about the picture. The defects in such reasoning are many; most importantly, our picture is only one representation of an idea. Another picture may be equally valid to represent the same idea, yet be very different in many other aspects. Students should be encouraged to draw their own pictures representing geometrical ideas. These various pictures can then be compared.

Notice that Postulate 1.3.5 does not really eliminate the possibility that two distinct lines can have *more* than two points in common; for if there *were* two points in common, then it *would* be true that the two lines have "either one or no points in common."

comment . . .

Although in everyday usage to say "there is one object" may sometimes imply "there is exactly one object" we do not assert this interpretation in mathematics. To say "there is one" really implies "there is one or more." However, if we bear the other postulates in mind we can eliminate this possibility. For if, indeed, there were two distinct lines which contain two or more points (say P and Q were two such points as in Fig. 1.3.7), then P and Q are distinct points which belong to two distinct lines and this violates Postulate 1.3.3.

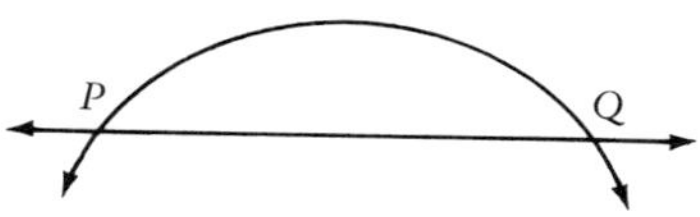

Figure 1.3.7

Thus, we have our first theorem.

1.3.6 Theorem

Two distinct lines cannot have two or more points in common.

At this time, we shall also be able to deduce another theorem from our postulates. Postulate 1.3.2 tells us there are an infinite number of points on a line. How many lines do you think there are through any point? Suppose A is a point of our space. Then, by Postulate 1.3.1, we know there is at least one line, say l. Now our postulates do not specify whether l contains A or not. So, our diagram will look like either Fig. 1.3.8 or 1.3.9. In Fig. 1.3.8, we know that l contains an infinite number of points.

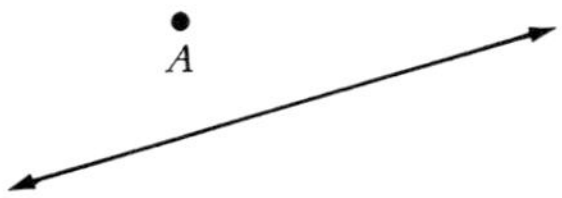

Figure 1.3.8

Figure 1.3.9

(Why?) Thus, if B is a point of l, then there is a line $\overleftrightarrow{AB}$, and if C is another point of l, there is another line $\overleftrightarrow{AC}$. (Which postulate tells us this?) Now $\overleftrightarrow{AC}$ and $\overleftrightarrow{AB}$ must be different lines for, if not, then A would be on the same line as B and C; that is, A would be on l which contradicts the assumption in Fig. 1.3.8. Likewise, for each of the infinite points of l, we would obtain a line through A. Thus, there would be an infinite number of lines through A. See Fig. 1.3.10. Now, if A and l are arranged as in Fig. 1.3.9, then we know that there is a point which is not on l, say B, by Postulate 1.3.4. Also, since l has an infinite number of points, there is one point on l distinct from A, say C. Now, since B and C are distinct, we know there is a line $\overleftrightarrow{BC}$. (Why?) Now $\overleftrightarrow{BC}$ contains an infinite number of points and each point of $\overleftrightarrow{BC}$ is on a line containing point A. Thus, once again we see there is an infinite number of lines through point A. Thus, we see that from our set of five postulates we can derive a theorem.

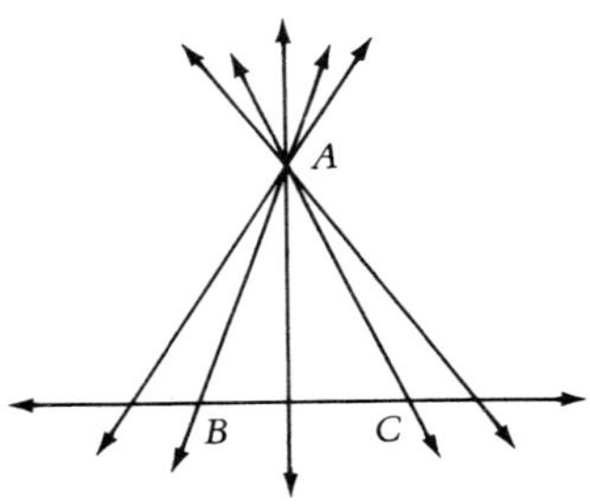

Figure 1.3.10

1.3.7 Theorem

There are an infinite number of lines through each point in space.

Exercise 1.3.1

1. In Fig. 1.3.11, name all the lines.
2. In Fig. 1.3.11, name a point not on each of the lines named in Exercise 1.
3. Draw diagrams showing line t and points P, Q, and R to fit each of the following conditions.

a. $t = \overleftrightarrow{PQ}, \quad R \notin t$ b. $t \neq \overleftrightarrow{PQ}, \quad P \in t$ and $R \in t$

c. $t \cap \overleftrightarrow{PQ} = \{R\}$ d. $\{P, Q\} \subset t, \quad R \notin t$

e. $t \cap \overleftrightarrow{PQ} = \phi$

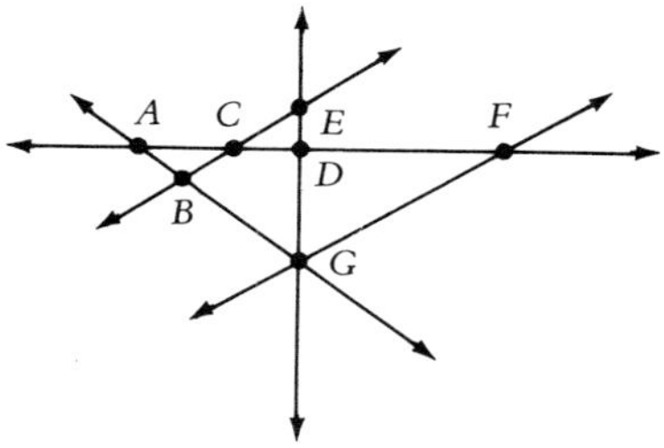

Figure 1.3.11

The next fundamental idea of our geometry is that of a plane. Any flat surface such as the face of a blackboard, the floor of a room, or a page of this book suggests the idea of a plane. Of course, these illustrations are inadequate not only because they possess features not common to the idea of a plane, but also because they are only finite in extent. As with a line, a plane is an infinite subset of space whose elements are points. We will usually denote a plane by a figure such as Fig. 1.3.12 and represent the entire plane by a script letter, in this case $\mathcal{P}$.

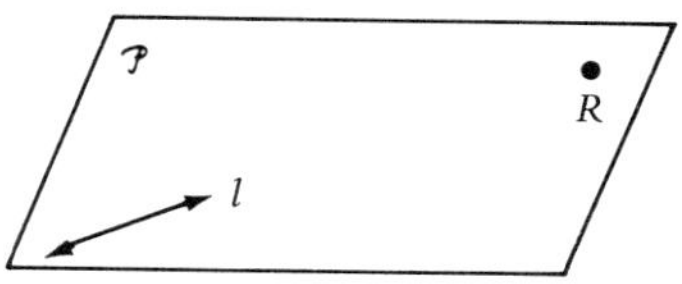

Figure 1.3.12

One of the points of the plane $\mathcal{P}$ pictured in Fig. 1.3.12 is denoted by the letter R. We can say that R is a point of the plane $\mathcal{P}$ in any of the following equivalent ways.

1. $R \in \mathcal{P}$.
2. $\mathcal{P}$ contains R.
3. R lies in the plane $\mathcal{P}$.
4. R is contained in $\mathcal{P}$.
5. The plane $\mathcal{P}$ passes through R.
6. The point R is incident with the plane $\mathcal{P}$.

Except for our intuition and drawings, we have not yet made any distinction between lines and planes since both are infinite subsets of space whose elements are points. The distinction will be made by adopting some additional postulates. There are several ways in which we might arrive at our postulates but the method we shall employ will point out a rather interesting relationship between points, lines, and planes. We shall reword each postulate we already have by replacing the word point by line and line by plane. In order to see how this is done, we will list each postulate and then the statement which is derived from it. Thus, we get the following.

1.3.1 Postulate

There is at least one line.

1.3.1′ Derived statement

There is at least one plane.

1.3.2 Postulate

Every line *contains an infinite number of* points.

1.3.2′ Derived statement

Every plane *contains an infinite number of* lines.

1.3.3. Postulate

There is exactly one line *through any two* points.

1.3.3′ Derived statement

There is exactly one plane *through any two* lines.

1.3.4 Postulate

Not all points *of space belong to the same* line.

1.3.4′ Derived statement

Not all lines *of space belong to the same* plane.

1.3.5 Postulate

Two distinct lines *have either one or no* points *in common.*

1.3.5′ Derived statement

Two distinct planes *have either one or no* lines *in common.*

If we consider the derived statements carefully, we see that in some cases we have obtained very reasonable statements about planes and lines. We shall adopt these as postulates. Thus, we get the following.

1.3.8 Postulate

There is at least one plane.

1.3.9 Postulate

Every plane contains an infinite number of lines.

Before we proceed any further, perhaps we should be more careful with what we mean by saying a line is in a plane. This becomes clarified if we give the following definition.

1.3.10 Definition

A line l is in a plane $\mathcal{P}$ if and only if all the points of l are in $\mathcal{P}$, that is, $l \subset \mathcal{P}$.

Now, if we consider situations in which l is not contained in $\mathcal{P}$, we must consider cases in which some points of l are not in $\mathcal{P}$. Figures 1.3.13–1.3.15 give some suggestions. Of course,

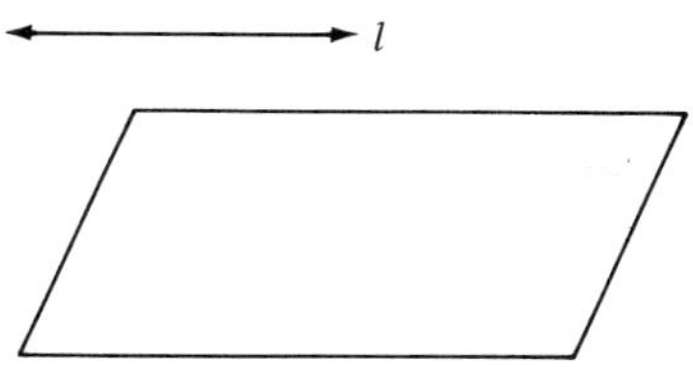

Figure 1.3.13
No points of l are in $\mathcal{P}$

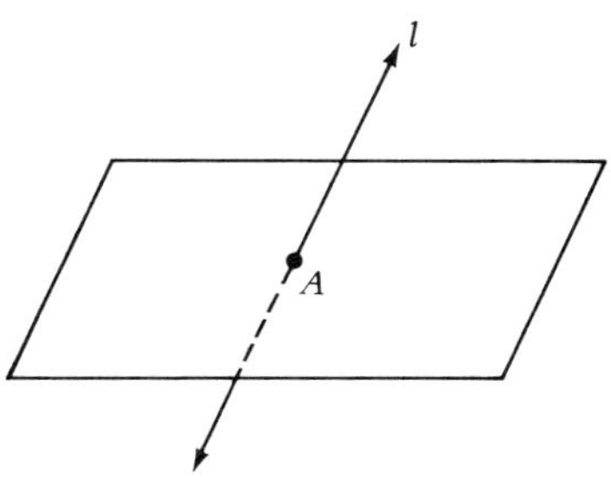

Figure 1.3.14
All but one point of l are not in $\mathcal{P}$

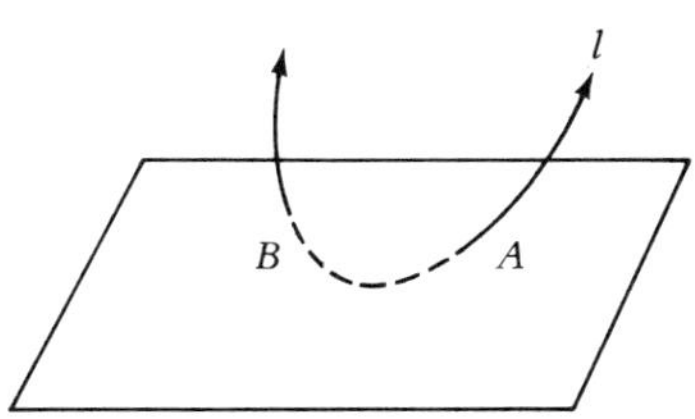

Figure 1.3.15
All but two points of l are not in $\mathcal{P}$

Fig. 1.3.15 does not really seem possible for it seems to go contrary to our interpretation of a "flat plane" and "straight line." In fact, this suggests another postulate.

1.3.11 Postulate

If two points of a line are in a plane, then every point of the line is in the plane.

Now if we consider the third derived statement, we no longer get a true statement. It is not always possible to pass a plane through any two distinct lines. As an example of a pair of lines that are not in the same plane, consider the lines of Fig. 1.3.18. The lines in this figure are called skew lines. Those in Fig. 1.3.16 are called intersecting lines. Because we will be using these concepts very frequently we shall give their definitions now.

1.3.12 Definition

Two distinct lines intersect if and only if they have one point in common (Fig. 1.3.16).

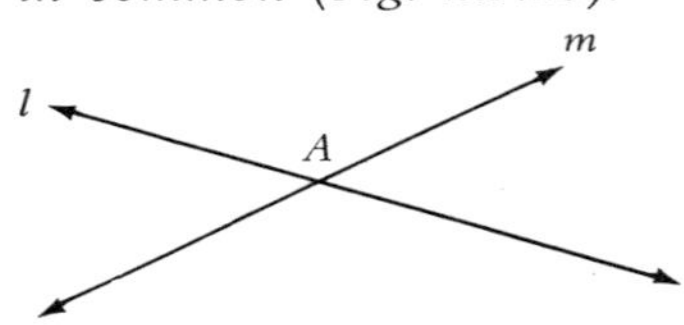

Figure 1.3.16
Intersecting lines

In Fig. 1.3.16 we can say any of the following.

1. l and m intersect at A.
2. The intersection of l and m is A.
3. $l \cap m = \{A\}$.

1.3.13 Definition

Two distinct lines are parallel if and only if they are in the same plane and do not intersect (Fig. 1.3.17).

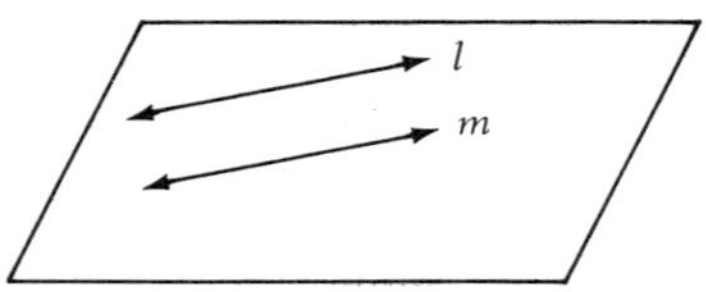

Figure 1.3.17
Parallel lines

1.3.14 Definition

Two distinct lines are skew if and only if they are not in the same plane and do not intersect (Fig. 1.3.18).

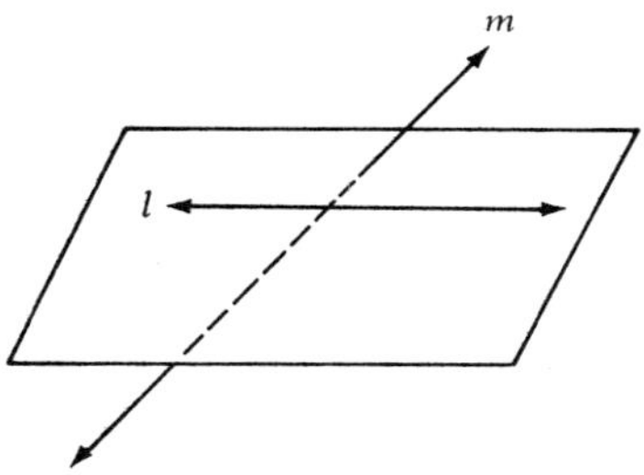

Figure 1.3.18
Skew lines

In Figs. 1.3.17 and 1.3.18 we can say all of the following.

1. l and m do not intersect.
2. $l \cap m = \phi$.
3. The intersection of l and m is empty.

As the previous examples suggest, it is not possible to pass a plane through a pair of lines when they are skew. However, it *is* true for any two intersecting lines and any two parallel lines. Thus, we obtain our next two postulates.

1.3.15 Postulate

There is exactly one plane through any two intersecting lines.

1.3.16 Postulate

There is exactly one plane through any two parallel lines.

Now, if we examine the fourth derived statement, we should realize that we have already been assuming it since we have acknowledged the existence of skew lines. Thus, we have our next postulate.

1.3.17 Postulate

Not all lines of space belong to the same plane.

The last derived statement is also true and it becomes our next postulate.

1.3.18 Postulate

Two distinct planes of space have either one or no lines in common.

So far, we have obtained statements which indicate the relationship between lines and planes. What about the relationship between points and planes? First, it should be clear that we wish to be able to pass an infinite number of planes through any point of space. In fact, we shall call this a postulate.

1.3.19 Postulate

There is an infinite number of planes through every point of space.

Now, suppose we select two points of space (say P and Q). How many planes can we pass through P and Q? Here, again, we find that there are an infinite number of planes. How about three points? It really depends on how the points are arranged. If they are collinear (that is, lie on the same line), then there are an infinite number of planes. However, if they are not collinear, then there is exactly one plane. This can be shown to follow from our postulates and it becomes our next theorem.

1.3.20 Theorem

There is exactly one plane which contains any three noncollinear points.

To see why this is true, let us suppose that A, B, and C are three noncollinear points. (See Figs. 1.3.19 and 1.3.20.) Then we know $\overleftrightarrow{AB}$ and $\overleftrightarrow{CB}$ are two distinct lines which intersect at B.

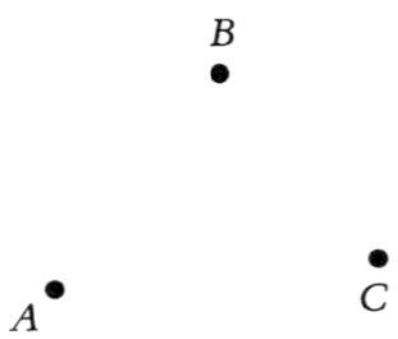

Figure 1.3.19

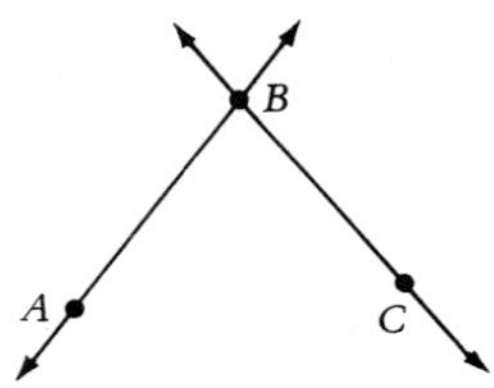

Figure 1.3.20

Thus, by Postulate 1.3.15, there is exactly one plane which contains $\overleftrightarrow{AB}$ and $\overleftrightarrow{BC}$. Thus, we know that there is at least one plane which contains points A, B, and C. Furthermore, we know there is no other plane containing these points for if there were, then this plane would contain $\overleftrightarrow{AB}$ and also $\overleftrightarrow{BC}$ by Postulate 1.3.11. Thus, we would have two planes containing $\overleftrightarrow{AB}$ and $\overleftrightarrow{BC}$ and this violates Postulate 1.3.15.

In a similar manner, we can justify the following theorem and this will be left as an exercise.

1.3.21 Theorem

There is exactly one plane which contains a line and a point not on the line.

Example 1

Consider three points in space: namely, P, Q, and R. Let us assume that the three points do not all lie in the same line. (See Fig. 1.3.21.) Then P, Q, and R determine a unique plane and we

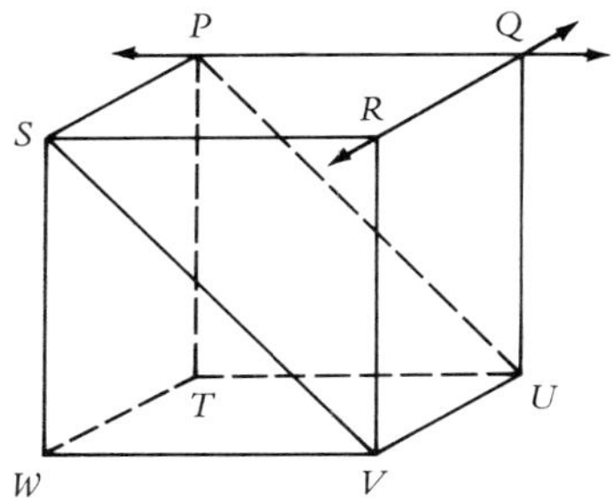

Figure 1.3.21

can designate it by $\overleftrightarrow{PQR}$. Also, if S lies in the same plane as P, Q, and R, then we could call the plane $\overleftrightarrow{PSR}$, or for that matter $\overleftrightarrow{PQS}$, $\overleftrightarrow{PRQ}$, $\overleftrightarrow{QPR}$, $\overleftrightarrow{QPS}$, and so on. Note that the order of the letters is unimportant.

Exercise 1.3.2

1. Frequently it is useful to suggest to students that the properties stated in Postulates 1.3.15, 1.3.16, and Theorems 1.3.20 and 1.3.21 have simple applications in the physical world. For example, using a tripod to hold a camera in a fixed position is an application of Theorem 1.3.20. Make a list of other applications of each of these properties.
2. Use the same technique of replacing the words point and line by line and plane, respectively, to derive statements from Theorems 1.3.6 and 1.3.7. Do you believe that these new statements are correct? Explain your thinking.
3. The front and top of the box pictured in Fig. 1.3.22 suggest planes $\mathcal{P}$ and $\mathcal{S}$, respectively.

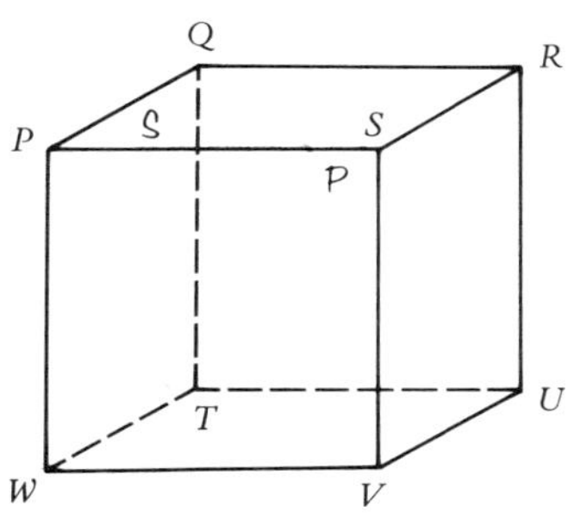

Figure 1.3.22

a. Name four points contained in $\mathcal{P}$.
b. Name four points contained in $\mathcal{S}$.
c. Name six lines lying in $\mathcal{S}$.
d. Name six lines lying in $\mathcal{P}$.
e. What is $\mathcal{P} \cap \mathcal{S}$?
f. Name plane $\mathcal{P}$ in two different ways.
g. Name plane $\mathcal{S}$ in two different ways.
h. Name all other planes in Fig. 1.3.22 suggested by the labeled points.
i. Name two lines that do not intersect but lie in a common plane.
j. Name two lines that do not intersect and do not lie in a common plane.
k. Name a line and a point that determine plane $\mathcal{P}$.
l. Name two intersecting lines that determine plane $\mathcal{S}$.
m. Name two parallel lines that determine plane $\mathcal{P}$.
n. Name a pair of skew lines in Fig. 1.3.22.

5. Let l, m, and n be three distinct lines in space such that l intersects m and l intersects n.
   a. Must m and n lie in the same plane?
   b. If there is a fourth line k which intersects all three lines l, m, and n, must m and n lie in the same plane?

1.4 DEFINITIONS

So far, we have used various phrases to describe certain incidence relationships between points, lines, planes, and space. Now that we have experienced many of them, it is time to assign names to some of the more common relationships.

1.4.1 Definition

Three or more points in space are said to be collinear *if and only if they all belong to the same line. If they do not all belong to the same line, they are said to be* noncollinear.

Example 1
See Fig. 1.4.1.

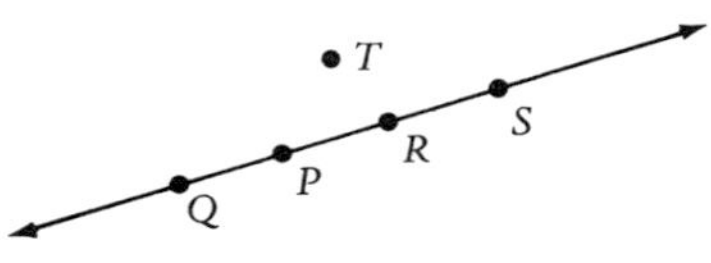

Figure 1.4.1
P, Q, R, and *S* are collinear; *P, R,* and *T* are noncollinear

1.4.2 Definition

Four or more points in space are said to be coplanar *if they all belong to the same plane. If they do not all belong to the same plane, they are said to be* noncoplanar.

Example 2

See Fig. 1.4.2.

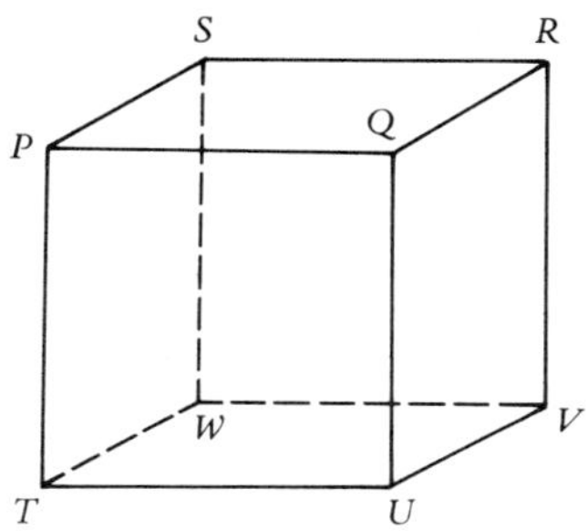

Figure 1.4.2
P, R, Q, and *S* are coplanar;
P, Q, R, and *T* are noncoplanar

1.4.3 Definition

Two or more geometrical objects (that is, subsets of space) are said to be coincident *if they consist of the same points. Otherwise, they are said to be* noncoincident.

Example 3

See Fig. 1.4.3.

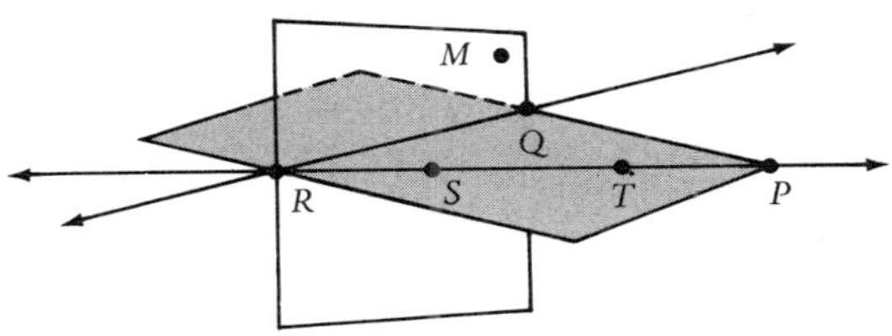

Figure 1.4.3
$\overleftrightarrow{ST}$ and $\overleftrightarrow{RP}$ are coincident; $\overleftrightarrow{RPQ}$ and $\overleftrightarrow{STQ}$ are coincident;
$\overleftrightarrow{RP}$ and $\overleftrightarrow{RQ}$ are noncoincident $\overleftrightarrow{MRQ}$ and $\overleftrightarrow{RPQ}$ are noncoincident

1.4.4 Definition

Two or more lines (or three or more planes) are said to be concurrent *if they have at least one point in common. Otherwise, they are said to be* nonconcurrent.

Example 4

See Fig. 1.4.4.

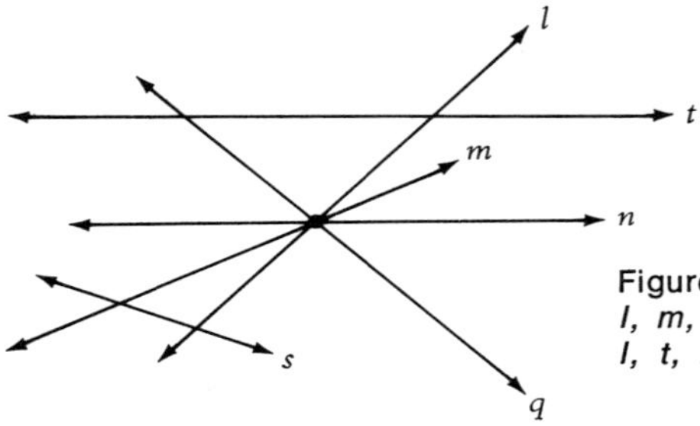

Figure 1.4.4
l, m, n, and *q* are concurrent;
l, t, m, and *s* are nonconcurrent

in our opinion . . .

It is appropriate that definitions of ideas follow the discussion of these ideas rather than the other way around. A definition is not an interpretation of a name, but rather the naming of an idea. It is much like naming a child Donald. The name Donald represents the child; the child does not represent Donald. It is risky, indeed, to start a discussion of an idea with young children by making a definition. Rather, it is important to formulate an idea understood by all which is finally formalized in a definition.

1.4.5 Definition

A line which does not lie in or intersect a plane is said to be parallel to the plane. Planes that have no points in common are called parallel planes.

Example 5
See Fig. 1.4.5.

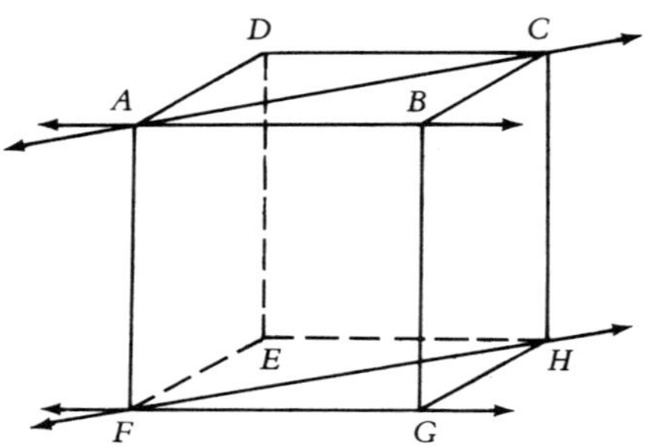

Figure 1.4.5
$\overleftrightarrow{AC}$ is parallel to $\overleftrightarrow{FH}$;
$\overleftrightarrow{AC}$ is parallel to $\overleftrightarrow{EFG}$;
$\overleftrightarrow{ABC}$ is parallel to $\overleftrightarrow{EFG}$

Note we will use the symbol "$||$" to mean "is parallel to." Thus, in our example, $\overleftrightarrow{AC} || \overleftrightarrow{FH}$, $\overleftrightarrow{AC} || \overleftrightarrow{EFG}$, and $\overleftrightarrow{ABC} || \overleftrightarrow{EFG}$.

comment . . .

Any classroom provides a very effective model of space with all of its important subsets. This is also true of the case when we coordinatize space in the next chapter. Sometimes a lot of pointing is very important to the understanding of an idea. It is important, however, never to allow the model to be confused with the idea.

A common mistake is to think that any two lines lying in parallel planes must be parallel to each other. In Fig. 1.4.6, $\overleftrightarrow{PQR}$ and $\overleftrightarrow{TUV}$ are to be considered parallel. Line $\overleftrightarrow{PR}$ lies in plane $\overleftrightarrow{PQR}$ and line $\overleftrightarrow{WV}$ lies in plane $\overleftrightarrow{TVU}$, but the line $\overleftrightarrow{PR}$ and $\overleftrightarrow{WV}$ are not parallel because they are noncoplanar. As we pointed out earlier, they are called skew lines.

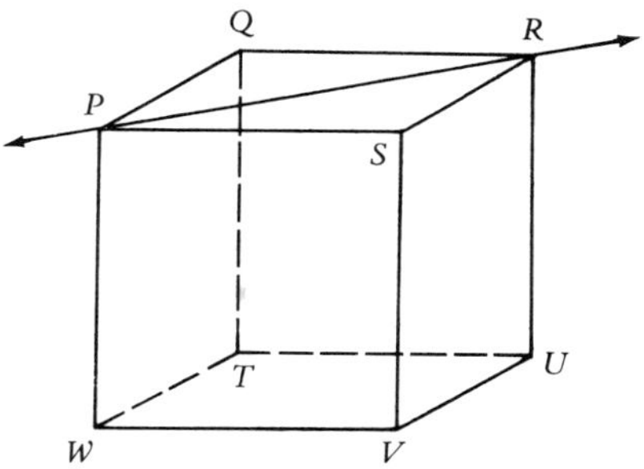

Figure 1.4.6

Exercise 1.4.1

1. Let l, m, and n be lines and $\mathcal{P}$, $\mathcal{S}$, and $\mathcal{R}$ be planes in space. Answer each of the following questions, justifying the answer by drawing a diagram.
   a. Let l and m be in the same plane, m and n in the same plane. Must l and n be in the same plane?
   b. Can l and m be coplanar, m and n be coplanar, l and n coplanar, but no plane exist containing all three lines l, m, and n?
   c. Let l, m, and n be concurrent. Must they be coplanar?
   d. Let $l \| m$, $m \| n$, $n \| l$. Must they be coplanar?
   e. If $l \cap m \neq \phi$ and $n \cap m \neq \phi$, must $l \cap n \neq \phi$?
   f. If $l \| \mathcal{P}$ and $l \| \mathcal{S}$, must $\mathcal{P} \| \mathcal{S}$?
   g. If $\mathcal{P} \cap \mathcal{S} = l$, $\mathcal{P} \cap \mathcal{R} = n$, and $l \| n$, must $\mathcal{S} \| \mathcal{R}$?
   h. If $\mathcal{P} \cap \mathcal{S} = l$, $\mathcal{S} \cap \mathcal{R} = m$, $\mathcal{P} \cap \mathcal{R} = n$, can l, m, and n be parallel?
   i. If $l \| m$, $m \| n$, can l be skew to n?

2. True or False. Base your answers on your intuition.
   a. Two skew lines determine a plane.
   b. If l is a line and $P \notin l$, then there exists a line m such that $P \in m$ and $l \| m$.
   c. If l is a line and $P \notin l$, then there is at most one line containing P that is parallel to l.
   d. If l is a line and $P \notin l$, then there exists a line m such that $P \in m$ and $l \cap m \neq \phi$.
   e. If l is a line and $P \notin l$, then there exists at most one line containing P which intersects l.
   f. If l is a line and $P \notin l$, then there is a line m such that $P \in m$ with m and l skew.
   g. If l is a line and $P \notin l$, then there is at most one line containing P such that it and l are skew.
   h. If $\mathcal{P}$ is a plane and $P \notin \mathcal{P}$, then there is a plane $\mathcal{S}$ with $P \in \mathcal{S}$ such that $\mathcal{P} \| \mathcal{S}$.
   i. If $\mathcal{P}$ is a plane and $P \notin \mathcal{P}$, then there is at most one plane containing P such that the two planes are parallel.
   j. Any two distinct lines intersect in at most one point.
   k. If P and Q are any two points in a plane $\mathcal{P}$, then $\overleftrightarrow{PQ} \subset \mathcal{P}$.
   l. For any line l and any plane $\mathcal{P}$, if $l \not\subset \mathcal{P}$, then $l \cap \mathcal{P}$ contains at most one point.

1.5 EXPERIMENTS

Experiment 2

We know that two points in a plane determine a unique line. Suppose we have three noncollinear points P, Q, and R in a plane. It is easy to see that these three points determine three lines (Fig. 1.5.1). Suppose next we have four points P, Q, R, and S in

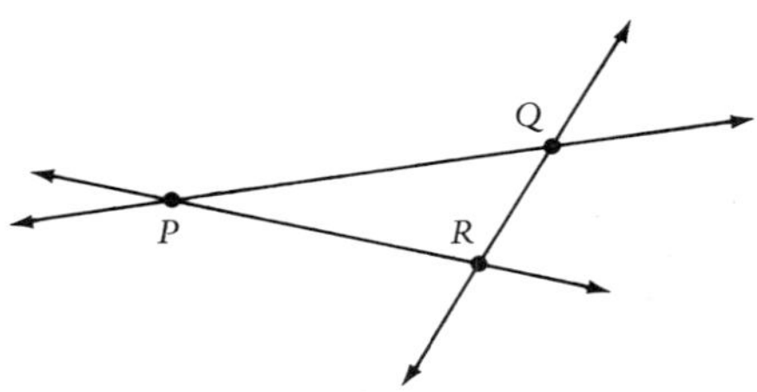

Figure 1.5.1

a plane, no three of which are collinear. By drawing a diagram, we discover that the number of lines determined by these four

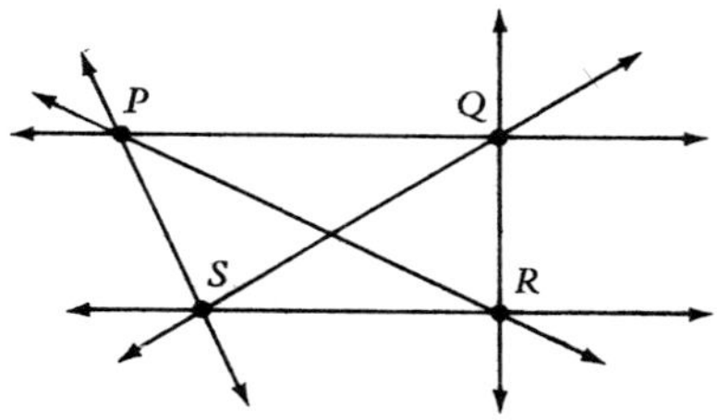

Figure 1.5.2

points is six (Fig. 1.5.2). Now complete the following table:

Number of points in a plane, no three of which are collinear	*Number of lines*
2	1
3	3
4	6
5	
6	
7	
8	
.	
.	
.	
n	

Did you notice that adding a new point to the system introduces a new line for each point already in the system? Can you discover the relationship between the number of lines at any step and the number of points in the preceding step and the number of lines at the preceding step? Can you discover a formula for the number of lines determined by **n** points in a plane, no three of which are collinear?

in our opinion . . .

The reason for the various experiments in this book is to set a good example. Every child is an empirical scientist. This can be observed in his play. Given some material with which he is unfamiliar, the child will make numerous experiments, notice some results of his investigation, experiment some more, formulate an opinion, and if anyone will listen, communicate his findings. This threefold approach to learning—namely, experimentation, thinking, and communication—is rather badly modified in many school situations where "real learning" is supposed to take place. All too often, experimentation is not allowed, thinking is discouraged, and communication is standardized.

Is it any wonder that so many children of an independent and curious nature fail to gain all the benefits of their mathematical study? Too frequently, they are not made sufficiently aware of the nature of the problem nor allowed to involve themselves enough to discover possible ways in which the problem might be approached. A child is creative and often fascinated by the things *he* learns. It is necessary that mathematics be presented as a creative and fascinating subject.

Geometry is a particularly appropriate branch of mathematics in which the student can be encouraged to experiment, to think, to communicate. In studying mathematics, one does not always start with a problem to be solved or a theorem to be proved. Instead, new materials and new concepts present many possibilities. In experimenting and thinking about new materials and new concepts, the student discovers the need for new vocabulary in order to be able to communicate. Once he has done this, it becomes possible for him to formulate a new problem.

For progress in learning, it is not sufficient that a student become adept at answering questions or solving problems. More important is to develop the ability to ask new questions and raise problems for solution. Perhaps it is fair to say that what the world needs to improve itself is not so much new answers as new questions.

In this process, the teacher must act the role of an intelligent guide and a constructive friendly critic. He must also have patience to really allow the student to learn. In many cases, the discovery method of learning is essentially a process in which the teacher has arranged things so that at least one student will arrive at an inescapable conclusion.

A better arrangement requires the teacher to trust his students, to encourage their curiosity, and to allow them to choose paths to knowledge not planned by the teacher. Certainly, the teacher must have goals, both mathematical and behavioral, but these must be *shared* with the students, *not* inflicted. The teacher will assemble the necessary materials or introduce the necessary concepts, making certain that all students understand the purpose and possibilities. Then he will allow these materials and concepts to be used by the students to allow for the growth and development of ideas.

Experiment 3

Given **n** points (**n** $>$ 4), no four of which are coplanar, how many planes are determined?

The approach suggested to complete these experiments is an example of *inductive reasoning;* that is, discovering a pattern from a few examples. Of course, this approach is quite fallible and in many cases may lead to faulty conjectures. Any formulas discovered in doing these experiments would require *deductive reasoning* to test their validity. That is, they must be proven. At the moment, we are interested only in discovering fairly reasonable conjectures; deductive reasoning will be delayed.

Exercise 1.5.1

1. Consider the following four rules of our geometry, hereafter called postulates. Determine the least number of points, lines, and planes that will satisfy Postulates 1–4. Draw a picture showing the relationships of these points, lines, and planes.
    1. Every line contains at least two points.
    2. For any two points, there exists exactly one line which contains them.
    3. There exist at least three noncollinear points.
    4. For any three noncollinear points, there exists exactly one plane which contains them.
2. The following are a few definitions as given by Euclid. Discuss these definitions and the various advantages and disadvantages of leaving concepts such as point, line, plane, and space undefined.
    a. A point is that which has no part.
    b. A line is breadth-less length.
    c. A straight line is a line which lies evenly with the points on itself.
    d. A plane surface is a surface which lies evenly with the straight line on itself.
    e. A boundary is that which is an extremity of anything.
    f. A figure is that which is contained by any boundary or boundaries.
3. Any checkerboard consists of 64 small squares. It also contains larger squares formed by four of the smaller squares that are adjacent. There are also larger squares formed by groups of small squares that contain a smaller square in the center. How many squares of all sizes are contained in a checkerboard? The most intelligent approach to this problem is by using inductive reasoning. First, determine the number of squares contained in each of the following special checkerboards.

Checkerboard	*Number of Squares*
Consisting of just one square 	1
Consisting of four squares 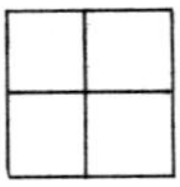	5
Consisting of nine squares 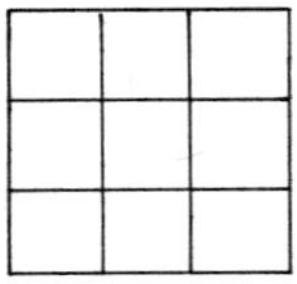	
Consisting of sixteen squares, etc. 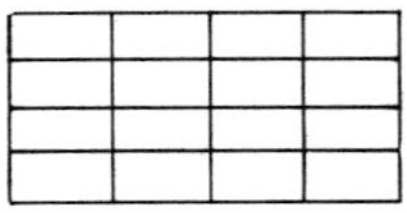	

1.6 SUMMARY

1. The physical idea of location or position suggests the mathematical idea of a point. A straight line is suggested by the edge of a ruler and a plane is suggested by a flat surface.
2. Space is an infinite set of points. Any subset of space, finite or infinite, which is nonempty is a geometric figure.
3. Two distinct points in space determine a unique line containing the two points.
4. Points that lie on the same line are said to be collinear; points that lie in the same plane are said to be coplanar.
5. Two or more lines are said to be concurrent if they contain a common point; three or more planes are said to be concurrent if they contain a common point.
6. If two lines intersect, they have a unique intersection point. Lines that are coplanar and do not intersect are said to be parallel. Lines that are noncoplanar are said to be skew.
7. Sets consisting of the same points in space are coincident.
8. Planes that have no points in common are said to be parallel.
9. If two points lie in a plane, the line determined by them lies in the plane.
10. A line which does not lie in a plane has either one point or no point in common with the plane.
11. A plane is determined uniquely by any of the following:
    a. three noncollinear points

b. a line and a point not on the line
c. two intersecting lines
d. two parallel lines

12. The set of points common to two planes is either a line or the empty set.

Exercise 1.6.1

Fill in the blanks.

1. The mathematical idea of point is suggested by the physical idea of ______ or position.
2. The set of all points is called ______.
3. Space consists of a(n) ______ number of points.
4. Through any two ______ points in space there is exactly ______ line(s).
5. The lines through any one point form a(n) ______ set.
6. Two distinct lines in space have either ______ point(s) or ______ points in common.
7. Two lines that intersect have exactly one ______ point.
8. If two distinct points lie in a plane, then the ______ determined by them lies in the ______.
9. The set of planes passing through any one line in space is a(n) ______ set.
10. ______ points not on the same line determine a plane.
11. A line and a point not on the line determine a ______.
12. Two intersecting ______ determine a plane.
13. A line that does not lie wholly in a plane has either ______ point(s) or ______ points in common with the plane.
14. In space, the set of points common to two planes is either a ______ or the ______ set.
15. Sets of points lying in the same plane are said to be ______.
16. Two sets consisting of the same points are called ______.
17. Two or more lines containing a given point are said to be ______.
18. ______ or more planes are said to be concurrent if they contain a given ______.
19. Two planes that have no points in common are called ______ planes.
20. Parallel lines are ______ lines that are not ______ and do not intersect.
21. A line that is not in a plane and that does not intersect the plane is said to be ______ to the plane.
22. Two parallel lines determine a ______.
23. Two distinct lines that are not coplanar are called ______ lines.
24. Three points not on the same ______ determine a ______.
25. If a line lies in a ______, then every ______ of the line lies in the plane.

COORDINATE SYSTEMS: AN INTRODUCTION

2

2.1 PROLOGUE

By now you should realize that geometry can be viewed as a study of sets of points, lines, and planes. Actually, if we examine the historical development of geometry, we find that from the time of Euclid, the only approach was synthetic. That is, all geometric ideas were developed without the use of measurement. It was not until René Descartes introduced coordinate geometry in 1637 that mathematicians began to realize that a treatment of measurement could allow one a greater degree of flexibility and conceptual power in considering geometric ideas.

Descartes' work showed that the two branches of mathematics, algebra and geometry, could be fused into a single, comprehensive theory. Such a thought was unprecedented for, prior to Descartes, algebra and geometry were considered as two separate and unrelated branches of mathematics.

The viewpoint that this book takes is that geometry is the study of sets of points and algebra provides a convenient vehicle for describing the relations on these sets of points. In order to do this, we shall want to develop a method of identifying each point in space in a unique way. As mentioned in Chapter 1, we shall want to interpret "space" in different ways. Sometimes, rather than considering the universe as space, we shall want to confine our discussion to points in a single plane or even points on a single line. Thus, we will find it convenient to have one-, two-, and three-dimensional coordinate systems to correspond to sets of points on a line, in a plane, and in the universe, respectively. The remainder of this chapter will be devoted to a development of these three types of coordinate systems.

2.2 A ONE-DIMENSIONAL COORDINATE SYSTEM

If all the points we are considering are assumed to be subsets of a single line, then we only need a one-dimensional coordinate system to identify them. Such a system is frequently referred to as the *real number line*. Quite simply, we arbitrarily select a point on the line and allow the number zero to correspond to it. We shall call this point the *origin*. By selecting a suitable unit of measure, we envision congruent segments (that is, segments of equal length called *unit segments*) being marked off consecutively in either direction. The endpoints of these segments correspond to the integers, with $+1, +2, +3, \ldots$ corresponding to the points to the right of zero and $-1, -2, -3, \ldots$ corresponding to those on the left.

comment . . .

Actually, the decision of using the positive numbers to identify points to the right of zero and the negatives to identify those on the left is a matter of convention, probably related to our habit of reading from left to right.

Now, we can associate each integer with exactly one point on the line. Each point is called the *graph* of the integer associated with it and each integer is called the *coordinate* of the point with which it is associated. Thus, in Fig. 2.2.1, each of the points of the line is the graph of one of the members of the set $\{\ldots, -3, -2, -1, 0, +1, +2, +3, \ldots\}$ and each member of the set $\{\ldots, -3, -2, -1, 0, +1, +2, +3, \ldots\}$ is the coordinate of some point of the line.

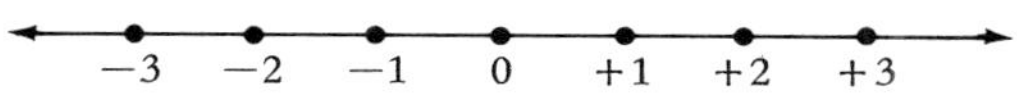

Figure 2.2.1

In a similar fashion, the rational number $\frac{1}{2}$ becomes the coordinate of the midpoint of the unit segment with endpoints whose coordinates are 0 and $+1$; $-\frac{3}{2}$ is the coordinate of the midpoint of the unit segment whose endpoints have coordinates -1 and -2, and so forth.

Likewise, all the rational numbers can be used as coordinates of points which divide suitable unit segments into a particular ratio.

An important question to consider at this point is the following: Once we have identified each rational number with a unique point of our line, are there any points of this line that are not yet associated with any number? If you have studied the real numbers in an earlier course, perhaps you recall that there are real numbers that are not rational. We are referring, of course, to the irrational numbers. Furthermore, the decimal representation of all irrational numbers is infinite and nonrepeating. Thus, the number $N = 1.54554555455554\ldots$ is a numeral representing an irrational number. We would like to know if there is a point on the number line which corresponds to this irrational number N. Perhaps we can reach an intuitive answer to the question if we consider the following sequence of inequalities. A careful examination of the decimal representation of N reveals that

$$\begin{array}{lcl}
1.5 & < N < & 1.6 \\
1.54 & < N < & 1.55 \\
1.545 & < N < & 1.546 \\
1.5455 & < N < & 1.5456 \\
1.54554 & < N < & 1.54555 \\
1.545545 & < N < & 1.545546 \\
\vdots & \vdots & \vdots \\
1.54554555455554 & < N < & 1.54554555455555 \\
\vdots & \vdots & \vdots
\end{array}$$

Observing that all the numbers on the extreme left and right of the above inequalities are rational numbers, we see that the irrational number N is being "squeezed" in between rational numbers which are getting closer and closer to one another. Now we know that each of these rational numbers corresponds to points on our line and the graphs of all the numbers which are on the left of these inequalities are getting closer and closer to all the graphs of the numbers on the right as Fig. 2.2.2 suggests.

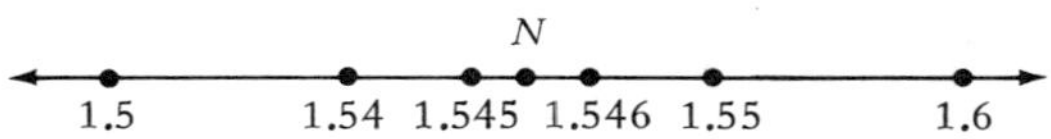

Figure 2.2.2

Thus, it seems that the graphs of these rational numbers are "squeezing in" on a point of the line which, although it is being located by this sequence of points with rational coordinates, it is not itself associated with any rational coordinate. We can call this point the graph of the number 1.54554555455554. . . . Now we can make the following statements about the graphs of N and its related sequence of rational numbers.

The graph of N is to the right of the graph of 1.5 and to the left of the graph of 1.6.

The graph of N is to the right of the graph of 1.54 and to the left of the graph of 1.55.

The graph of N is to the right of the graph of 1.545 and to the left of the graph of 1.546.

The graph of N is to the right of the graph of 1.5455 and to the left of the graph of 1.5456.

The graph of N is to the right of the graph of 1.54554 and to the left of the graph of 1.54555.

The graph of N is to the right of the graph of 1.545545 and to the left of the graph of 1.545546.

$$\vdots$$

The graph of N is to the right of the graph of 1.54554555455554 and to the left of the graph of 1.54554555455555.

$$\vdots$$

The important point here is that given any irrational number, we can always find two rational numbers whose graphs are on either side of the irrational number. Furthermore, these two rational numbers can be made to be as close to each other as we wish. This process allows us to locate the graph for an irrational number as accurately as we wish.

Thus, we are now able to identify each real number with some point of our line. The next question which naturally arises is: Are there any other points of the line which are not yet associated with a coordinate? The answer is no. However, the argument supporting this assertion is difficult to follow. Intuitively, it goes something like this: If P is any point on our line, then to the left and right of P there are points which have real coordinates which are as close to P as we wish. Thus, we can construct two sequences of reals which "squeeze in" on some real number. This real number must correspond to the point P and thus P has a real number coordinate.

Thus, we have established a one-to-one correspondence between the set of points on the line and the set of real numbers. Furthermore, it is asserted that under this correspondence, the following is true: Point P is to the left of point Q if and only if the coordinate of P is less than the coordinate of Q. This line, together with the real numbers associated with it, is called a *real number line.* This real number line will serve as our one-dimensional coordinate system. With respect to this coordinate system, each point P of the real number line is the graph of some real number r and this real number r is called the coordinate of the point P.

Example 1

Indicate on the real number line the graphs of the points A, B, C, and D whose coordinates are $A(+2)$, $B(-8/3)$, $C(.101001000100001\ldots)$, and $D(-\sqrt{3})$.

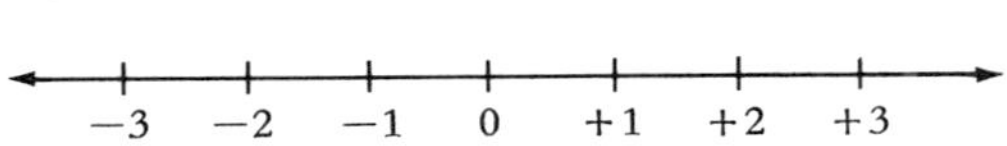

Figure 2.2.3

First, we sketch a real number line. (See Fig. 2.2.3.) Then we locate as accurately as possible the points which correspond to the four coordinates given above. We indicate each graph by a large dot. Thus, the four graphs are as shown in Fig. 2.2.4.

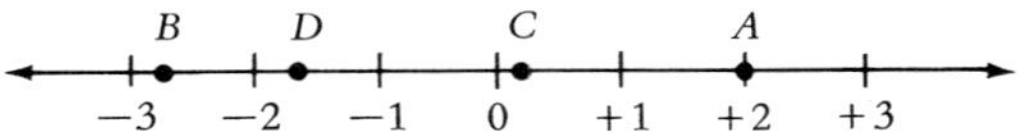

Figure 2.2.4

Example 2

Estimate rational numbers to approximate the coordinates of each of the graphs in Fig. 2.2.5. One set of rationals might be: $A(.5)$, $B(+3.2)$, $C(-2.3)$, $D(-3.9)$, and $E(+.8)$.

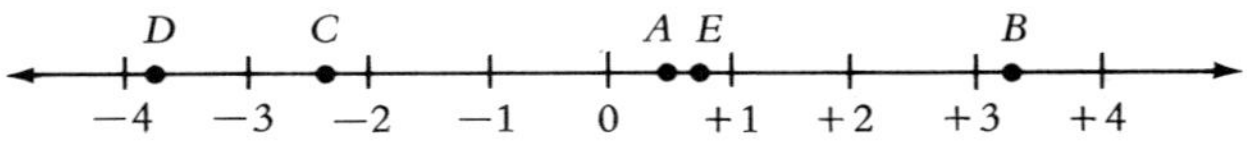

Figure 2.2.5

in our opinion . . .

Most of the current mathematics textbooks and curricula instill a negative attitude toward the use and concept of irrational numbers. As if their name isn't prejudicial enough, we are consistently exposed to subtle reminders that the irrationals ought to be avoided. For example, rarely, if ever, is a student asked to graph points with irrational coordinates. Seldom, if ever, is a student asked to solve simple linear equations with irrational coefficients. Most measurements are given as rational in verbal problems. Such obvious omissions of the use of irrationals ultimately lead the student to the following *incorrect* conclusions.

1. Irrationals cannot be graphed—or at least they are extremely more difficult to graph than are rationals.
2. There are relatively few irrationals when compared to rationals, by virtue of the fact that they are so rarely encountered in assignments and problems.

In response to conclusion 1, we suggest that graphing irrationals is no more difficult than graphing the rational .5. Just as we must estimate the position of the point corresponding to π, so must we approximate the location of the point whose coordinate is .5. However, the student believes that it is more difficult because irrationals have infinite decimal representations, and we are therefore forced to "round it off" and use a rational to approximate it. Although this is true, we hasten to suggest that in graphing such rationals as 49999/100000, .69723876, $-3.21\overline{5215}$, −1.00000000001, we find ourselves forced to approximate them as .5, .7, −3.2, and −1, respectively.

comment . . .

The numeral $-3.215\overline{215}$ refers to the infinite repeating decimal $-3.215215215215\ldots$ The bar is used to indicate the digits which repeat.

in our opinion . . .

Thus, we see that we also estimate most rationals by "simpler rationals" to graph them. We are doing nothing more when we graph the irrationals.

In response to conclusion 2, we propose that the curricula should make concentrated efforts to include more irrational numbers in the exercises given to students. We are building into the curriculum an obvious bias against the irrational number. We do this, yet we cannot understand why students find it difficult to learn and retain important ideas about the real number system.

Exercise 2.2.1

1. Estimate and label the location of the graphs of each of the following points whose coordinates are given below.
   a. $A(-4.8)$ b. $C(+\pi/2)$
   c. $E(-.687)$ d. $B(+5\ ^{21}/_{40})$
   e. $D(-2\sqrt{2})$ f. $F(-5.23223222322223\ldots)$
2. Use rational numbers to approximate the coordinates of each of the points whose graphs are indicated in Fig. 2.2.6.

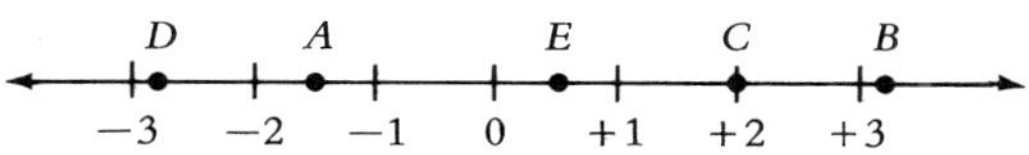

Figure 2.2.6

3. Use irrational numbers to approximate the coordinates of each of the points whose graphs are indicated in Problem 2.

2.3 GRAPHS OF SETS OF REAL NUMBERS

We are frequently interested in graphing sets of real numbers instead of individual reals. We often are able to do this by graphing representatives of the set and then inferring from this what the entire graph would look like.

Example 1

Determine the graph of the set $\{x \mid -3 < x \leqq \sqrt{2}\}$. Thus, we want to graph all the points that correspond to all the real numbers greater than -3 but less than or equal to positive $\sqrt{2}$.

We recognize that our graph is bounded by -3 on the left and $\sqrt{2}$ on the right. Furthermore, the left boundary is not included as part of the graph while the right boundary is. Thus, our graph looks like Fig. 2.3.1.

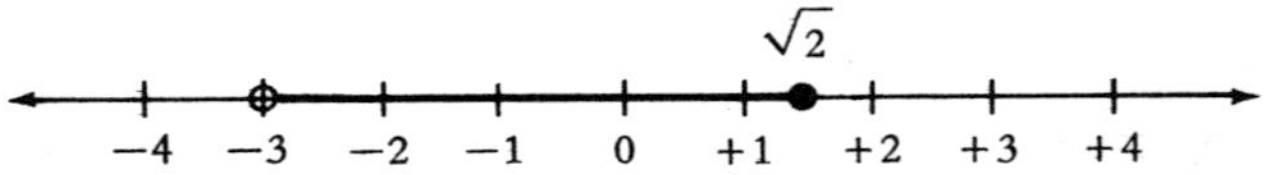

Figure 2.3.1

comment . . .

We will use the solid • to indicate that the point is included in the graph and the open circle ○ to indicate that the point is not included.

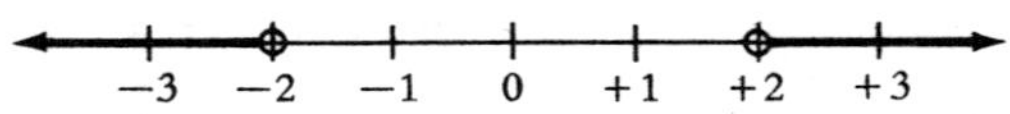

Figure 2.3.2

Example 2

Describe the set of real numbers that corresponds to the graph in Fig. 2.3.2. We could describe the set of coordinates corresponding to this set in several acceptable ways, such as:

1. The set of all reals less than -2 or greater than $+2$.
2. $\{x \in R | x < -2\} \cup \{x \in R | x > +2\}$, R stands for the set of real numbers.
3. $\{x \in R | x^2 > 4\}$.

It may not appear obvious to the reader that 2 and 3 are equivalent ways of describing the same set of reals. The reader should study both sets and see if he can convince himself that they do describe the same subset of real numbers.

Exercise 2.3.1

1. Determine the graphs of the following subsets of real numbers.
   a. $\{x \in R | x \geqq +4\}$
   b. $\{x \in R | -\pi \leqq x < +2.1\}$
   c. $\{x \in R | x = 2 \cdot n, n \in N\}$, N stands for the set of natural numbers.
   d. $\{x \in R | 2x + 3 = 3(x + 2)\}$
   e. $\{x \in R | 5x - 2 \leqq 2x + 8\}$
   f. $\{x \in R | x \in I\}$, I stands for the set of integers.
   g. $\{x \in R | \pi x + 3 \leqq -2\}$
   h. $\{x \in R | x^2 < +3\}$
   i. $\{x \in R | x^3 > +5\}$
   j. $\{x \in R | x^2 = +3\pi\}$
   k. $\{x \in R | x^2 = -4\}$
   l. $\{x \in R | x^2 > -4\}$
   m. $\{x \in R | x^2 < -4\}$
   n. $\{x \in R | x^2 \geqq .\bar{1}\}$
2. For each graph in Fig. 2.3.3, describe the set of coordinates indicated, using set-builder notation.

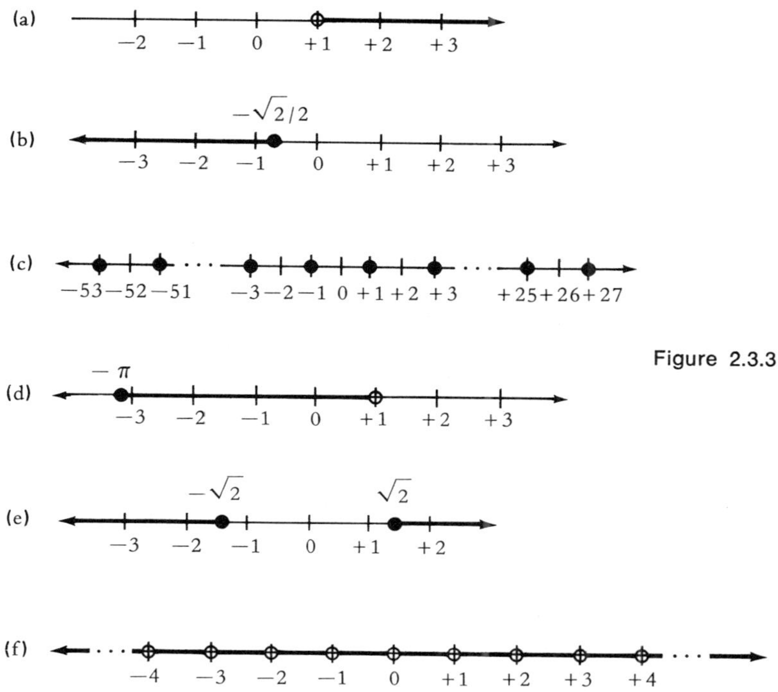

Figure 2.3.3

2.4 A TWO-DIMENSIONAL COORDINATE SYSTEM

Before we proceed much further, it seems appropriate to review some fundamental ideas concerning *ordered pairs*. If we have two sets A and B, then the *cross product* $A \times B$ is defined as follows.

2.4.1 Definition

If A and B are sets, then $A \times B$ is the set of all ordered pairs such that the first component is an element of A and the second component is an element of B. $A \times B = \{(x, y) | x \in A, y \in B\}$.

Example 1
If set $A = \{1, 2, 3\}$ and $B = \{2, 4\}$, then $A \times B = \{(1, 2), (2, 2), (3, 2), (1, 4), (2, 4), (3, 4)\}$.

2.4.2 Definition

A relation *from set A to B is any subset of $A \times B$.*

Thus, for Example 1, $\{(1, 2), (2, 4)\}$ is a relation from set A

to set B. Likewise, $\{(1, 2), (2, 2), (3, 2)\}$ is a relation from set A to set B.

2.4.3 Definition

If R is a relation from the set A to the set B, then by the domain *of R we mean the set of all elements which are first components of the ordered pairs of R and the* range *of R is the set of all elements which are the second components of the ordered pairs of R.*

Thus, in the previous example, the domain of the relation $\{(1, 2), (2, 4)\}$ is the set $\{1, 2\}$ and the range is the set $\{2, 4\}$. Likewise, the relation $\{(1, 2), (2, 2), (3, 2)\}$ has a domain of $\{1, 2, 3\}$ and a range of $\{2\}$.

2.4.4 Definition

A function from set A to set B is a relation R from A to B which satisfies the following two conditions.

1. *The domain of R is the set A.*
2. *Each member of the domain of R is paired with exactly one member of the range.*

Thus, in the previous example, the relation $\{(1, 2), (2, 4)\}$ is not a function from A to B since the first condition is not satisfied. Also, the relation $\{(1, 2), (2, 4), (3, 2), (2, 2)\}$ is not a function because condition 2 is not satisfied. On the other hand, $\{(1, 2), (2, 2), (3, 2)\}$ is a function from A to B.

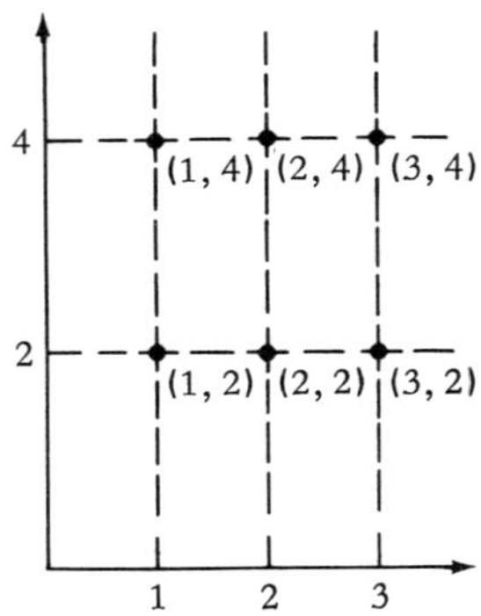

Figure 2.4.1

Sometimes it becomes difficult or even impossible to list the members of a cross product. Thus, frequently a *lattice* is used to assist us. For $A \times B$ of Example 1, the corresponding lattice would be as shown in Fig. 2.4.1. (*Note:* The dashed lines are only there to assist the reader in determining how to identify the points. Actually, the lattice looks like Fig. 2.4.2.)

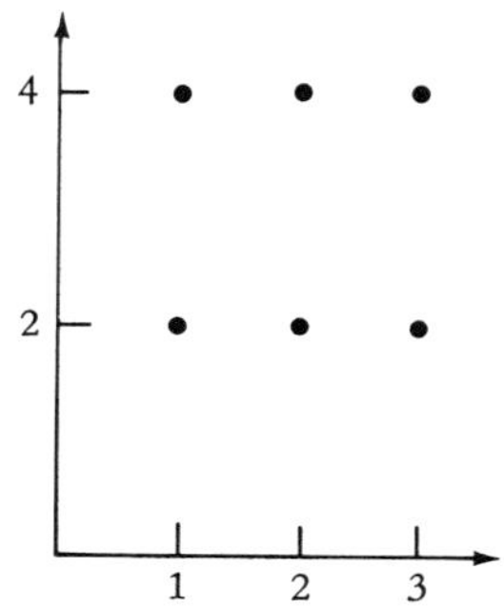

Figure 2.4.2

The vertical and horizontal lines are called *axes* and by convention we indicate the members of the set A (members of the first set) on the horizontal axis and the members of the set B (members of the second set) on the vertical one.

The set of points on a lattice of $A \times B$ which corresponds to the ordered pairs of a given relation from A to B is called the *graph* of the relation. For the lattice in Fig. 2.4.2, the graphs of the relations $\{(1, 2), (2, 4)\}$ and $\{(1, 2), (2, 2), (3, 2)\}$ are shown in Figs. 2.4.3 and 2.4.4.

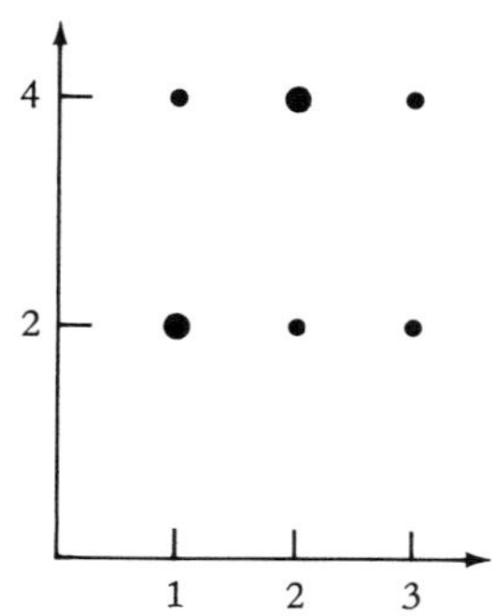

Figure 2.4.3
Graph of $\{(1, 2), (2, 4)\}$

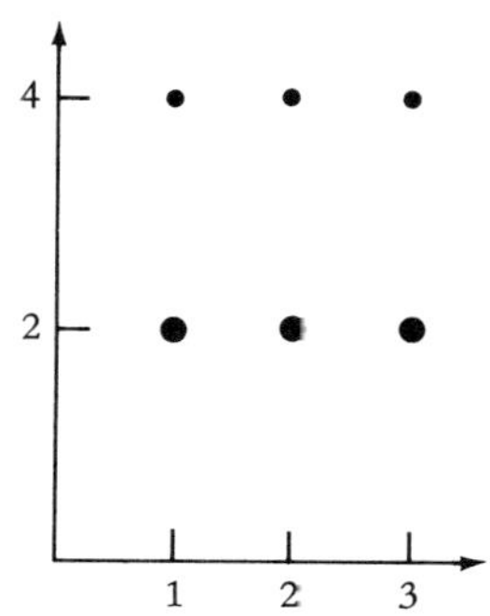

Figure 2.4.4
Graph of $\{(1, 2), (2, 2), (3, 2)\}$

Exercise 2.4.1

In each of the following problems, let $A = \{1, 3, 5, 7\}$ and $B = \{2, 4, 6, 8\}$.

1. List the members of $A \times B$.
2. Draw the lattice for $A \times B$.

3. Graph the following relations from set A to set B and determine if they are functions.
   a. $\{(1, 2), (3, 6), (7, 4), (3, 8)\}$
   b. $\{(1, 2), (3, 4), (5, 6), (7, 8)\}$
   c. $\{(3, 2), (5, 1), (7, 1)\}$
   d. $\{(7, 6), (5, 4), (3, 2), (5, 6), (1, 2)\}$
   e. The set of all ordered pairs from $A \times B$ where the first member is one more than the second member; that is, $\{(x, y) | x = y + 1\}$.
   f. The set of all ordered pairs from $A \times B$ where the sum of the first and second members is 9; that is, $\{(x, y) | x + y = 9\}$.
   g. $\{(x, y) | x = 3\}$
   h. $\{(x, y) | y = 4\}$
   i. From the examples above, how do you distinguish between the graphs of those relations that are not functions and those that are?

Now suppose we consider the cross product of $N \times N$ and its lattice, where N is the set of natural numbers. Since $N \times N$ is the set of all ordered pairs of natural numbers,

$$N \times N = \{(1, 1), (1, 2), (1, 3), \ldots, (2, 1), (2, 2), (2, 3), \ldots, (3, 1), (3, 2), (3, 3), \ldots\}$$

You may have noticed that it was difficult to list the members of $N \times N$. The partial lattice for $N \times N$ may be helpful in indicating this set. (See Fig. 2.4.5.)

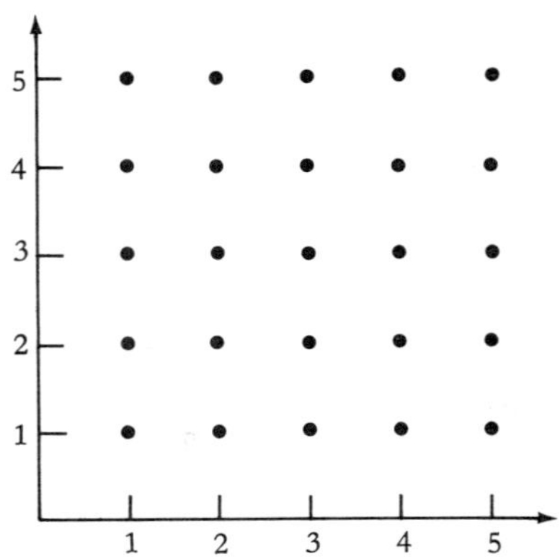

Figure 2.4.5

For $W \times W$ we could alter the lattice for $N \times N$ by placing 0 at the point of intersection. Here W is the set of whole numbers (Fig. 2.4.6).

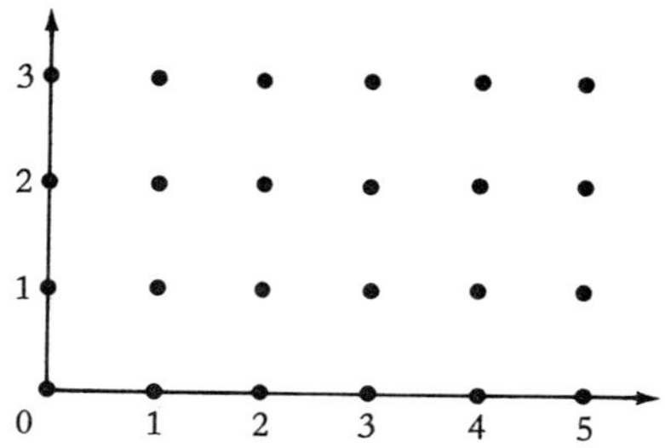

Figure 2.4.6

Notice that there are lattice points on the axes as well. The point of intersection of the two axes is called the *origin* and has (0, 0) as its ordered pair.

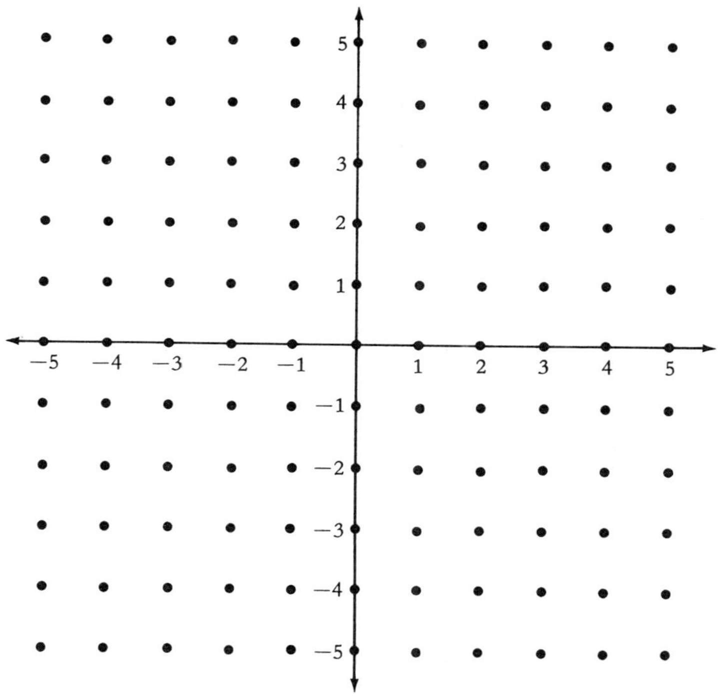

Figure 2.4.7

A further alteration of the lattice for $W \times W$ will produce $I \times I$ by extending each axis into its opposite direction to form lines. Here I is the set of integers. (See Fig. 2.4.7.)

Suppose we define $I_{1/2}$ to be the set of all integers and all halves of integers: $I_{1/2} = \{x|x = \frac{1}{2}\, y, y \in I\}$. Then $I_{1/2} = \{\ldots, -3, -\frac{5}{2}, -2, -\frac{3}{2}, -1, -\frac{1}{2}, 0, \frac{1}{2}, 1, \frac{3}{2}, 2, \frac{5}{2}, 3, \ldots\}$. The lattice for $I_{1/2} \times I_{1/2}$ would be as shown in Fig. 2.4.8.

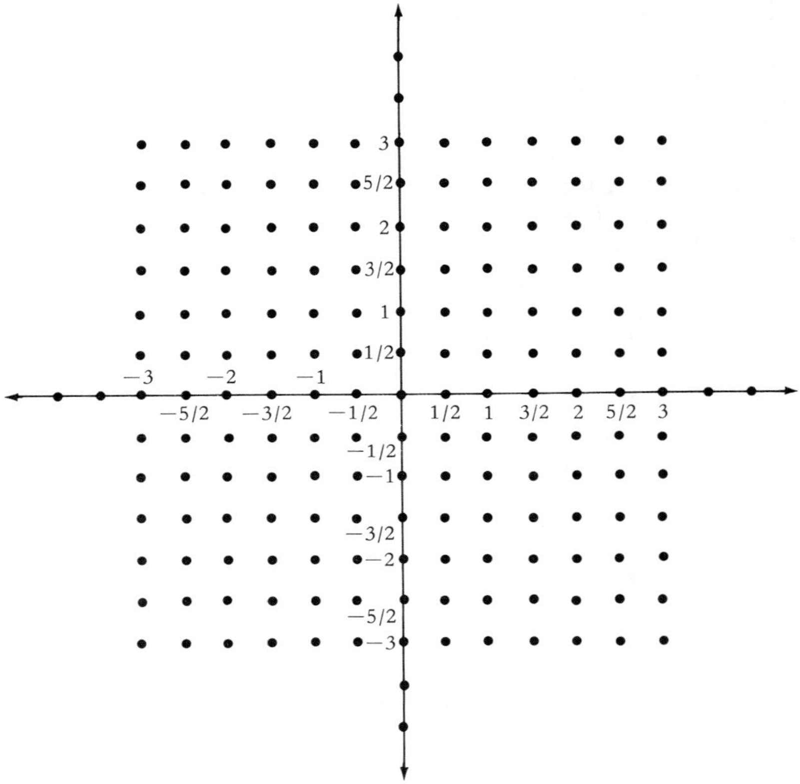

Figure 2.4.8

Can you imagine what $Q \times Q$ would look like where Q is the set of rationals? Perhaps Fig. 2.4.9 may be of help to you. Although the points would be densely distributed in the plane, we know that there are many gaps, ordered pairs where at least one member is irrational.

Now if we look at $R \times R$, where R is the set of real numbers, we find that finally every point of the plane is included, as in Fig. 2.4.10. That is, for every point in the plane there is an ordered pair of real numbers and for every ordered pair of real numbers there is a point in the plane. In other words, there is a one-to-one correspondence between the points in the plane and the ordered pairs of $R \times R$.

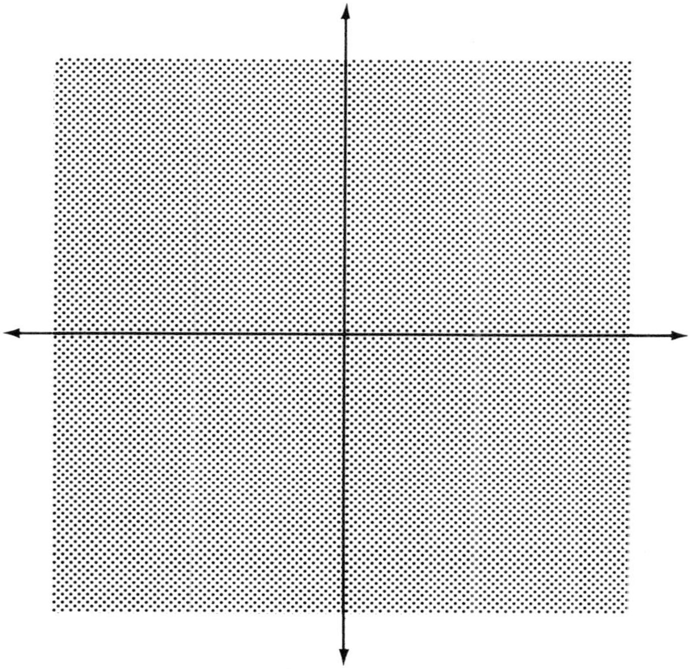

Figure 2.4.9

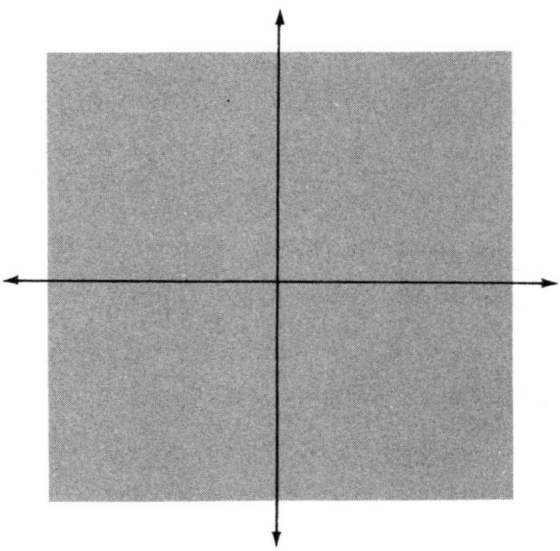

Figure 2.4.10

Example 2

Graph the members of the following relations from R to R.

$$\{(0, 5), (-1, 2), (2, -\tfrac{3}{4}), (-1.1, -3.2), (\sqrt{5}, \pi/2)\}$$

See Fig. 2.4.11.

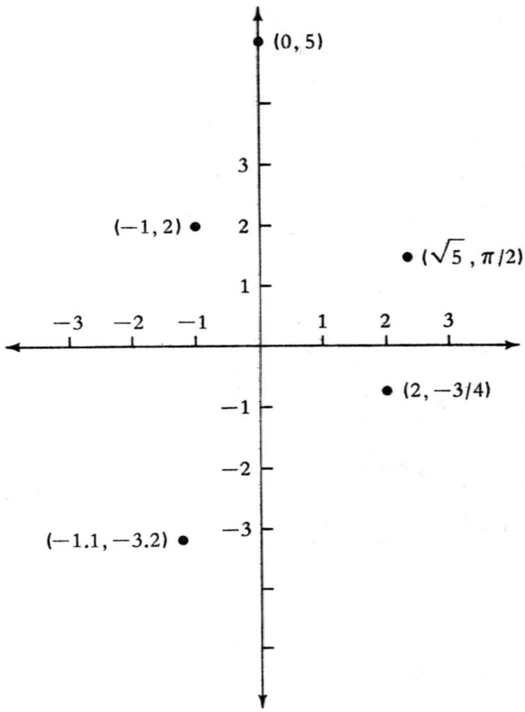

Figure 2.4.11

Example 3

Determine the graph of $\{(x, y)|x + y = 5, x \varepsilon I, y \varepsilon I\}$.

We seek the graph of the set of all ordered pairs of integers for which the sum of the first and second members is 5. This set is

$$\{\ldots, (-5, 10), (-4, 9), (-3, 8), (-2, 7), (-1, 6), (0, 5), (1, 4), (2, 3), (3, 2), (4, 1), (5, 0), (6, -1), (7, -2), (8, -3), (9, -4), (10, -5), \ldots\}$$

The graph for this set is Fig. 2.4.12 (not all points are labeled).

Example 4

Use the results of Example 3 to determine the graph of $\{(x, y)| x + y = 5, x \varepsilon R, y \varepsilon R\}$.

Now we wish to graph the set of ordered pairs of all real numbers for which the sum of the first and second members is 5. In addition to the integers listed in Example 3, we can obtain some of the following:

$$\{(-2\tfrac{1}{2}, 7\tfrac{1}{2}), (8\tfrac{2}{3}, -3\tfrac{2}{3}), (1\tfrac{1}{7}, 3\tfrac{6}{7}), (6 + \sqrt{2}, -1 - \sqrt{2}), (\pi, 5 - \pi)\}$$

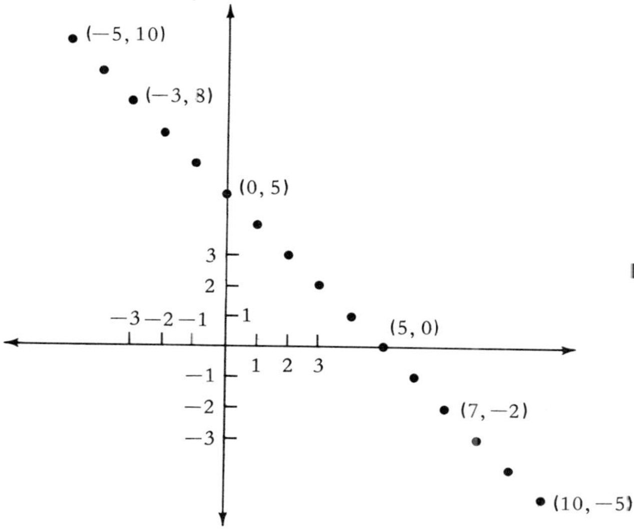

Figure 2.4.12

When we graph these along with some of the integers, our graph becomes as shown in Fig. 2.4.13.

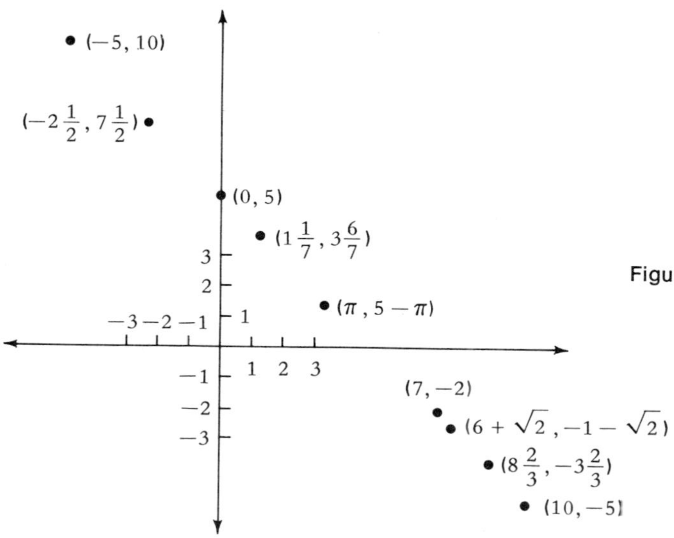

Figure 2.4.13

Now we have not yet graphed *all* the ordered pairs of reals in our relations. But from the representative graphs, so far it appears as if all the ordered pairs are collinear. Graphing more points appears to be filling in the gaps. Thus, our guess is that the graph will be a straight line (Fig. 2.4.14).

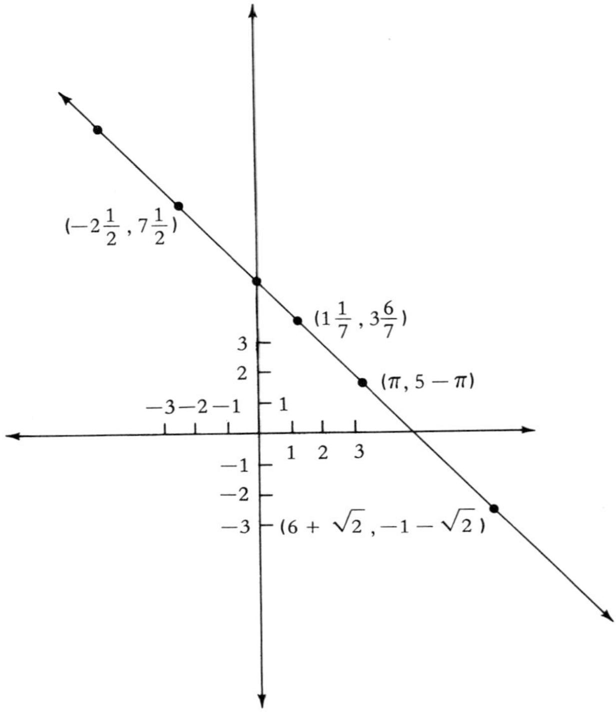

Figure 2.4.14

Exercise 2.4.2

1. Graph the members of the following relations from R to R. Label the graphs of each ordered pair.
   a. $\{(7, 1), (-3, 5), (5, 0), (-1, -7), (\frac{1}{2}, -\frac{3}{4}), (\pi, 1), (-\sqrt{3}, -\frac{1}{2}), (0, 6), (-2, -8), (.6, 3.2)\}$
   b. $\{(0, 0), (1, 1), (2, 2), (3, 3), (4, 4), (5, 5), (-1, -1), (-2, -2), (-3, -3), (-4, -4)\}$
   c. $\{(0, 0), (3, 1\frac{1}{2}), (-5, -2\frac{1}{2}), (-1, -\frac{1}{2}), (-4, -2), (2, 1), (-2, -1), (1, \frac{1}{2}), (-3, -1\frac{1}{2})\}$
   d. $\{(8, 0), (5, 3), (4, 4), (-2, 10), (0, 8), (2, 6), (7, 1), (6, 2), (3, 5), (2, 6), (1, 7), (-1, 9), (9, -1), (10, -2)\}$
   e. $\{(1, 3), (2, \frac{3}{2}), (3, 1), (4, \frac{3}{4}), (-1, -3), (-2, -\frac{3}{2}), (-3, -1), (-4, -\frac{3}{4})\}$
   f. $\{(0, 0), (1, 1), (2, 4), (3, 9), (4, 16), (-1, 1), (-2, 4), (-3, 9), (-4, 16)\}$
   g. $\{(5, 0), (4, 3), (3, 4), (2, \sqrt{21}), (1, \sqrt{24}), (0, 5), (-1, \sqrt{24}), (-2, \sqrt{21}), (-3, 4), (-4, 3), (-5, 0)\}$

h. $\{(-1, -2), (0, 1), (1, 4), (2, 7), (3, 10), (4, 13), (5, 16), (6, 19), (7, 22)\}$

2. a. List all the members of the relations which are described below.
   b. Graph each relation.
   c. Indicate what you think the graph of each relation would look like if you extended the domain and range to all real numbers which make the description meaningful.
   d. Which of these relations are functions from R to R under this extension?
      1. The set of all ordered pairs such that the sum of the first and second members is 5, where the domain is the set of all integers between -2 and 5, inclusive.
      2. The set of all ordered pairs such that the first member is one more than the second member and the domain is the natural numbers less than 10.
      3. The set of all ordered pairs such that the second member is the cube of the first member and the domain is the integers between -2 and 2, inclusive.
      4. The set of all ordered pairs such that the second member is equal to the first member if the first member is even, and the second member is the additive inverse of the first member if the first member is odd, and the domain is the set of whole numbers less than 10.
      5. The set of all ordered pairs such that the first member is 3 more than 2 times the second member and the range is the integers between -3 and 3, inclusive.
      6. The set of all ordered pairs such that the second member is less than or equal to the first and both the domain and range are the set of whole numbers less than 5.
3. Use complete sentences to thoroughly describe the relations in Exercise 1, parts b–h.
4. For each relation below:
   a. Verbally describe the relation indicated.
   b. List the members.
   c. Graph the relation.
   d. Indicate what you think the graph would look like if you extended the domain and range to all real numbers which make the description meaningful.
   e. Which of the relations are functions from R to R under this extension?
      1. $\{(x, y) | x - y = 4, x \varepsilon I, -3 \leqq x \leqq 2\}$
      2. $\{(x, y) | x = 3y, x \varepsilon W, x < 7\}$
      3. $\{(x, y) | y = x^2 + 1, x \varepsilon I, -2 \leqq x \leqq 2\}$
      4. $\{(x, y) | 2x + y = 5, y \varepsilon W, y < 10\}$
      5. $\{(x, y) | xy = 12, x \varepsilon I, -6 \leqq x \leqq 6, x \neq 0\}$

6. $\left\{\begin{array}{ll}(x, y) \mid x + y = 4, \text{ if } x \text{ is even} & \\ & x \in W, x \leqq 8 \\ \quad x - y = 4, \text{ if } x \text{ is odd} & \end{array}\right\}$

7. $\{(x, y) \mid x + y \geqq 5, x \in W, x < 5, y \in W, y < 4\}$

5. Use set-builder notation to describe the relations in Problems 1 and 2.

2.5 A THREE-DIMENSIONAL COORDINATE SYSTEM

In the last section we saw how we can identify each point of a given plane with a unique set of ordered pairs of real numbers. We note that the lattice for this cross product of $R \times R$ is called the *Cartesian plane.* If we now consider the problem of identifying each point in the universe, we must drop the restriction that all the points be in the same plane. Thus, we will extend our cross product $R \times R$ of ordered pairs to $R \times R \times R$ of ordered triples. Perhaps it might be useful to consider this space as a large book. In the two-dimensional case, we were looking at one page of the book. In the three-dimensional case, we must also identify the page as well. Thus, the third member of the triple identifies the "page" or plane in space and then the first and second members identify the point on that "page" or plane. For an additional intuitive model of a three-dimensional coordinate system, consider the following experiment.

Experiment 2

An enterprising real estate agent built an apartment complex which consisted of a series of three-story dwellings built on five parallel streets which ran east and west. There were seven buildings on each street. Within a building, all the rooms on a particular story comprised a single apartment. The agent realized that each apartment could be positively identified through a series of three numbers; the first designating the street on which the building was located, the second the building on that street, and the third the floor level. Thus, an apartment with code number (2, 5, 3) corresponded to the apartment which was on the third story of the fifth building on the second street.

1. Draw a sketch of the complex.
2. Where is apartment (1, 5, 2) located?
3. What would be true about the code numbers for all the apartments on street number 5?
4. What would be true about the code numbers of all second-story apartments?

5. What would be true about the code numbers for all the apartments which are in the seventh building on the fourth street?
6. Suppose the real estate agent designated for each street a street supervisor who was in charge of all the apartments on a particular street. How many numbers of the code would each street supervisor have to pay attention to?
7. Suppose, further, that each street supervisor hired a building supervisor for each building on his street. How many numbers in the apartment code would each building supervisor have to worry about?
8. How many apartments are there in the complex?
9. How many apartments is each street supervisor responsible for? Each building supervisor?
10. What is the code number of the apartment which is farthest from the apartment (1, 1, 1)?
11. The agent decided to give air conditioning to all the apartments for which the first two numbers of the code added up to 8. List the apartments which received air conditioning.
12. The agent also decided to raise the rent of all the middle level apartments for which the first and second numbers of the code were equal. List these apartments.

Now, let us formally extend cross product to three sets.

2.5.1 Definition

If A, B, and C are sets, then $A \times B \times C$ is the set of all ordered triples *where the first member is an element of A, the second member is an element of B, and the third is an element of C; that is,* $A \times B \times C = \{(x, y, z) | x \varepsilon A, y \varepsilon B, z \varepsilon C\}$.

Example 1
Let $A = \{1, 2, 3\}$, $B = \{2, 6\}$, $C = \{3, 4, 5\}$. Then

$$A \times B \times C = \left\{ \begin{matrix} (1, 2, 3), (1, 2, 4), (1, 2, 5) \\ (1, 6, 3), (1, 6, 4), (1, 6, 5) \\ (2, 2, 3), (2, 2, 4), (2, 2, 5) \\ (2, 6, 3), (2, 6, 4), (2, 6, 5) \\ (3, 2, 3), (3, 2, 4), (3, 2, 5) \\ (3, 6, 3), (3, 6, 4), (3, 6, 5) \end{matrix} \right\}$$

As before, we can construct a lattice for this set of ordered triples, but now we need three axes and the lattice must be in three dimensions. Since C contains three members, we can think of three duplicates of the lattice for $A \times B$ arranged in layers, one above the other. The bottom layer can correspond to identifying the third member 3 with each member of $A \times B$, the second layer with the number 4, and the third layer with the number 5. The members of set C can be indicated on a third axis drawn perpendicularly to the planes determined by the axes for $A \times B$.

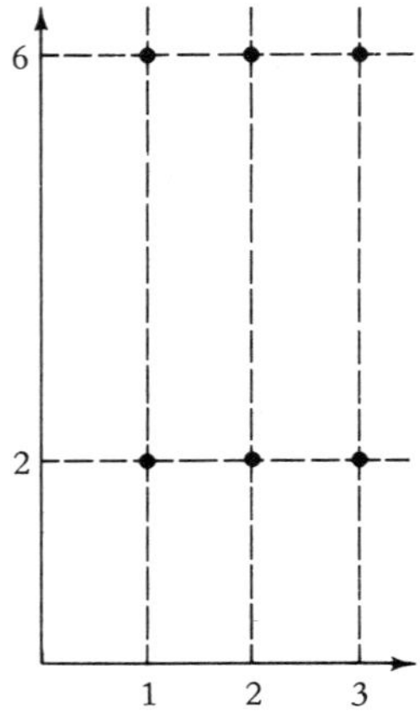

Figure 2.5.1

In Fig. 2.5.1, you will see the lattice for $A \times B$. Then, since we want to be able to see three duplicates of this lattice arranged in layers one above the other, we shall alter the angle between the axes to look like perpendicular lines viewed from an angle rather than from directly above. (See Fig. 2.5.2.)

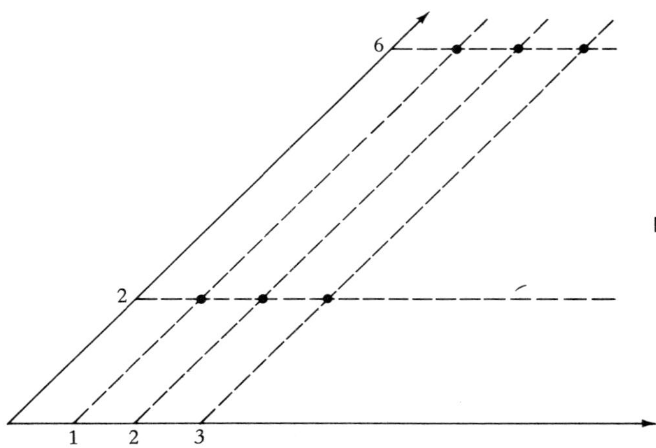

Figure 2.5.2

In Fig. 2.5.3 you will see that the third axis has been drawn perpendicularly to the first two axes. The numbers corresponding to set C are indicated on this axis.

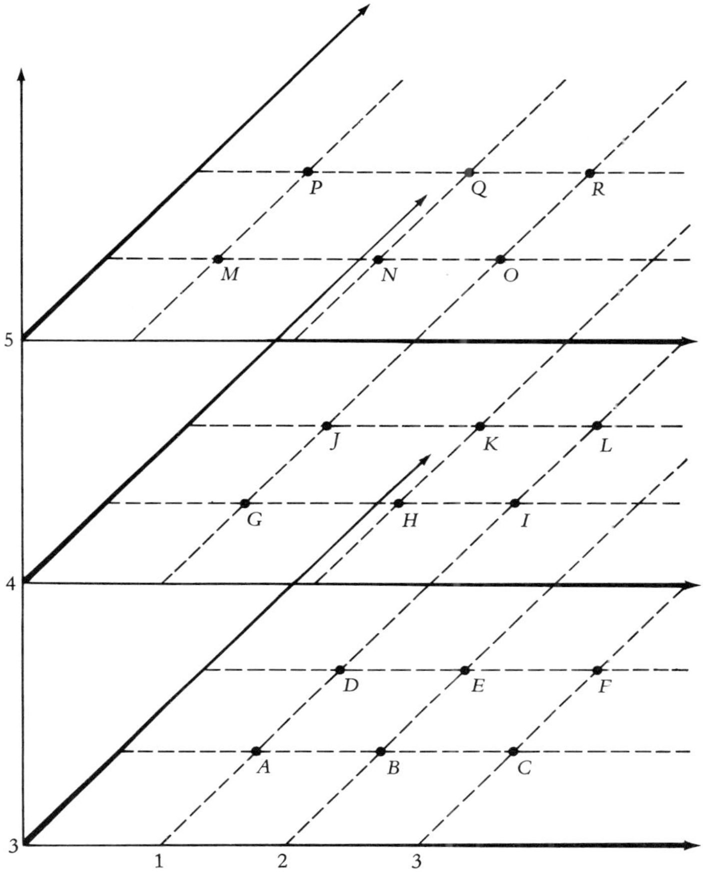

Figure 2.5.3

If you look closely at the set of ordered triples listed for $A \times B \times C$, you will notice that the first two members correspond to the ordered pairs of $A \times B$. Each of these is then ordered with each element of C.

Exercise 2.5.1

Refer to Fig. 2.5.3 in answering each of the following questions.

1. Give the letter of the point of the lattice which corresponds to each ordered triple below:
   a. (2, 6, 3) b. (3, 2, 4)
   c. (1, 6, 4) d. (2, 6, 5)
   e. (1, 2, 3)
2. Give the ordered triple which corresponds to each of the following letters:

a. *F* b. *O*
c. *G* d. *R*
e. *B* f. *N*

3. Frequently, the corner of a room makes a good model for a three-dimensional coordinate system. Sometimes you can use the tiles in the floor or the concrete blocks in the wall as your units. Settle on one vertex of your room as the origin and establish a unit on each axis. Then estimate the coordinates of several objects in your room.

In general, a three-dimensional coordinate system has three axes intersecting at a single point called the *origin* with each axis perpendicular to the plane of the other two.

Earlier in the chapter, we referred to a subset of a cross product as being a relation. So, too, will we call any subset of ordered triples a relation, and to distinguish between them we

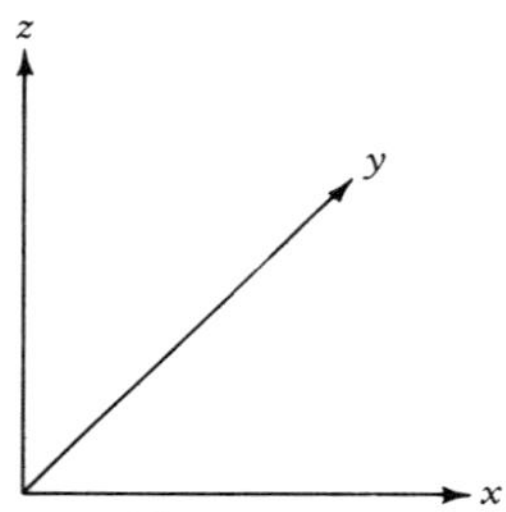

Figure 2.5.4

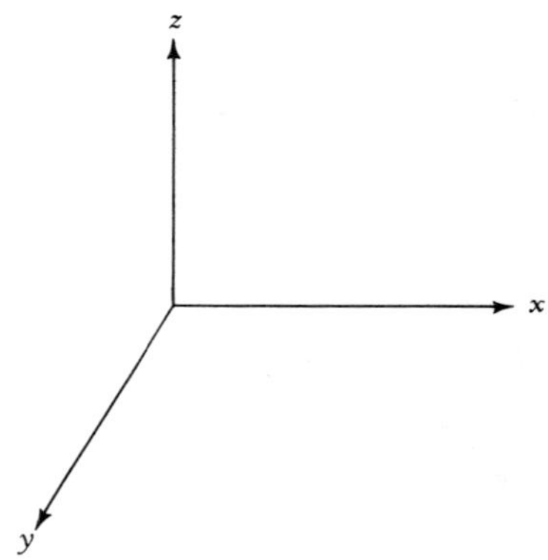

Figure 2.5.5

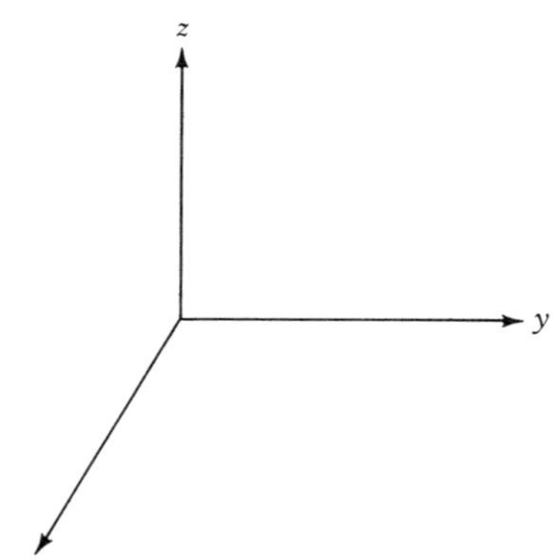

Figure 2.5.6

shall call the subsets of ordered pairs *binary* relations and the subsets of ordered triples *ternary* relations. Likewise, the set of points of the lattice which corresponds to the ternary relation is called the *graph* of the relation. (We can also call subsets of $R \times R \times R$ ternary relations.)

If we now consider the three-dimensional lattice for $R \times R \times R$, we have three axes each of which corresponds to a real number line. These axes are, by tradition, called the x-, y-, and z-axes. Various texts choose to draw and label these axes in different ways. Some of the more conventional selections are given in Figs. 2.5.4 to 2.5.6. In this text, we shall use the arrangement in Fig. 2.5.6.

We shall now proceed to examine the graphs of certain ternary relations of $R \times R \times R$.

Example 2

Consider the graph of $\{(x, y, z)|x = 3\}$. First, let us consider the two-dimensional counterpart $\{(x, y)|x = 3\}$. This is a line parallel to the y-axis (Fig. 2.5.7).

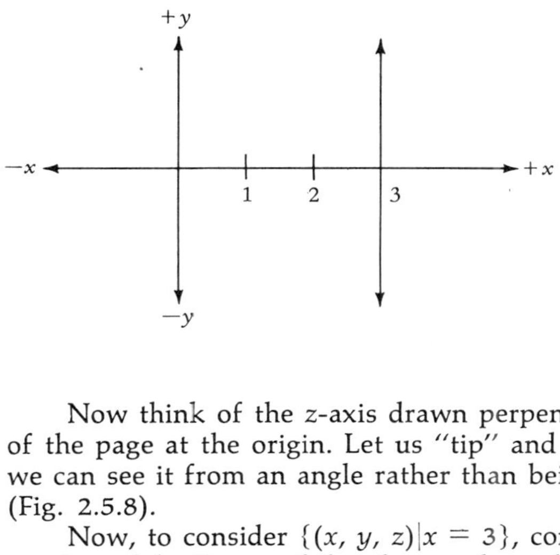

Figure 2.5.7
Graph of $\{(x, y)|x = 3\}$

Now think of the z-axis drawn perpendicular to the plane of the page at the origin. Let us "tip" and rotate the plane so we can see it from an angle rather than being directly above it (Fig. 2.5.8).

Now, to consider $\{(x, y, z)|x = 3\}$, consider an infinite number of duplicates of the plane and graph of Fig. 2.5.8, arranged in layers. The graph of $\{(x, y, z)|x = 3\}$ becomes "layers of lines," thus producing a plane as in Fig. 2.5.9.

Exercise 2.5.2

1. Use your graphs to some of the problems of Exercise 2.4.2 to help you decide what the following graphs look like over the

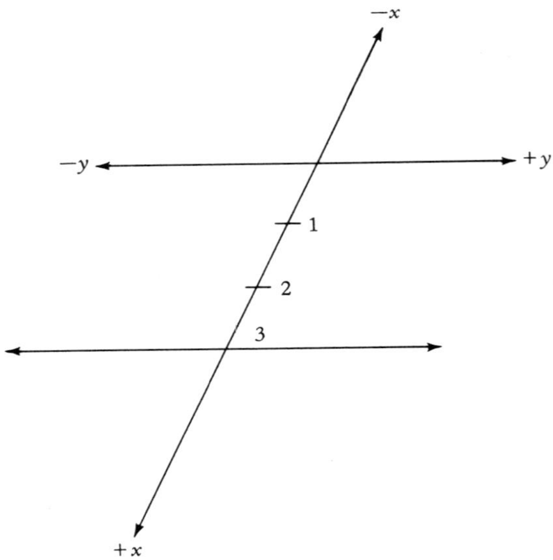

Figure 2.5.8

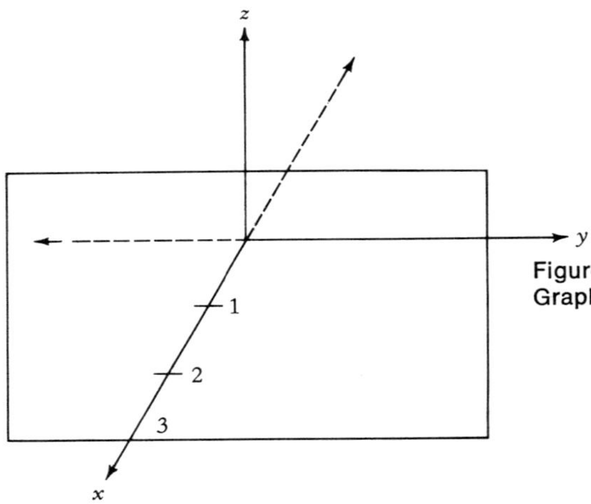

Figure 2.5.9
Graph of $\{(x, y, z)|x = 3\}$

set of reals. Either draw a picture or describe your result. The numbers in parentheses refer to previous graphs from Exercise 2.4.2 which may help you.

a. $\{(x, y, z)|x^2 + y^2 = 25, 0 \leqq z \leqq 3\}$ (1-g)
b. $\{(x, y, z)|x - y = 4, -3 \leqq x \leqq 2, -4 \leqq z \leqq 4\}$ (4-e-1)
c. $\{(x, y, z)|x = 3y\}$ (4-e-2)
d. $\{(x, y, z)|x + y \geqq 5, x < 5, y < 4, 1 \leqq z \leqq 4\}$ (4-e-7)

2.6 SOME FINAL OBSERVATIONS

We have seen that we can properly identify each point in space with a unique ordered triple by means of some fixed set of coordinate axes. Also, we have seen that the coordinate plane becomes a special case of this three-dimensional case where all the points are assumed to be in the same plane, and we use ordered pairs to identify these points. We also considered an even more specialized case by not only assuming the points to be in the same plane but also on the same line in that plane. In this case, we needed only to use a singleton of real numbers and the lattice was the familiar real number line. If we now consider the graphs of the following sets: $\{x|x = 3\}$, $\{(x, y)|x = 3\}$, and $\{(x, y, z)|x = 3\}$ you should become aware that they are quite different from one another.

Thus, if we encounter the equation $x = 3$ it does not by itself determine a graph. It depends on other restrictions that are placed on it. Are we to consider all points on a given line in a given plane, all points on a given plane, or all points in the universe? In the first case, $x = 3$ determines a point, in the second a line, and in the third a plane. Strictly speaking, then, it is not really accurate to talk about the graph of an equation (or inequality). What we really mean is graph the set of all reals or ordered pairs of reals or ordered triples of reals which satisfy the given condition. Thus, we should not refer to graphs of equations or inequalities unless the dimension is apparent from the context.

2.7 SUMMARY

1. The one-dimensional coordinate system displays a one-to-one correspondence between all the points of a line and all the real numbers.
2. The two-dimensional coordinate system displays a one-to-one correspondence between all the points in a plane and the set of all ordered pairs of real numbers.
3. The three-dimensional coordinate system displays a one-to-one correspondence between all the points in space and the set of all ordered triples of real numbers.
4. A subset of R is called a *unary* relation on R, a subset of $R \times R$ is called a *binary* relation on $R \times R$, and a subset of $R \times R \times R$ is called a *ternary* relation on $R \times R \times R$.
5. A relation $\mathbf{T}$ from R to R is a function from R to R if and only if (a) the domain of $\mathbf{T} = R$ and (b) for each element of the domain of $\mathbf{T}$ there is exactly one member of the range of $\mathbf{T}$ paired with it.

6. A relation from set A to set B is not a function if there is a vertical line which intersects its graph at more than one point.

Exercise 2.7.1

Fill in the blanks.

1. Each part below refers to the one-dimensional coordinate system.
   a. Each point in the coordinate system is called the ______ of the number with which it is matched, and the number is the ______ of the point.
   b. The graph of *0* is called the ______.
   c. The segment whose endpoints have coordinates 0 and 1 is called a ______ segment.
   d. The set of points corresponding to a set of numbers is called the ______ of the set.
2. Each part below refers to a two-dimensional coordinate system.
   a. By means of a rectangular (Cartesian) coordinate system it is possible to set up a ______ correspondence between the set of all ______ pairs of real numbers and the set of all ______ in a plane.
   b. The components of the ordered pair (x, y) are called ______.
   c. The x-component of a point tells its distance from the ______ axis; its distance from the x-axis is given by the ______ coordinate.
   d. The coordinates of the origin are ______.
   e. A relation is a set of ordered ______ of numbers.
   f. The set of all first components of a relation is called the ______ of the relation and the set of all second components is the ______.
   g. If each element of the domain of a relation is matched with only one element in its ______ the relation is a ______.
3. Each part refers to a three-dimensional coordinate system.
   a. By means of a rectangular (Cartesian) coordinate system it is possible to set up a ______ correspondence between the set of all ordered ______ of real numbers and the set of all points in ______.
   b. The components of the ordered triple (x, y, z) are called ______.
   c. The x-coordinate of a point tells its distance from the ______ plane; the distance from the x-y plane is given by the ______ coordinate.
   d. The coordinates of the origin are ______.
   e. In a three-dimensional coordinate system, the set of all points on the x-axis can be described as all ordered triples of the form ______. The y-axis is the set of all ordered triples of the form ______. The z-axis is the set of all ordered triples of the form ______.

GEOMETRY OF POSITION

3

3.1 PROLOGUE

Not all questions of geometry are dependent upon measurement. Certain very important geometrical properties are solely functions of the general position of the lines and points of a figure. In this chapter and the next, we investigate some of these properties such as: betweenness, separation, interiorness, exteriorness, and convexity. These concepts, while relatively intuitive, especially when we have a picture, a drawing, or some physical object to which we can point, cannot be considered at all obvious. It is perhaps the mistaken notion that these concepts were trivial that accounts for the fact that Euclid in his *Elements* failed to consider them, making many of his arguments invalid. Our problem will be to describe these ideas in a general way, building on our intuition drawn from special situations.

These ideas of position play an important role in a branch of mathematics known as Topology. Topology owes its origins both to geometry and analysis. It is a branch of mathematics that is more qualitative than quantitative, describing spatial features in many new ways. The following describes what might be called the first truly topological problem.

In the eighteenth century in the town of Königsberg, there were seven bridges over the river Pregel. Four of the bridges led from opposite banks to an island called Kneiphof. One bridge connected Kneiphof with another island and two bridges joined this island with the mainland. See Fig. 3.1.1. Strollers enjoyed taking walks over these bridges and a question which occurred to them was: "How can one plan a walk so as to cross each of the seven bridges without recrossing any of them?" Repeated trials led to the belief that this was impossible but it remained for the great Swiss mathematician Leonard Euler (1707–1783) to prove this in 1736. As often is the case in mathematics, the significance of Euler's methods far outweighed the problem that had motivated them.

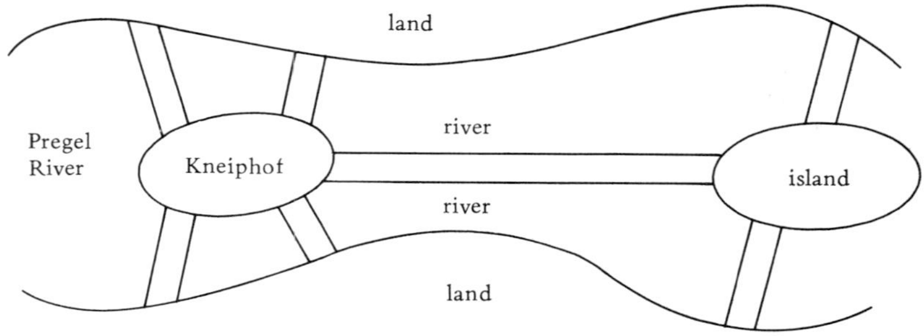

Figure 3.1.1

Euler reduced the problem by replacing the land by points and the bridges by arcs connecting these points. Once this simplification had been effected, the problem was reduced to answering the question: "Can Fig. 3.1.2 be drawn with one continuous motion of the pencil, without lifting it from the paper?" This is the equivalent of physically traversing the seven bridges in one journey.

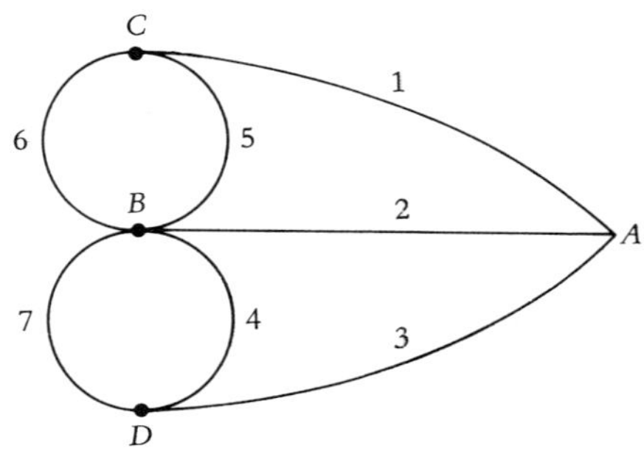

Figure 3.1.2

In Chapter 1 we observed that point and line can be interpreted in many different ways. In particular, a point can represent a land mass and a line can represent a bridge. In Fig. 3.1.2, point B represents the island Kneiphof, A the other island, and C and D the two mainland areas. The numbered lines represent the seven bridges. The points A, B, C, and D will be called *vertices*. Note that at vertex A, three lines terminate; at vertex B, there are five; at vertex C and D, there are three lines each.

The picture in Fig. 3.1.2 is called a *network* and the problem of the Königsberg bridges is to traverse this network. A network is *traveled* (or traversed, or traced) by passing through all the lines exactly once. A network is simply a figure consisting of a finite number of points called vertices and a number of lines (or arcs). The vertices are the endpoints of the arcs, and no two arcs have a common point, except perhaps a common vertex. A vertex is called *odd* or *even*, according to whether the number of arcs forming the vertex is odd or even.

Euler's investigation of networks led to the conclusion that the Königsberg bridges cannot be traversed. He also discovered how one can determine if a particular network can be traversed by merely considering the nature of the vertices and the arcs that terminate at them. He also discovered a technique for determining how many distinct journeys are necessary to traverse any given network.

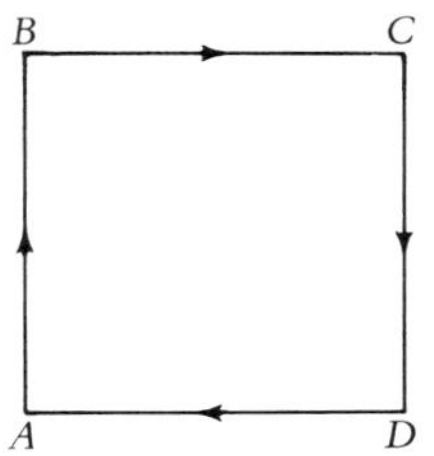

Figure 3.1.3

Consider the network in Fig. 3.1.3 which can be traversed, starting at A and following the arrows until we return to A. Notice that we have four even vertices and no odd vertices. On the other hand, the network in Fig. 3.1.4 cannot be traversed where the starting and ending point are the same. This can be determined by trial and error and the observation that the various paths we can attempt starting at A or C are of the same nature and this is also the case of the vertices B and D. Here we see that we have two even vertices, namely, A and C; and two odd vertices, namely B and D. Although it is not possible to traverse this network and return to the same point, it is possible to make a journey which goes through every vertex and travels each arc but once. Such a trip would start at B, move to C, to D, to A, back to B, and then end at D; symbolically, $BCDABD$.

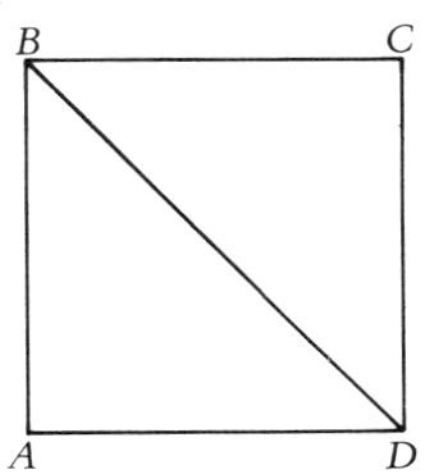

Figure 3.1.4

As an example of a network that cannot be traversed in either of the above ways, we have the schematic Königsberg bridge network which has four odd vertices and no even vertices.

Experiment 1

Figure 3.1.5 consists of numerous networks and a table to be completed. Attempt to discover if the networks can be traversed in a single journey beginning and ending at the same point. If not, how many journeys would be required to traverse the network? Can you draw with a single continuous stroke of your pencil those networks which can be traversed by a single journey?

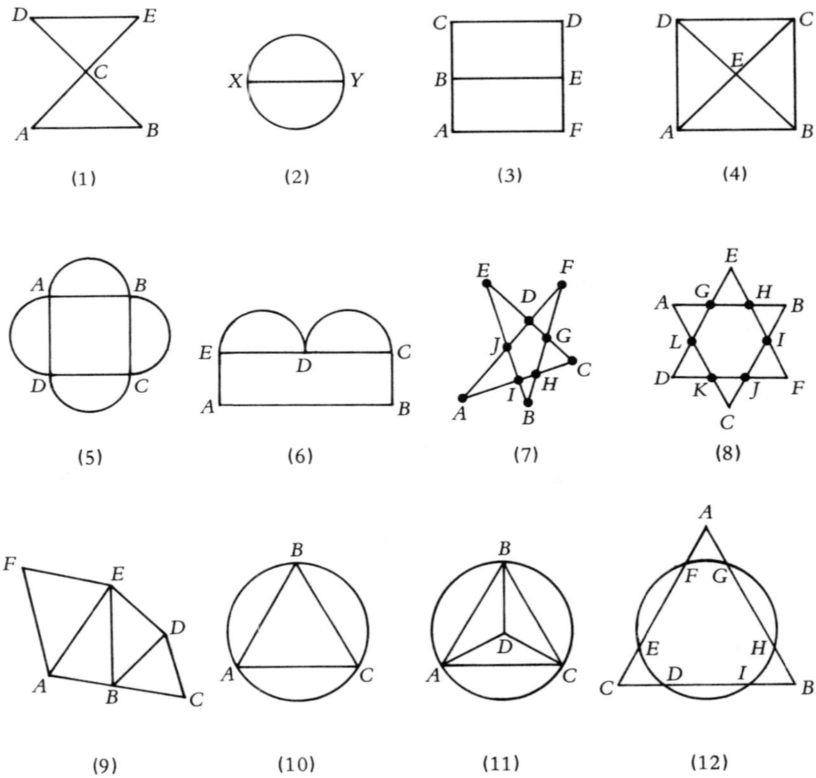

Figure 3.1.5

Does it matter where you start? Fill in the table. Can you make any observations or predictions about the networks from the data concerning the vertices?

Network	*Even vertices*	*Odd vertices*	*Network*	*Even vertices*	*Odd vertices*
1			7		
2			8		
3			9		
4			10		
5			11		
6			12		

comment . . .

Students may be asked to construct networks that can be traversed in a single journey. These networks can then be exchanged among the students.

Exercise 3.1.1

1. *Water, Gas, and Electricity*
   We wish to lay water, gas, and electricity connections to three houses, *A*, *B*, and *C*, without any pipes crossing one another. Either find a solution or offer an explanation why it cannot be done. (See Fig. 3.1.6.)

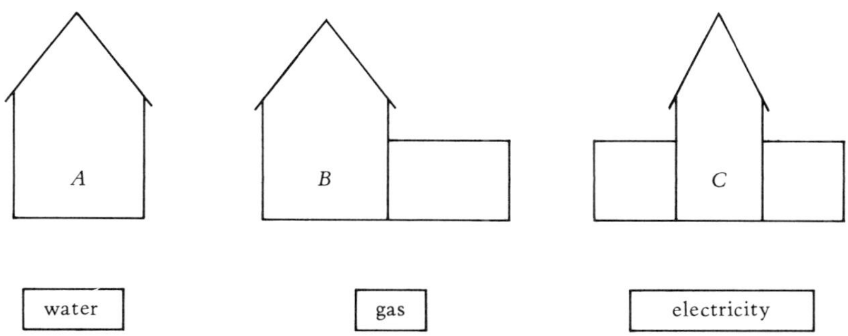

Figure 3.1.6

2. The two graphs of Fig. 3.1.7 can be traversed. In each case, starting with vertex *A*, list the order of the vertices that allows the figures to be traversed.
3. Each of the graphs in Fig. 3.1.8 shows the floor plan of a house. Can you take a trip through every door of this house without passing through any door more than once?

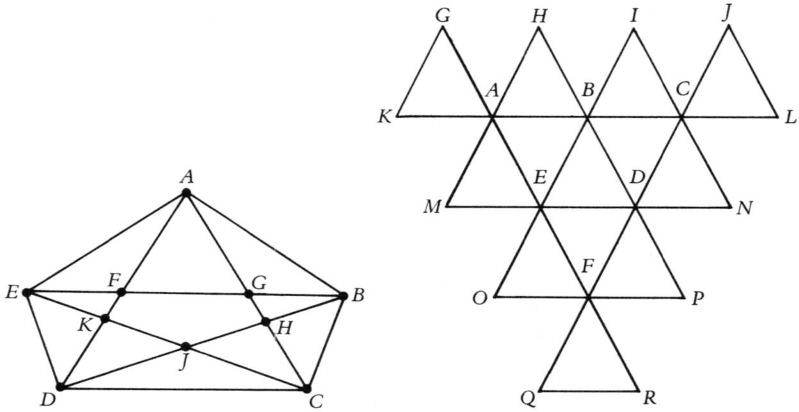

Figure 3.1.7

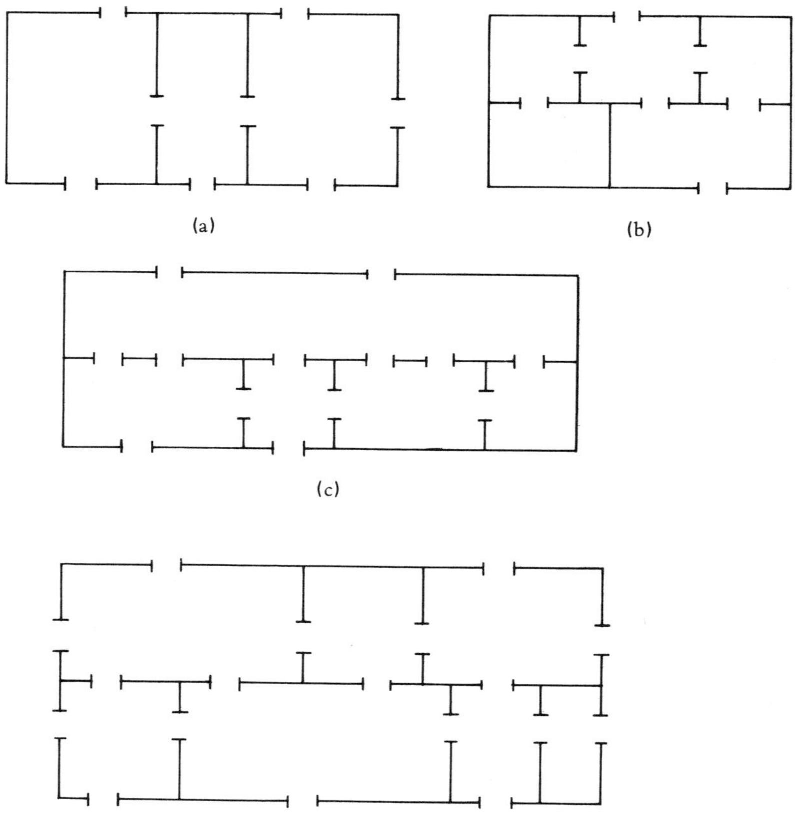

Figure 3.1.8

3.2 SEPARATION IN SPACE

Many geometric figures of special interest and importance have boundaries or frontiers that are special examples of networks. Later, when we investigate the measurement of these figures, we will note the prominent role played by these boundaries. They will separate space into numerous regions, some of which we shall desire to measure. Before reaching this point, we must investigate the notion of separation itself.

comment . . .

Many of the geometric figures now to be described were encountered in the previous chapter as graphs of certain sets of real numbers. There our interest was to find some convenient way to visualize these sets and in doing so perhaps be able to observe some important properties of these sets. In this chapter, we wish to describe these geometric figures and indicate their nature by describing the properties we will demand that they have.

Starting with the simplest kind of separation, let l be any line and P some point of the line (Fig. 3.2.1). The point P separates the line l into three disjoint subsets of l. The first subset consists of the point P alone. The other two subsets are infinite subsets of l called *half-lines* of l with *boundary* P. We can make the following observations.

Figure 3.2.1

1. Each point of l with the exception of P belongs to exactly one half-line.
2. The point P does not belong to either half-line of l.

comment . . .

If we coordinatize the line l and the coordinate of P is 3, how could you describe the coordinates of all the points in one of the half-lines? What about the other half-line?

3.2.1 Postulate

Any point of any line in space separates the line into two half-lines and their common boundary, the point.

In Fig. 3.2.1, there appears to be two sides of the line, the side to the left of the point P and the side to the right of P.

Unfortunately, the concepts of left and right are not geometric concepts. Still, we wish to discuss the idea of two sides of the line l, each of the sides being a half-line.

We will say that points Q and R are on the *same side* of P if Q and R belong to the same half-line. Points Q and S are said to be on *oppcsite sides* of point P if they do not belong to the same half-line (Fig. 3.2.2).

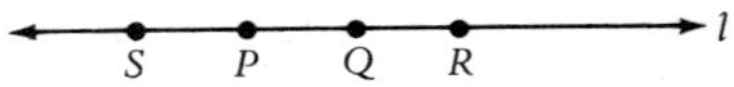

Figure 3.2.2

comment . . .

If we coordinatize the line l, we could also describe what it means to say Q and R are on the same side of P, in terms of the coordinates of P, Q, and R. What would our description be?

There is a special name for the union of a half-line and its boundary; namely, a *ray*. Another description is that the ray $\overrightarrow{PQ}$ (read ray P, Q) is the half-line with endpoint P, and Q any other point of the half-line. Note in symbolizing a ray $\overrightarrow{PQ}$ (Fig. 3.2.2), the first letter, P, names the endpoint of the ray, the second letter any point of the half-line. Thus, in Fig. 3.2.2 the ray $\overrightarrow{PQ}$ is the same as the ray $\overrightarrow{PR}$. We would say that $\overrightarrow{PQ}$ and $\overrightarrow{PR}$ are coincident. The ray $\overrightarrow{PS}$ is a ray opposite to $\overrightarrow{PQ}$. In general, *opposite rays* are rays which are subsets of the same line having the same endpoint, but which do not share any other points.

comment . . .

Rays or half-lines are parallel if they are subsets of parallel lines. Rays or half-lines are said to be skew if they are subsets of skew lines.

Figure 3.2.3 pictures a line l with points P, Q, and R, elements of the line. Note that Q and R of l are on opposite half-lines

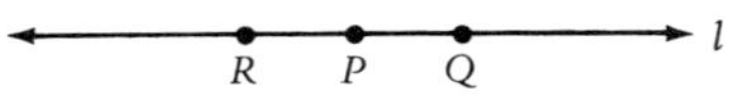

Figure 3.2.3

with a boundary of P. In this situation, we say that P is *between* Q and R (equivalently, P is between R and Q).

comment . . .

If we coordinatize l, what could you say about the coordinates of P, Q, and R if P is between Q and R?

3.2.2 Definition

In general, a point P *is* between *points* Q *and* R *if and only if (a)* P, Q, *and* R *are collinear and (b)* Q *and* R *are on opposite half-lines with boundary* P. *If these conditions are satisfied, we use the symbolic form* $R - P - Q$ *(or* $Q - P - R$*) to indicate that* P *is between* R *and* Q.

3.2.3 Postulate

Given three collinear points P, Q, *and* R *exactly one is between the other two.*

That is, exactly one of the following relationships holds: $P - Q - R$, $P - R - Q$, $R - P - Q$. (There is no reason to consider the relationships $R - Q - P$, $Q - R - P$, $Q - P - R$ as they are respectively the same as the three mentioned above.)

3.2.4 Postulate

If A *and* B *are any two points on a line, then there is a point* C *on this line such that* $A - B - C$. *In addition, there is a point* D *on the line such that* $A - D - B$.

In other words, given two points A, B in space, there is a point C on the line $\overleftrightarrow{AB}$ such that B is between A and C. Also, there is a point D on the line $\overleftrightarrow{AB}$ such that D is between A and B (Fig. 3.2.4).

Figure 3.2.4

As a consequence, we can say that the points on a line are dense. That is, there are no gaps in the line. If we take a negative point of view, Fig. 3.2.5 would *not* represent any line we have in mind.

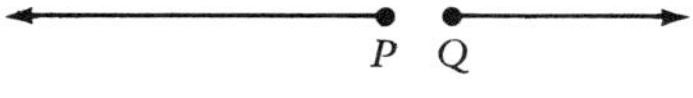

Figure 3.2.5

in our opinion . . .

This is similar and related to our understanding of the real number system. Between any two real numbers, there is another real number. Actually, between any two real numbers, there are an infinite number of real numbers. Similarly, between any two points on a line, there are an infinite number of points.

This density property is shared by the set of rational numbers. That is, between any two rational numbers, there exists another rational number. The set of integers does not have the density property. The real numbers have a property not shared by the rationals; namely, the completeness property. This is precisely the property we described in Chapter 2 when we established a one-to-one correspondence between the set of real numbers and the set of points in a Euclidean plane.

Some students might feel that the study of betweenness is "much ado about nothing." But this attitude only betrays the disposition to continue thinking about some physical object rather than the geometric idea. This disposition is hardly uncommon, since Euclid himself seemed to ignore this concept. In doing so, he made it impossible to derive all the results of his geometry that he desired.

Perhaps the student should look at the notion of betweenness being included because we do wish to have a geometry consistent with our intuition. The betweenness properties herein described guarantee the existence of an infinite number of points on a line and that a line is not terminated at any point. Further, they assure that the order of points on a line is serial, that is they satisfy Postulate 3.2.3: of three collinear points, exactly one is between the others.

Exercise 3.2.1

Suppose space is our earth, considered as a sphere. Let the lines of this space be the great circles of the earth. Investigate the betweenness properties of points on great circles. What incidence postulates exist among points, circles, and this space?

Suppose we now consider two points P and Q in space (Fig. 3.2.6). These two points determine a line $\overleftrightarrow{PQ}$ and a ray $\overrightarrow{PQ}$. Each of these is infinite in extent, the line being infinite in both senses of direction while the ray originates at point P and extends infinitely in one of the senses of direction. There is an additional important region of the line determined by P and Q, that being the infinite subset of points between and including P and Q. We call this subset a *line segment;* in particular, the line segment P, Q denoted by $\overline{PQ}$. There is only one segment with endpoints P and Q. Note that $\overline{QP}$ also symbolizes $\overline{PQ}$.

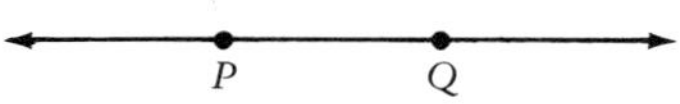

Figure 3.2.6

comment . . .

Segments are parallel to each other if they are subsets of parallel lines. Segments that are subsets of skew lines are said to be skew to each other.

Now that we have considered how a point separates a line, the next step is to consider how a line in a plane separates the plane. In Fig. 3.2.7, the line l separates the plane $\mathcal{P}$ into three disjoint subsets. The first subset is the line l itself. Each of the other two subsets is an infinite set of points which will be called half-planes of $\mathcal{P}$ with boundary l. The boundary of the two half-planes is sometimes referred to as the *edge* of the half-plane. As in the case of the half-lines, we speak of these two half-planes as *opposite* half-planes with edge l.

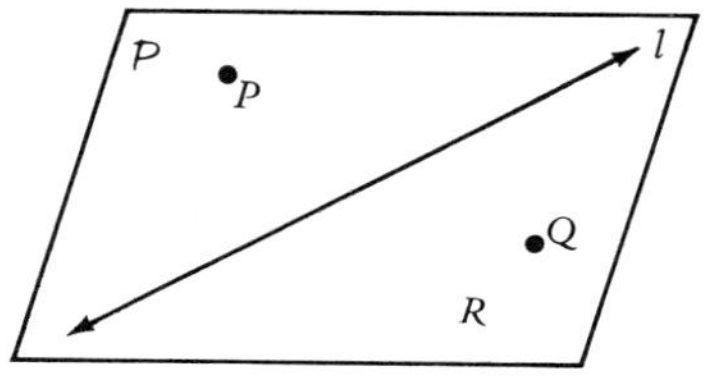

Figure 3.2.7

We make the following observations: (i) Each point of $\mathcal{P}$ with the exception of the points of the line l belongs to exactly one half-plane, (ii) the line l does not belong to either half-plane. If we use the same technique of Chapter 1, replacing the words point and line in Postulate 3.2.1 by line and plane, respectively, we obtain our next postulate.

3.2.5 Postulate

Any line of any plane in space $\mathcal{P}$ separates the plane into two half-planes and their common boundary, the line.

In Fig. 3.2.7, point P is on one side of the line l and so we call the half-plane containing P the P-side of l. The point Q is on the other side of l (that is, the opposite half-plane) and so the half-plane containing Q is named the Q-side of l. Note that point R is on the Q-side of the plane.

An interesting property of half-planes is that if Q and R are two points of the half-plane, then the line segment $\overline{QR}$ is completely contained in the half-plane. This is called the property of *convexity*. Mathematically, sets with this property seem to be easier to deal with than those not possessing the property. This is particularly true when we attempt measurement. For this and other reasons, we will spend most of our time studying sets which are said to be convex.

3.2.6 Definition

A set of points S is said to be convex *if whenever any two points X and Y belong to S, then every point of the line segment $\overline{XY}$ belongs to S.*

It is easy to see that line segments, rays, and lines are examples of convex sets. Figure 3.2.8 shows some sets that at least intuitively appear to be convex. The set we have in mind in each case is the bounded region and its boundary.

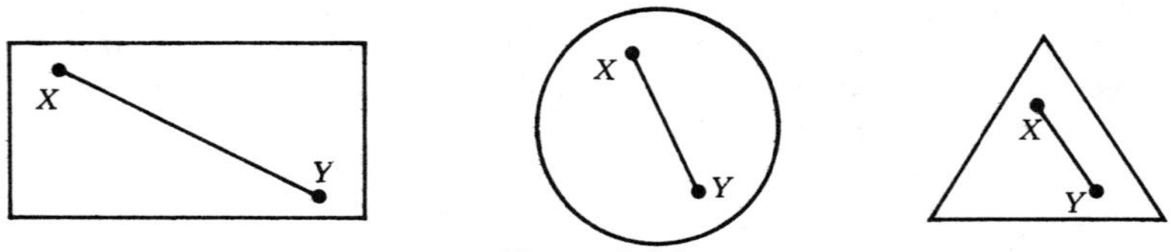

Figure 3.2.8

The regions in Fig. 3.2.9 represent subsets of space that are not convex, showing X and Y in the bounded regions, while $\overline{XY}$ is not completely contained in the set.

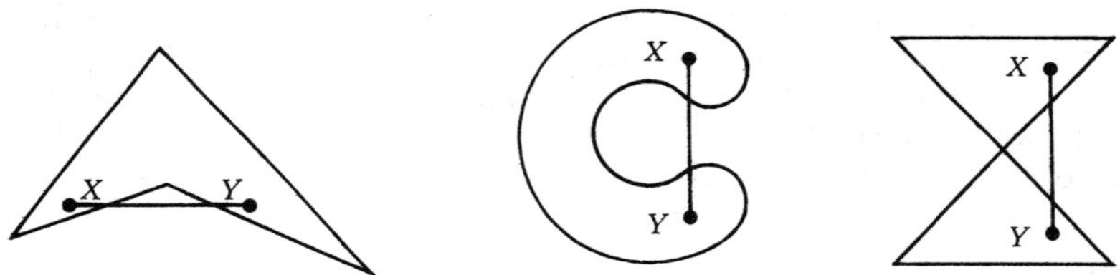

Figure 3.2.9

Example 1

Show that the diagram in Fig. 3.2.10 is not convex and sketch the smallest convex set of which it is a subset (called a *convex*

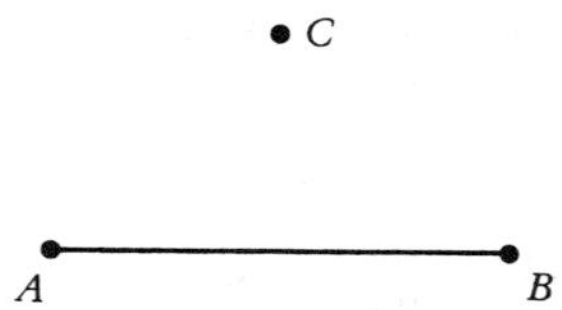

Figure 3.2.10

superset). This diagram is not convex since any segment joining point C with any point of $\overline{AB}$ contains points not on the diagram. The convex superset would be in Fig. 3.2.11.

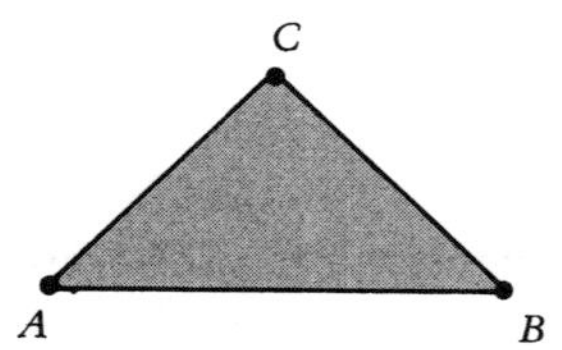

Figure 3.2.11

Exercise 3.2.2

Consider each of the following sets of points in space. Decide if they are themselves convex. If not, determine the smallest convex set of which they are subsets.

Set	*Convex superset*
1. A set consisting of one point	1. ______
2. A set consisting of two points	2. ______
3. A set of three noncollinear points	3. ______
4. A plane	4. ______
5. Two intersecting line segments	5. ______
6. Two nonintersecting line segments	6. ______
7. A line and a point not on the line	7. ______
8. Two parallel lines	8. ______
9. Two intersecting lines	9. ______

comment . . .

The idea of convexity of sets gives a way to decide whether two points of a plane are in the same half-plane or in opposite half-planes. In Fig. 3.2.12, the fact that P and Q are in opposite half-planes with edge l can be described by saying that the line segment $\overline{PQ}$ contains a point of l; namely, X. On the other hand, the fact that R and Q are in the same half-plane with edge l means that $\overline{RQ}$ has no points in common with l.

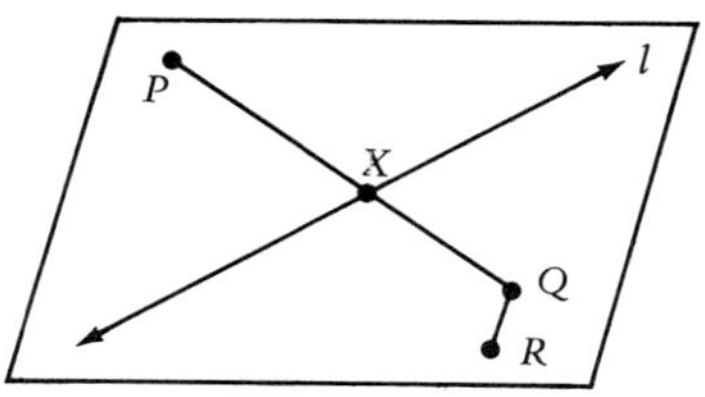

Figure 3.2.12

Now turning to the separation of space by a plane, we find that in a completely analogous manner to the separation of a line by a point and of a plane by a line, a plane separates space into three distinct subsets. The first is the plane itself and the other two are disjoint infinite subsets of space called *half-spaces* with the plane as boundary.

3.2.7 Postulate

Any plane in space separates space into three disjoint regions, two half-spaces and their common boundary, the plane.

In Fig. 3.2.13, point P is on one side of plane $\mathcal{P}$ which is referred to as the P-side and point R is on the opposite side of $\mathcal{P}$, which will be referred to as the R-side. Note that $\overleftrightarrow{PR}$ intersects $\mathcal{P}$ in the point Q. In general, two points P and R are on opposite sides of a plane $\mathcal{P}$ if there is a point of $\mathcal{P}$ between P and R. To say that points P and S are on the same side of $\mathcal{P}$ means that there is no point of $\mathcal{P}$ between P and S.

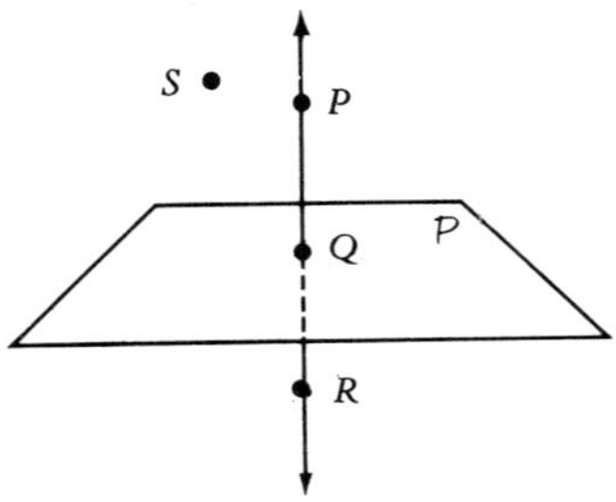

Figure 3.2.13

in our opinion . . .

Notice that mathematics often re-asks questions in similar situations and uses similar terminology to describe similar ideas. Thus, we have half-line, half-plane, and half-space. This is another reason why students must learn to ask good questions, not merely to parrot answers. The same question in mathematics is often asked in a spiraling fashion, getting more and more general and abstract. Understanding the question at some level of abstraction will aid the student in answering the question at levels of greater abstraction.

Exercise 3.2.3

1. How many rays are determined by any three noncollinear points in space? How many rays are determined by any n points in space, no three of which are collinear?
2. Draw a diagram to picture each of the following figures.
   a. two rays whose union is a line
   b. two rays whose intersection is a point and whose union is not a line
   c. a diagram showing three different points A, B, and C on a line such that $A - B - C$. Choose a point D between B and C. Is D between A and C? Is D between A and B?
3. If $\overline{PQ}$ and $\overline{RS}$ are parallel segments, how are the lines $\overleftrightarrow{PQ}$ and $\overleftrightarrow{RS}$ related? If $\overline{PQ}$ and $\overline{RS}$ are neither parallel nor skew segments, how are the lines $\overleftrightarrow{PQ}$ and $\overleftrightarrow{RS}$ related?
4. Which of the shaded regions in Fig. 3.2.14 are convex?

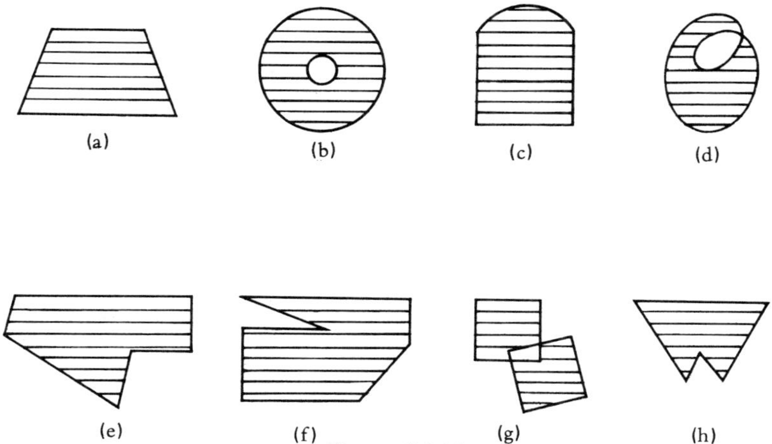

Figure 3.2.14

5. Give an argument showing that the intersection of two convex sets is a convex set. Give an example to show that the union of two convex sets need not be convex. Explain what you would have to do to show that a set is not a convex set.

3.3 ANGLES

In the next chapter, we will discuss separation of space by special sets known as polygons and polyhedrons. Before we can do this, we must first consider the concept of an angle.

Our idea of an angle is illustrated in Fig. 3.3.1.

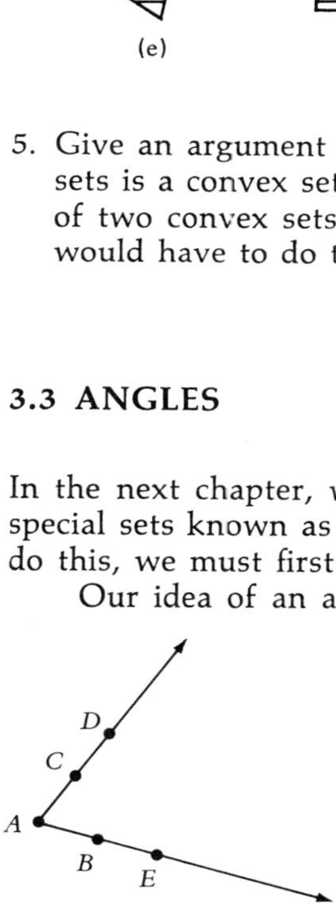

Figure 3.3.1

3.3.1 Definition

An angle *is the union of two rays originating from a common point.*

In Fig. 3.3.1, the angle is $\overrightarrow{AC} \cup \overrightarrow{AB}$ such that $\overrightarrow{AC} \cap \overrightarrow{AB} = \{A\}$. The rays are the *sides* of the angle and the common endpoint of the rays is called the *vertex* of the angle. We will use the symbol $\angle BAC$, read "angle B, A, C," to represent the angle pictured in Fig. 3.3.1.

Note that the middle letter of the three letters used in naming the angle is the vertex of the angle. Other names for $\angle BAC$ are $\angle CAB$, $\angle DAE$, $\angle EAD$, $\angle CAE$, and so on.

Any angle lies in a unique plane and separates the plane into three distinct regions. The first region is the set of points of the angle itself. The other two regions are infinite subsets of the plane which are called the *interior* of the angle and the *exterior* of the angle. In Fig. 3.3.2, we illustrate the region which we mean to be the interior of the angle and the region which is meant to be the exterior of the angle.

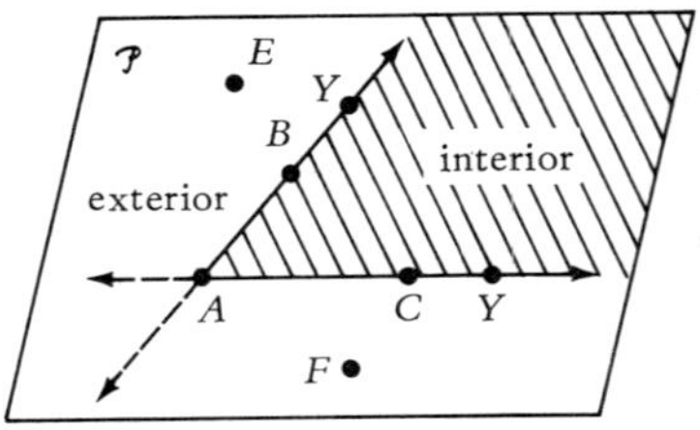

Figure 3.3.2

As long as we have a picture to point to, it is fairly clear which region is to be considered the interior of the angle. But since an angle is an idea defined in terms of sets, no picture will be satisfactory, and we will have to make the idea of interior of an angle more precise.

We say that the interior of $\angle BAC$ is the intersection of the half-plane with edge $\overleftrightarrow{AC}$ that contains B with the half-plane with edge $\overleftrightarrow{AB}$ that contains C. Note in Fig. 3.3.3. we have marked the half-plane with edge $\overleftrightarrow{AC}$ containing B, using vertical lines. We have marked the half-plane with edge $\overleftrightarrow{AB}$ containing C with horizontal lines. The region which has crossing vertical and horizontal lines is the interior of $\angle BAC$. We then agree that the exterior of $\angle BAC$ is any part of the plane which does not contain any points of $\angle BAC$ or its interior.

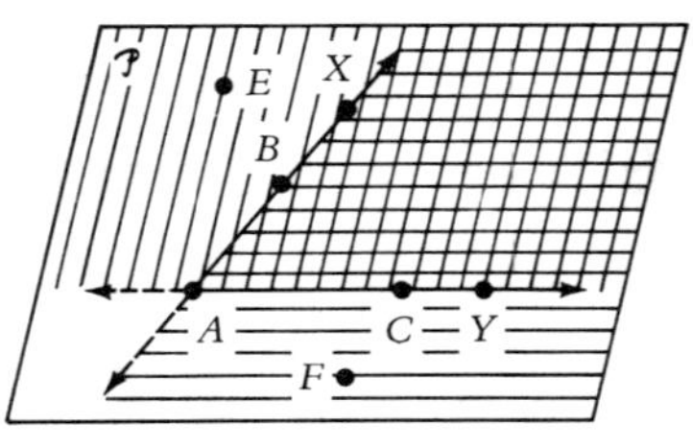

Figure 3.3.3

Our description of the interior of $\angle BAC$ does not depend on the points B and C with which we have arbitrarily chosen to label the angle. It does depend on the rays $\overrightarrow{AB}$, $\overrightarrow{AC}$ and their common endpoint A. Thus, in our diagram, the interior of $\angle BAC$ is the same as the interior of $\angle XAY$, since $\angle BAC = \angle XAY$. We say that the interior of $\angle BAC$ is coincident with the interior of $\angle XAY$. Note that the point D is in the interior of $\angle BAC$ while points E and F are in the exterior of $\angle BAC$.

There is one type of angle where it does not make any sense to speak of the interior or exterior of the angle. This is

Figure 3.3.4

the angle illustrated in Fig. 3.3.4, formed by two opposite rays $\overrightarrow{AB}$ and $\overrightarrow{AC}$. We call such an angle a *straight angle*. This is because the sides of the angle form a straight line.

comment . . .

Our definition of an angle as the union of two rays with a common endpoint does not rule out that the two rays coincide. Such an angle, called a *zero-angle*, will play a valuable role in mathematics. As with a straight angle, it has neither interior nor exterior.

in our opinion . . .

Students often think of angles as finite in extent; that is, they forget that the sides of angles are rays which are infinite in length. This leads to confusion when we wish to measure angles. Students will often say that $\angle BAC$ in Fig. 3.3.5 is bigger than

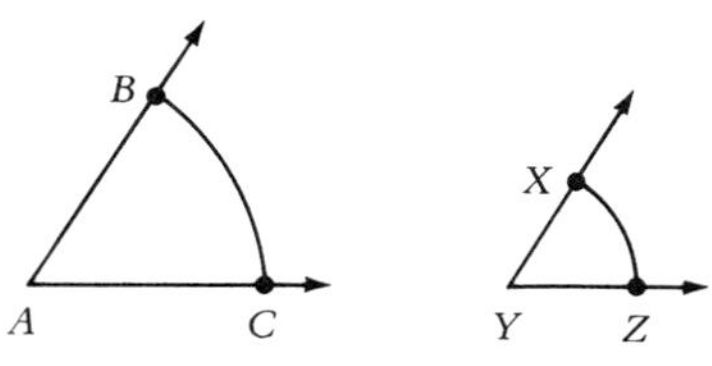

Figure 3.3.5

$\angle XYZ$. The question is what are we referring to when we speak of bigger? Essentially, the confusion is similar to that raised if we ask which is bigger in Fig. 3.3.6. As we can see here, the

Figure 3.3.6

question is ambiguous. Do we mean which is the bigger symbol or which is the bigger value? With our angles, $\angle BAC$ has a bigger *representation* than $\angle XYZ$, but this does not mean that $\angle BAC$ is bigger than $\angle XYZ$. We will return to this discussion when we investigate measurement.

In Chapter 1, we introduced the notion of intersecting lines. In Fig. 3.3.7, lines l and m intersect in a point P and form four angles. The angles are $\angle APB$, $\angle BPC$, $\angle CPD$, and $\angle DPA$. Of course, the straight angles, $\angle APC$ and $\angle BPD$, are also formed by the lines l and m. When we refer to the angles formed by two intersecting lines, we shall mean those nonzero angles that are not straight angles.

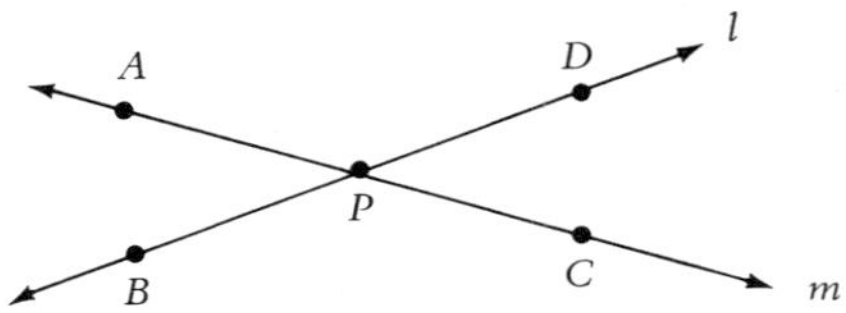

Figure 3.3.7

We notice that $\angle APB$ and $\angle BPC$ have a common side, namely $\overrightarrow{PB}$. It is also clear that these two angles have no other points in common and that their interiors have no points in common. We call two such angles *adjacent angles.*

3.3.2 Definition

Adjacent angles *are two coplanar angles that have a common vertex and a common ray, but whose interiors have no points in common.*

Example 1
In Fig. 3.3.8, $\angle SPR$ and $\angle RPQ$ are adjacent angles; $\angle SPR$ and $\angle SPQ$ are not adjacent angles since their interiors have points in common.

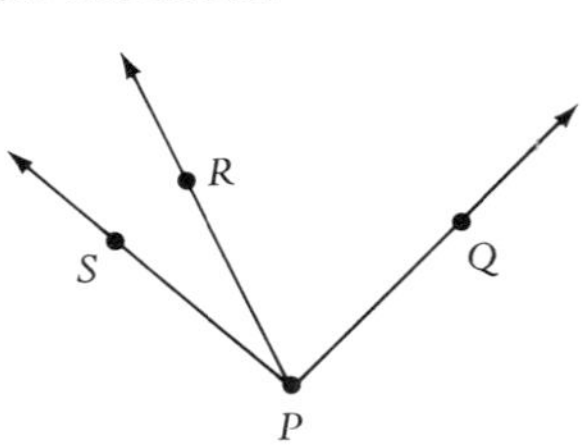

Figure 3.3.8

Example 2
In Fig. 3.3.9, $\angle ABE$ and $\angle ECD$ are not adjacent angles because they do not have a common vertex. $\angle XYZ$ and $\angle YZW$ are

not adjacent angles as they do not have a common ray, even though they have a common segment.

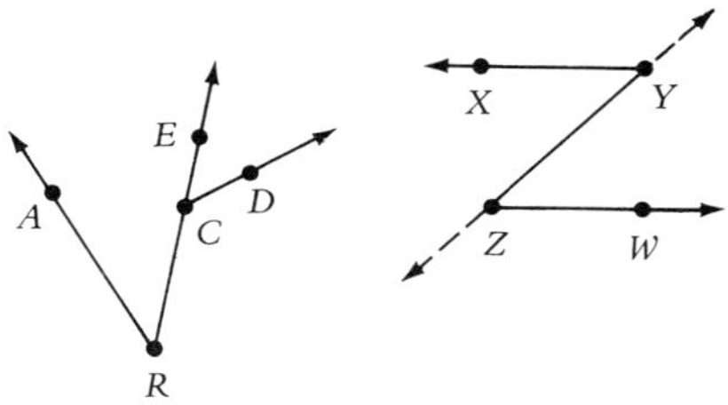

Figure 3.3.9

Returning to the picture of two intersecting lines (Fig. 3.3.7), we note that the angles $\angle APD$ and $\angle BPC$ do *not* form a pair of adjacent angles. They do share a common point, namely the vertex P. Two angles formed by two intersecting lines with this property are called *vertical angles.* More precisely, we have the following.

3.3.3 Definition

If $A - P - C$, $B - P - D$, and $\overleftrightarrow{PA} \neq \overleftrightarrow{PB}$, then $\angle APD$ and $\angle BPC$ *form a pair of vertical angles.*

in our opinion . . .

In interpreting the notion of vertical angles, we used various symbols to represent the necessary conditions. Students should become able to interpret symbols. The danger with symbols all too often is that students merely read symbols without attaching any significance to them. It is something akin to reading a French menu but still having no idea what food is available. Mathematics is a language and, like any language, has certain idioms that must be learned. Students do not automatically translate ideas into symbols and symbols into ideas; they must be consciously taught these skills. If not, they will remain mathematical illiterates.

Exercise 3.3.1

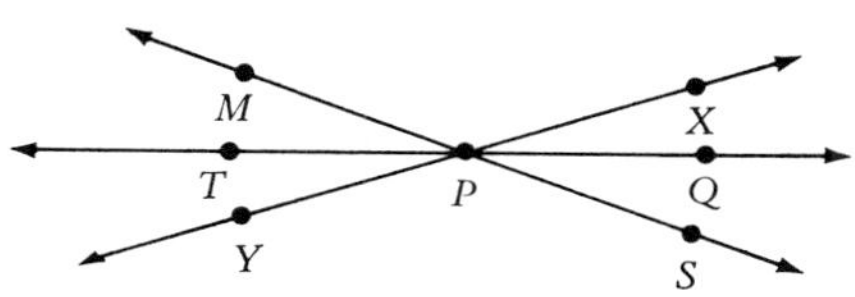

Figure 3.3.10

1. Refer to Fig. 3.3.10 to answer the questions.
   a. Name four angles that are not straight angles.
   b. Name two pairs of adjacent angles.

c. Name two straight angles.
d. Name a point in the interior of $\angle XPS$.
e. Name all the points pictured in the exterior of $\angle XPS$.
f. Name the sides of $\angle XPS$.
g. Name a pair of nonadjacent angles that are not vertical angles.
h. Name a pair of vertical angles.

2. a. Draw a picture showing three noncollinear points P, Q, and S and $\angle PQS$.
   b. Draw $\overline{PS}$, choose any point X *such that* $P - X - S$, and draw $\overrightarrow{XQ}$. Does Q lie on $\overline{PS}$?
   c. Refer to part b. Name two adjacent angles having $\overrightarrow{XQ}$ as one side. Do these angles form a pair of vertical angles?
3. Make a definition describing the fact that $\overrightarrow{OB}$ lies between $\overrightarrow{OA}$ and $\overrightarrow{OC}$. (Use Fig. 3.3.11.)

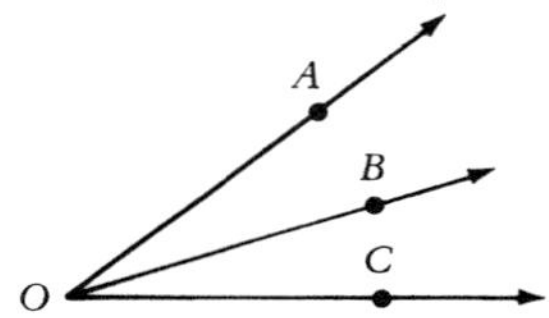

Figure 3.3.11

3.4 SUMMARY

1. Every point P on a line separates the line into three distinct subsets: the set consisting of P alone and the infinite sets called half-lines. The point P is called the endpoint of each half-line.
2. A half-line together with its endpoint is called a ray.
3. Collinear rays having only their endpoints in common are called opposite rays.
4. A point B is between A and C if A, B, and C are collinear and $\overrightarrow{BA}$ and $\overrightarrow{BC}$ are opposite rays. Between any two points in space, there is another point. The points of a line are dense and serial.
5. A figure consisting of two points P and Q and all points between P and Q is called a line segment with endpoints P and Q.
6. A network is a figure consisting of a finite number of points called vertices and a number of arcs. The vertices are the endpoints of the arcs, and no two arcs have a common point, except perhaps a common vertex. A vertex is called odd or even, according to whether the number of arcs forming the vertex is odd or even.

7. A subset S of space is said to be convex if whenever any two points P and Q belong to S, all points of the segment $\overline{PQ}$ belong to S.
8. Every line l in a plane separates the plane into three distinct subsets: the set consisting of the line l and two subsets called half-planes. The line l is called the edge of each of its half-planes.
9. Every plane $\mathcal{P}$ in space separates space into three distinct subsets: the set consisting of the plane $\mathcal{P}$ and two subsets called half-spaces.
10. An angle is the union of two rays with the same endpoint. The rays are called the sides of the angle and their common endpoint is called the vertex of the angle. If the sides of an angle are opposite rays, the angle is a straight angle.
11. The interior of a nonstraight angle $\angle ABC$ is the intersection of the A-side of $\overleftrightarrow{BC}$ with the C-side of $\overleftrightarrow{BA}$. If a point of the plane of $\angle ABC$ is not an interior point and does not lie on the sides of an angle, it is called an exterior point.
12. Adjacent angles are two coplanar angles having a common vertex and a common ray but whose interiors have no points in common.
13. Vertical angles are two nonadjacent angles formed by two intersecting straight lines.

Exercise 3.4.1

1. True or False:
   a. Every half-plane is a convex set.
   b. A set containing just one point is a convex set.
   c. A ray lying in a plane separates the plane into three distinct regions.
   d. An angle is the union of two rays with a common endpoint.
   e. An angle is the intersection of two rays with a common endpoint.
   f. If A, B, C are three distinct and noncollinear points, l is a line not passing through any of them and l contains a point of $\overline{AB}$, then l also contains a point of $\overline{AC}$ or $\overline{BC}$.
   g. The interior of an angle is a convex set.
   h. If a point D lies in the interior of $\angle DAC$, then every point of $\overrightarrow{AD}$ lies in the interior.
   i. If D is an interior point of $\angle BAC$, then $\overrightarrow{AD}$ intersects $\overline{BC}$.
   j. If A, B, C are three points of space, exactly one is between the other two.
2. Refer to the line l in Fig. 3.4.1 in answering the following questions.
   a. Name two points on the same side of R.
   b. Name two points on opposite sides of R.

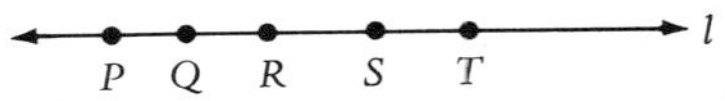

Figure 3.4.1

c. Give two other names for the set $\overrightarrow{QR}$.
d. What is the relationship between $\overrightarrow{RS}$ and $\overrightarrow{RP}$?
e. What is the endpoint of $\overrightarrow{QT}$?

3. Give arguments to support the following statements. All points and lines are lying in one plane.
   a. If P and Q are on opposite sides of the line l and X and Q are on opposite sides of the line l, then P and X are on the same side of l.
   b. If P and Q are on opposite sides of the line l, and Q and X are on the same side of l, then P and X are on opposite sides of l.
   c. If a ray has its endpoint on a line, but does not lie on the line, then all points of the ray except for the endpoint are on the same side of the line.
   d. Every side of a triangle lies, except for its endpoints, in the interior of the opposite angle.
   e. If $A - C - D$, $B - F - C$, and $A - F - G$ (all in the same plane) and A, B, and C are not collinear, then G is in the interior of $\angle BCD$.
   f. If D lies in the interior of $\angle BAC$, then $\overrightarrow{AD} - \{A\}$ (that is, the half-line A, D) lies in the interior of $\angle BAC$.
   g. If D lies in the interior of $\angle BAC$ and $G - A - D$, then $\overrightarrow{AG} - \{A\}$ lies in the side of $\overleftrightarrow{AC}$ that does not contain B.
   h. If D is in the interior of $\angle BAC$, and $F - A - C$, then F and B are on the same side of $\overleftrightarrow{AD}$.
   i. If D is in the interior of $\angle BAC$, then $\overrightarrow{AD}$ intersects $\overline{BC}$.

CURVES, SURFACES, AND SEPARATION

4

4.1 PROLOGUE

Most everyone over the age of five would recognize each of the diagrams in Fig. 4.1.1. In fact, children in kindergarten are frequently presented with these figures to prepare them for reading. Of course, these figures are given the names triangle, square, and circle. The kindergarten child is expected to note the difference in these figures and learn their names. Learning to distinguish between these figures will assist the children in learning to distinguish between the various letters of the alphabet.

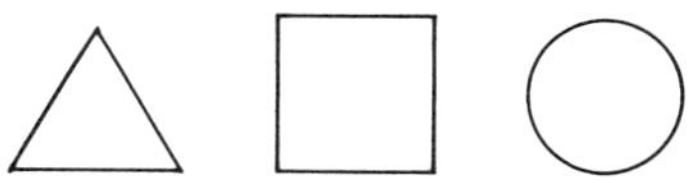

Figure 4.1.1

We will be interested in these figures for other reasons. We will wish to consider not only their differences but also their similarities. For instance, each is a plane figure which divides the plane into three distinct regions and the bounded regions are each convex. We will also be interested in the measurement of these figures and the regions they enclose, particularly their perimeters and areas.

Since many of our ideas of geometry have their origin in nature, everyone has some degree of geometric sense. Long before any formal study of the concepts of geometry, we were aware of location, lengths, surfaces, regions, areas, and volumes. One only has to observe a child at play to note his geometric sense developing. The preschooler explores his world and makes many sophisticated geometric discoveries in his play. He fits together pots and pans, builds elaborate structures with boxes and blocks, draws various pictures of curves with crayons, and cuts out numerous regions with scissors and paper. That adults have not decided to give names to these curves and regions takes nothing away from the child's geometric discoveries.

It is because we wish to communicate such geometric discoveries and to generate new information from them in the most efficient manner that we study these geometric properties in a more abstract setting. In this way, we strip away the irrelevant and concentrate more fully on the concepts that interest us. In this way, we do not rely on individual interpretations and arbitrary decisions. Thus, before we can progress far in our study, we must learn to make observations about geometric figures and classify them.

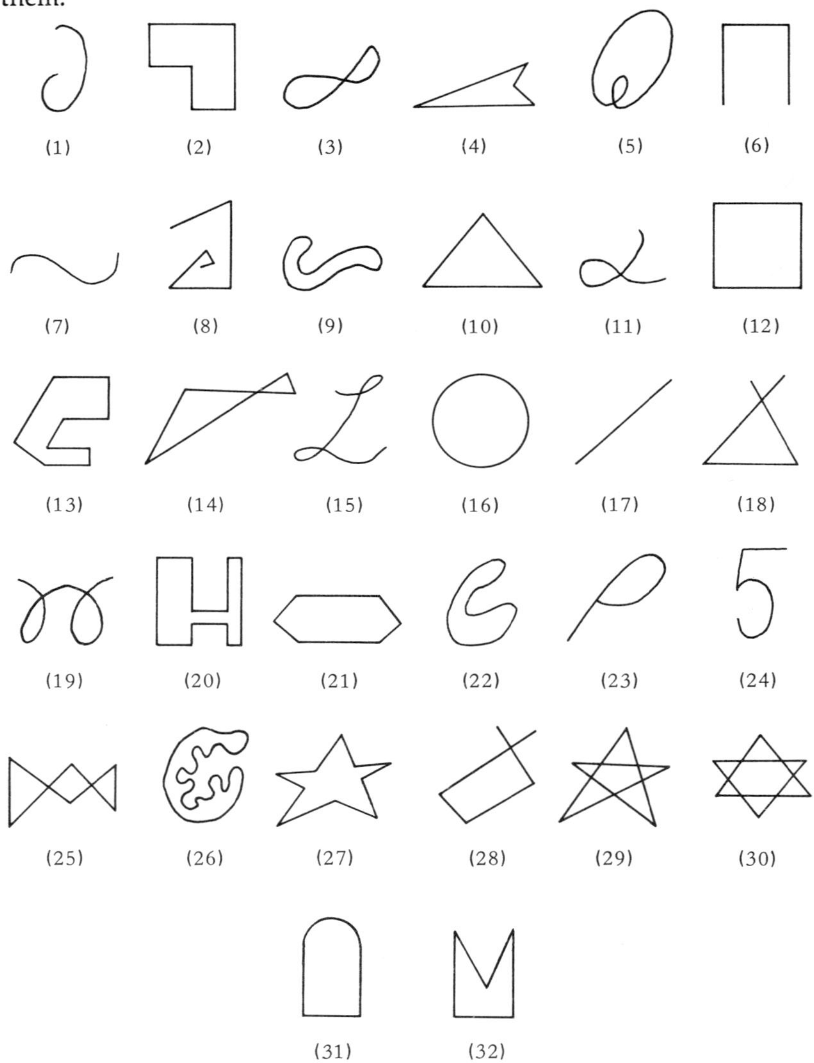

Figure 4.1.2

Experiment 1

The geometric shapes in Fig. 4.1.2 are all different in certain senses. We could attempt to describe how they differ but to do this most successfully we must desribe ways in which they are the same. In doing this experiment, we are engaged in what in mathematics is known as the *classification problem*. There are various ways in which these shapes can be classified. For instance, (10), (12), and (16) share properties. Describe them. Do any of the other figures share these properties? Are any of the other figures lacking in these properties? Why? As another example, (1), (7), and (24) share common features. Describe them. What other figures share these properties? What figures do not? Why? Figures (11) and (18) have properties in common. Describe them. Make as many other classifications as possible.

Experiment 2

We can make four interesting surfaces from a strip of stretchable material as suggested by the strip in Fig. 4.1.3. The first surface lies completely in a plane and is formed by stretching the strip and pasting side $\overline{AB}$ to side $\overline{DC}$, matching A and D, B and C.

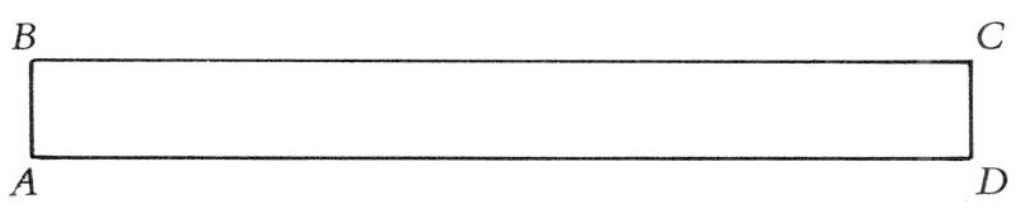

Figure 4.1.3

The resulting surface, known as an *annulus*, looks something like Fig. 4.1.4. The boundary of the strip separates the plane into three distinct regions. What are they? How many regions does the annulus separate the plane into? Can you describe them? Are there any other differences or similarities that you can note about the two surfaces?

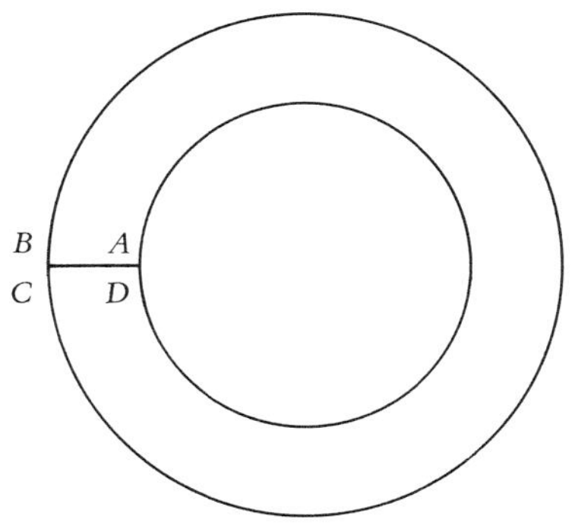

Figure 4.1.4

The second surface that can be formed using such a strip is known as a *cylinder*. It is also formed by pasting together sides $\overline{AB}$ and $\overline{DC}$ with A and D, B and C matched, only this surface is not formed in a plane. (See Fig. 4.1.5.)

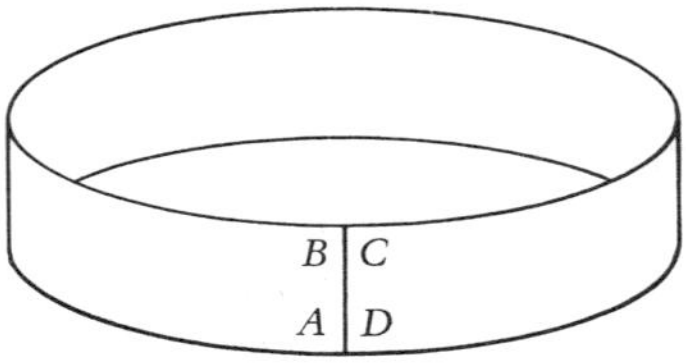

Figure 4.1.5

Using a strip of paper, construct a cylinder. Are there any other differences or similarities that you can note about the annulus and the cylinder?

The third surface can be constructed by first pasting together sides $\overline{BC}$ and $\overline{AD}$ matching B and A, C and D, to make a cylinder (Fig. 4.1.6). Next, paste together the two circular regions giving

Figure 4.1.6

what is known as a *torus* (donut-shape). Attempt to construct a torus using a strip of paper. Your surface will look something like Fig. 4.1.7. Now the original strip of paper and a circular region have common properties. Can you suggest some of them? Can you describe how a torus and a sphere differ? How they are the same?

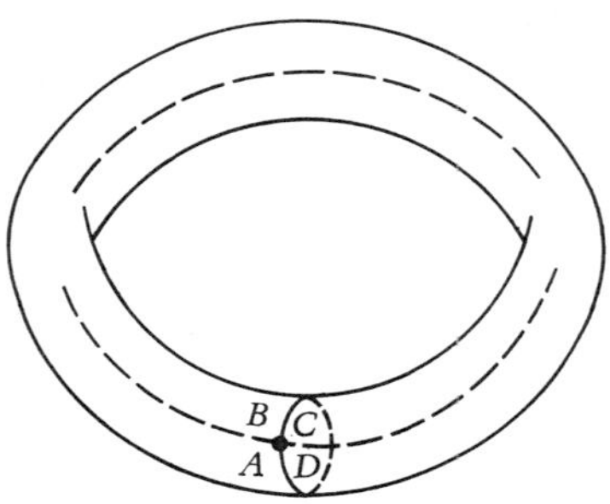

Figure 4.1.7

The next surface is called a *Möbius strip* and is formed by making a twist in the strip and pasting the sides $\overline{AB}$ and $\overline{CD}$ together, matching A and C, B and D. The resulting surface looks something like Fig. 4.1.8. Using two strips of paper, construct two Möbius strips. Label a point on the Möbius strip X and starting at X draw a continuous curve around the strip until you return to X. What do you observe? What can you conclude?

Cut the Möbius strip with scissors along a line in the middle of the strip. What do you observe? What can you conclude? Using the other Möbius strip, cut it along a line about $\frac{1}{4}$ of the way from an edge. What do you observe? What can you conclude? Describe an edge of a Möbius strip.

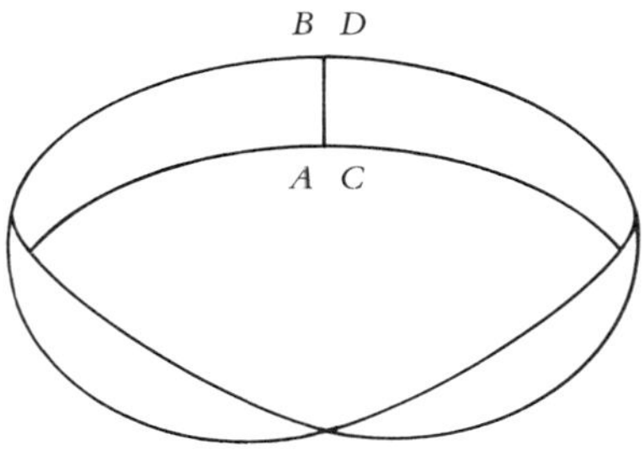

Figure 4.1.8

Next, form a double Möbius strip by placing two strips of paper together, giving both of them a half-twist and pasting their ends together (Fig. 4.1.9). Put your finger between the bands and run it all the way around until you return to the point at which you started. What do you observe? Make a mark on one of the strips between the two strips and draw a curve around the strip until you return to the original mark. What do you observe? What can you conclude? Open the double Möbius strip. What do you observe? What can you conclude?

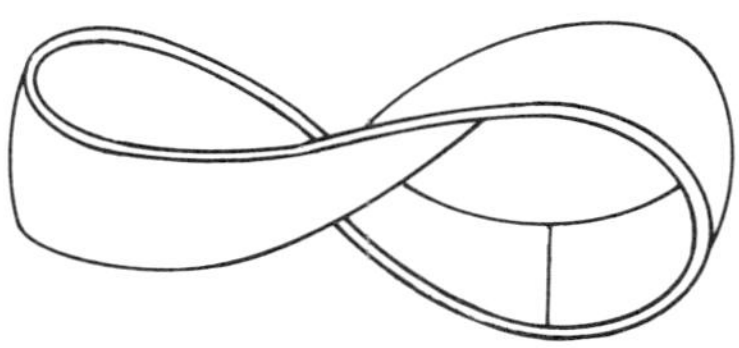

Figure 4.1.9

Exercise 4.1.1

An unsolved mathematical problem is the so-called *Four Color Problem:* "How many colors are necessary to color a map with any number of countries so that no two countries which adjoin on a boundary shall have the same color?" (See Figure 4.1.10.) Mathematicians have never found a plane map for which four colors would not suffice. At the same time, no one has been able to prove that four colors would be sufficient for any possible map. Color the maps of Fig. 4.1.10 to satisfy the condition with four or fewer colors. Make your own map to show that four colors are necessary and five colors are sufficient.

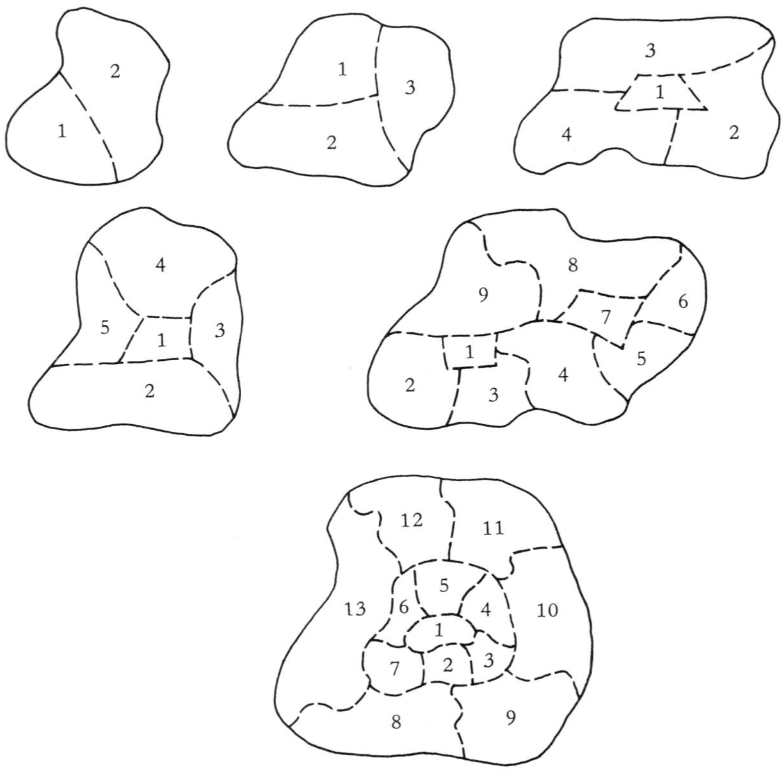

Figure 4.1.10

4.2 CURVES AND SURFACES

In the two experiments, we have examined two types of geometrical figures called *curves* and *surfaces*. In mathematics, a curve is most easily considered as the path of a point moving continuously with respect to both position and direction. It is not possible to give a precise definition of a curve without a knowledge of continuous functions. A curve may also be thought of as an infinite set of points and will include such geometric figures as rays and lines. When we speak of a curve, we will ordinarily consider a geometric figure lying wholly in one plane, and we will use a phrase like *space curve* or *twisted curve* for one which does not lie wholly in one plane.

There is no agreed definition of a surface that we can adopt. We may think of it as the boundary of a solid body or as the locus of an infinite set of points selected in some way from the infinite points of space, having a certain amount of continuity.

The plane is the simplest example of a surface. Since our aim in this book is to limit our study to certain "nice" surfaces, a precise definition will be unnecessary. We will be mostly concerned with the boundaries of certain geometric solids. With the exception of the cone, the cylinder, and the sphere, all will be formed by fitting together flat pieces of planes.

in our opinion . . .

All geometric figures are a union of curves, surfaces, and points. How these are fitted together predetermines such things as length, angle measure, perimeter, area, and volume. It also motivates the various classifications of geometric figures such as congruence and similarity. Curves and surfaces can be studied in non-Euclidean settings as well. Rigidity is the main characteristic of the shapes of elementary geometry. All that Euclid could tell about lines, circles, spheres, cones, and so on, is lost under the slightest deformation of such figures. The completely arbitrary curves and surfaces are the objects of topology and are studied with respect to its roughest intuitive properties, such as interior, exterior, connectivity, and so on.

Much of the geometry of a plane is devoted to the study of relations among subsets of a plane that are called *simple closed curves*. Among these are included circles, triangles, quadrilaterals; in fact, all polygons that you are familiar with. But simple closed curves include many other subsets of the plane and thus any definition of a simple closed curve would be extremely complicated. Thus, we will not formally define simple closed curves but will attempt to intuitively characterize what we have in mind.

The prototype of a simple closed curve is the circle, such as in Fig. 4.2.1. We have labeled four points of this circle by X, Y, Z, and W.

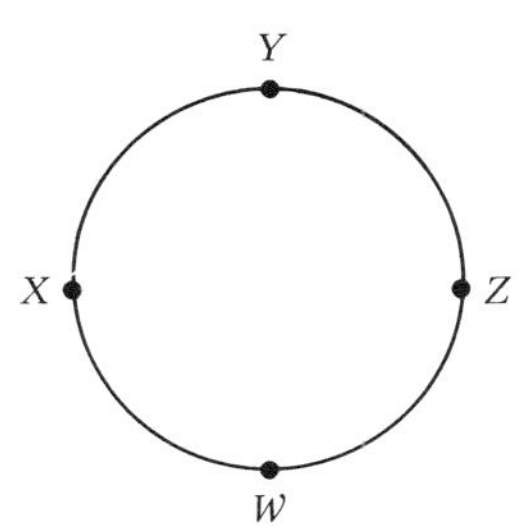

Figure 4.2.1

We will now allow certain distortions of the circle. For instance, we are allowed to stretch, shrink, and bend the path of the circle. None of these distortions will affect the order of the points X, Y, Z, and W on the circle. Nor will they affect the fact that the circle divides the plane into three disjoint regions: its interior, its exterior, and the circle itself. The result of these allowable distortions is a simple closed curve. To put it another way, a curve is a simple closed curve if it can be stretched, shrunk, and bent into the shape of a circle. Certain distortions of the curve are not allowed. For instance, we are not allowed to cut or tear the path nor in stretching and bending it are we allowed to have any two distinct points of it coincide.

In Fig. 4.2.2, we indicate some allowable distortions of the circle which are simple closed curves and some unallowable distortions. These latter curves do not represent simple closed curves.

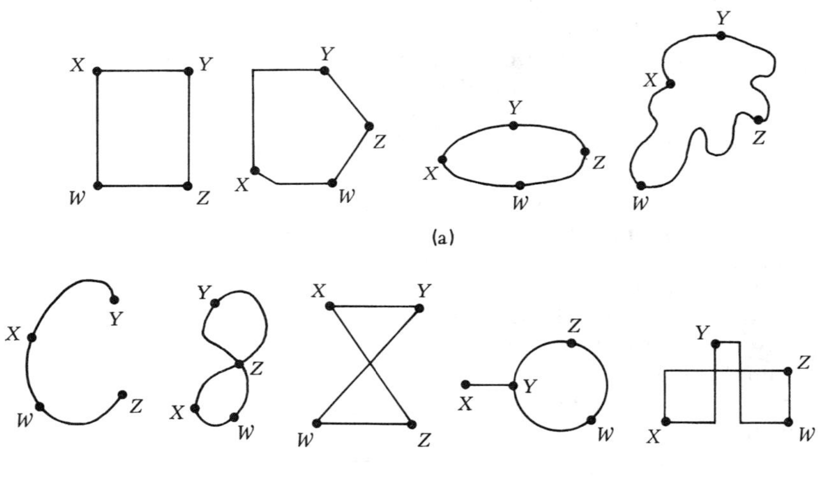

Figure 4.2.2 (a) Allowable; (b) Unallowable

We might suggest that to define a simple closed curve we should first define a simple curve, then a closed curve, and finally a simple closed curve. This makes sense and, in fact, we could designate that a simple curve is any allowable distortion of a line segment with the added stipulation that no two points, except the endpoints of the segment, can be made to coincide. The interpretation of a closed curve in this manner is less satisfactory since starting with some well-known closed curve and performing allowable distortions will not result in all figures that we might ordinarily think of as closed.

The point is that you can look at a curve and ask the question, "Is it possible to change the shape of this curve by allowable distortions and result in a circle?" If the answer is yes, you are looking at a simple closed curve. Similarly, you can look at a curve and ask if allowing a simple cut in the curve and then using other allowable distortions can permit the curve to acquire the shape of a line segment? If the answer is yes, you are looking at a simple curve.

Now, consider Fig. 4.2.3. Something tells us it should be considered closed, but what distortions will result in it being transformed into a more elementary closed curve, such as a circle, a triangle, a square, and so on?

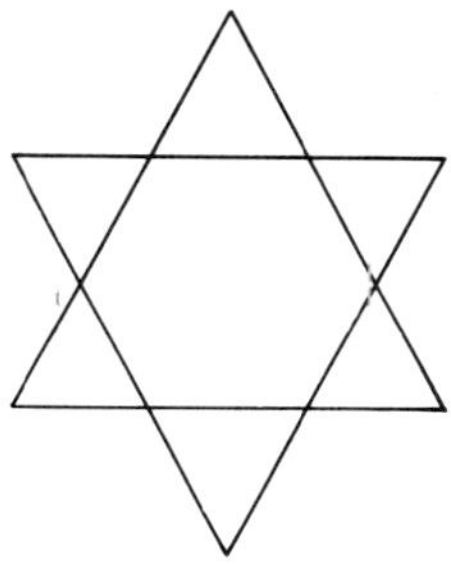

Figure 4.2.3

The reader will probably recognize that this is a network that cannot be traversed. In Chapter 3, we encountered other examples of networks that cannot be traversed, yet their shapes suggest that they should be considered closed.

The purpose of this discussion is not meant to confuse the reader, but to emphasize the role played by definitions in any discipline. A good definition must include everything that we want and exclude everything we do not want. As shown above, this can often prove a most difficult undertaking. Fortunately, for our purposes, we will be primarily concerned with simple closed curves and a characterization of such curves is sufficient for everything we hope to do.

Exercise 4.2.1

In Experiment 1, which curves represent simple closed curves? Why are the remaining curves not simple closed curves?

comment . . .

Students can be asked to describe their various doodles as simple curves, closed curves, and simple closed curves.

The most important property of a simple closed curve is that it separates the plane into exactly three distinct sets of points. These are the points which make up the curve which, in turn, acts as a boundary of two other sets of points of the plane. One of these sets is called the interior or inside of the curve, while the other is called the exterior or outside of the curve. The plane is the union of these three sets. The union of the interior of a simple closed curve and the curve itself is called a *region*. The set of points on and between two or more simple closed curves is also referred to as a region.

Example 1
The shaded areas in the pictures of Fig. 4.2.4 are regions.

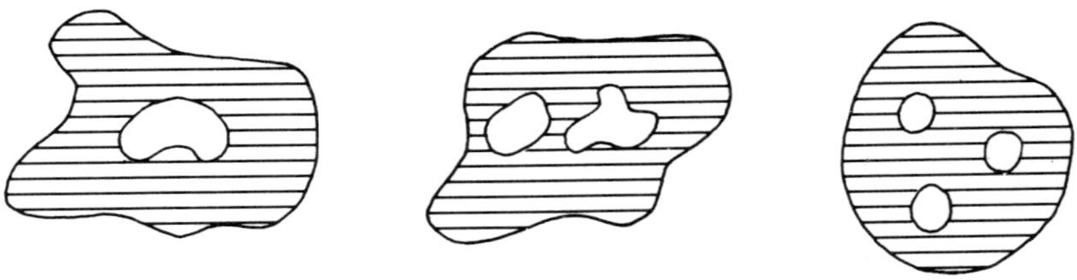

Figure 4.2.4

comment . . .

It is interesting to note that a simple closed curve may not appear all that simple and another curve which seems quite worthy of the name is not simple. For instance, curve (a) (in Fig. 4.2.5) is a simple closed curve and (b) is not.

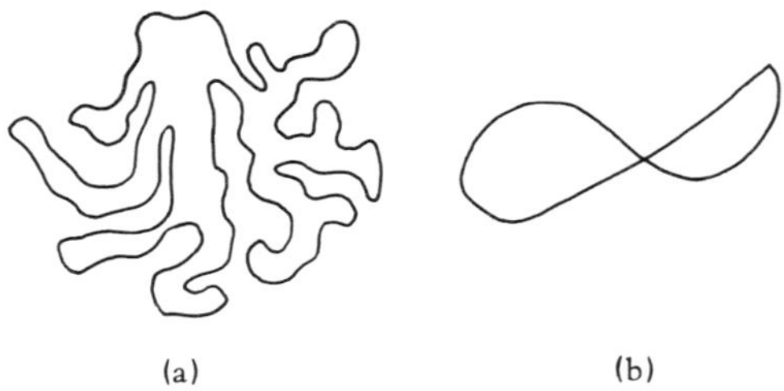

Figure 4.2.5

If X is a point in the interior of a simple closed curve α and Y any point in the exterior of α (Fig. 4.2.6), then the intersection of α with any curve containing X and Y contains at least one point of α.

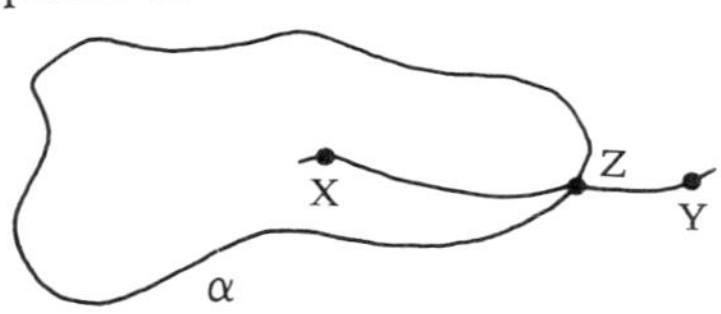

Figure 4.2.6

If X and Y (Fig. 4.2.7) are both interior points or both exterior points, then there is a curve containing X and Y which has no points in common with α.

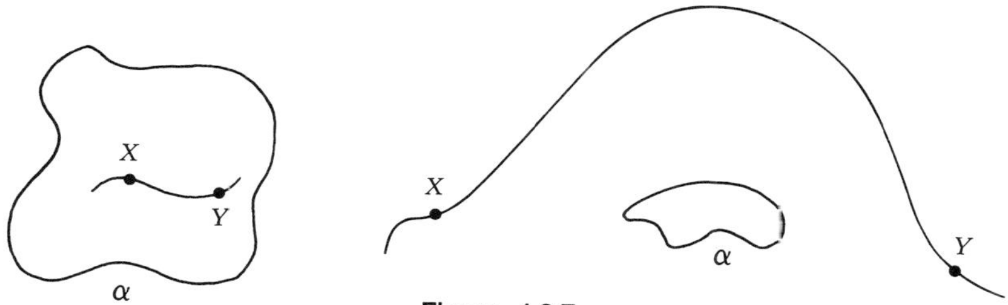

Figure 4.2.7

4.2.1 Definition

A simple closed curve is called convex *if and only if the line segment joining any two points of its interior is itself a subset of the interior.*

In other words, a simple closed curve is convex if and only if the set of points which is the union of the curve and its interior is a convex set.

comment . . .

This is an unfortunate case where a word, convex, is used to describe two related but different concepts. We could introduce a new name to describe a simple closed curve whose interior region is convex, but such curves are modified with enough adjectives as it is.

Example 2

See Figs. 4.2.8 and 4.2.9.

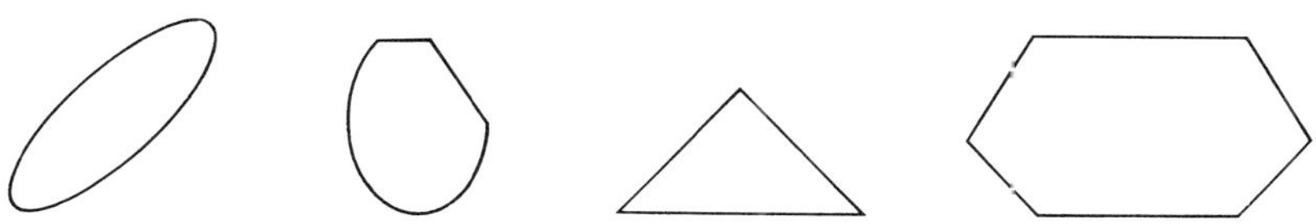

(a) (b) (c) (d)

Figure 4.2.8 Convex curves

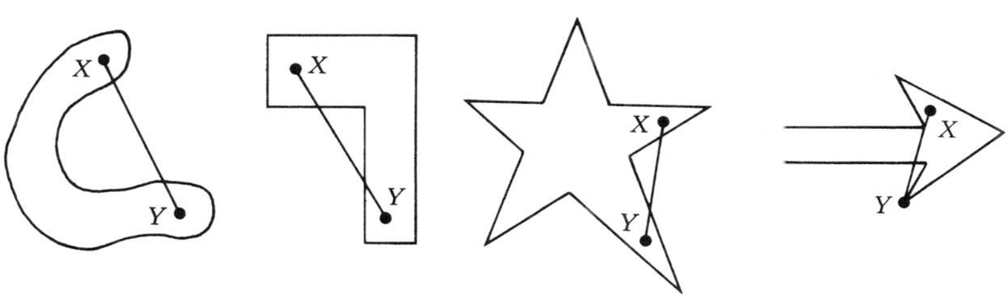

(a) (b) (c) (d)

Figure 4.2.9 Nonconvex curves

Exercise 4.2.1

1. Determine which of the curves in Fig. 4.2.10 are simple closed curves.

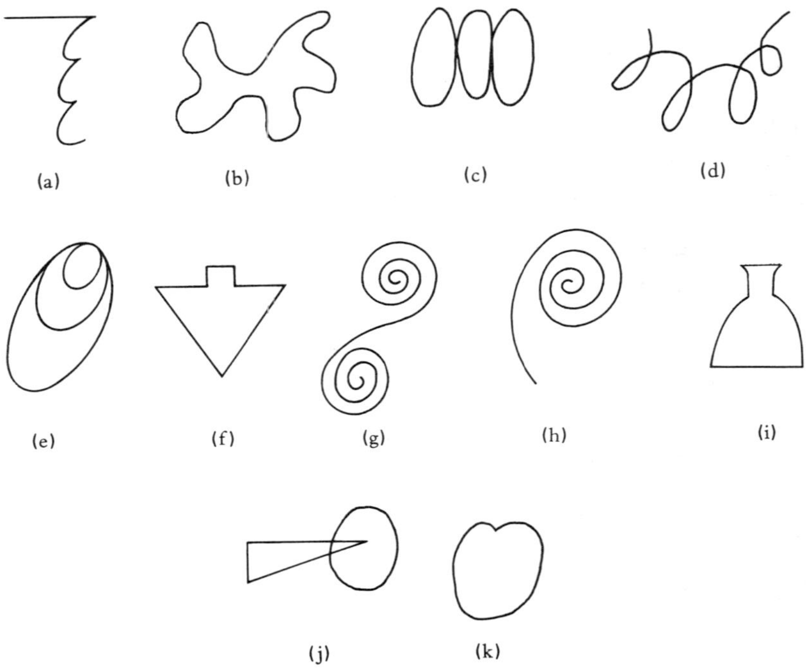

Figure 4.2.10

2. Look at the letters of the alphabet. Which letters represent simple closed curves? Does your answer depend on the form of script used in writing the letters? Answer the same questions for the numerals. Are any of the boundaries of the states of the United States not a simple closed curve?

4.3 POLYGONS

A particularly important family of space curves consists of simple curves formed by taking the union of line segments. We call such curves *plane polygonal paths* (Fig. 4.3.1). To be more precise, this family may be defined as follows.

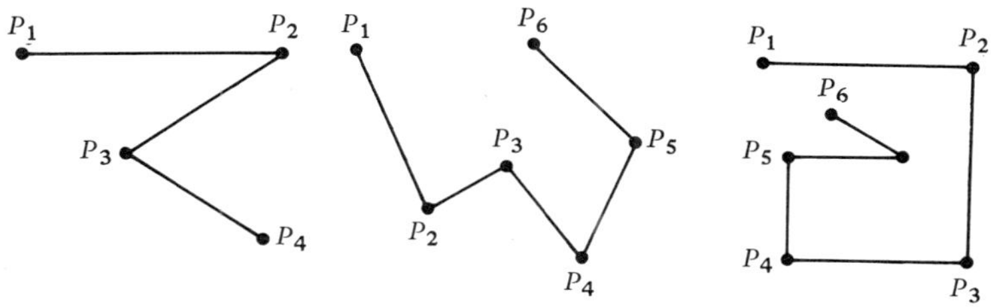

Figure 4.3.1

4.3.1 Definition

A plane polygonal path *is any simple curve which is the union of line segments:* $\overline{P_1P_2} \cup \overline{P_2P_3} \cup \cdots \cup \overline{P_{n-1}P_n}$ *where* $P_1, P_2, \ldots, P_n$ *are n points for which no three consecutive points are collinear.*

comment . . .

There are also polygonal paths which are not in the plane and they are defined in a similar fashion.

A polygonal path which is a simple closed curve is called a *polygon.* The line segments forming the polygon are called the *sides* of the polygon. When two sides intersect, the point of intersection is called a *vertex* and two such sides are called *adjacent sides.* The endpoints of any side of a polygon are called *adjacent vertices.* A line segment joining any two nonadjacent vertices is called a *diagonal* of the polygon.

4.3.2 Definition

More precisely, let $P_1, P_2, \ldots, P_n$ be n points in a plane with $n \geqq 3$ *and no three successive points collinear. A* polygon *is a union of line segments* $\overline{P_1P_2}, \overline{P_2P_3}, \ldots, \overline{P_{n-1}P_n}$, *and* $\overline{P_nP_1}$ *where these points are the only points of intersection of the given line segments. The segments* $\overline{P_1P_2}, \ldots, \overline{P_nP_1}$ *are the* sides *of the polygon and the points* $P_1, P_2, \ldots, P_n$ *are the* vertices *of the polygon.*

A polygon is named by its vertices $P_1P_2 \cdots P_n$. If $n = 3$, the polygon is a *triangle* and if $n = 4$, it is called *quadrilateral.* Table 4.3.1 contains names of other polygons

comment . . .

We have chosen to define a polygon as a simple closed curve. In doing so, Fig. 4.3.2 fails to qualify as a polygon. An interesting

Table 4.3.1

Number of sides	*Name of polygon*
5	Pentagon
6	Hexagon
7	Heptagon
8	Octagon
9	Nonagon
10	Decagon
.	.
.	.
.	.
n	n-gon

classroom experience is to have the students tell the teacher what is a quadrilateral. The teacher should take the students' instructions seriously and attempt to draw a picture meeting their definition. No doubt, their first attempt will result in a figure that in no way looks like a quadrilateral. In polishing their definition, the teacher should ask whether Fig. 4.3.2 should be considered a quadrilateral. There is no correct answer to this question. It really boils down to making a definition which will include all figures that we wish to consider as quadrilaterals, and this may or may not include Fig. 4.3.2, and to exclude all figures that we do not wish to consider as quadrilaterals.

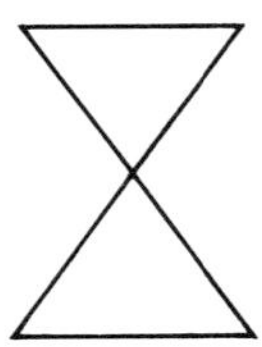

Figure 4.3.2

Our description of a *triangle* as a three-sided polygon is not sufficient to aid in classifying triangles as to size and shape. The properties that all triangles share are:

1. they lie in a plane
2. they are the union of three line segments
3. each line segment intersects each of the two other line segments in exactly one point, namely an endpoint of the segment.

The triangle in Fig. 4.3.3 will be symbolized $\triangle ABC$, read "triangle A, B, C" where A, B, and C are the three vertices of the triangle. Thus, $\triangle ABC = \overline{AB} \cup \overline{BC} \cup \overline{AC}$ such that $\overline{AB} \cap \overline{AC} = \{A\}$, $\overline{AC} \cap \overline{BC} = \{C\}$, and $\overline{BC} \cap \overline{AB} = \{B\}$. The triangle has three angles associated with it: $\angle ABC$, $\angle BCA$, and $\angle ACB$.

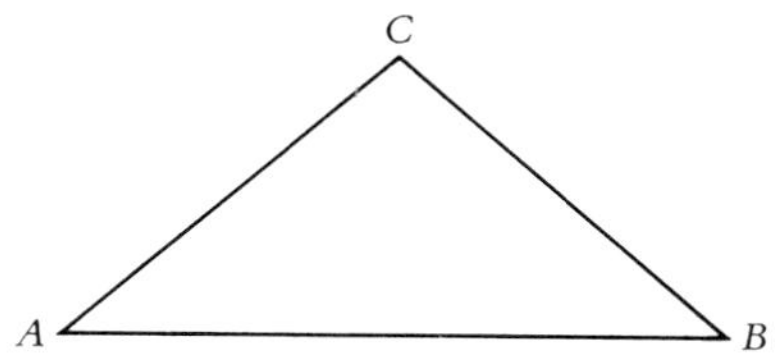

Figure 4.3.3

Notice, however, that $\triangle ABC$ does not contain these angles because the sides of an angle are rays and the sides of a triangle are segments.

The *interior* of $\triangle ABC$ is the intersection of the interiors of its three angles. We note that any triangle is a convex curve. Polygons with four or more sides may or may not be convex curves. For instance, the polygons in Fig. 4.3.4 are not convex curves. We call such polygons *concave* while those polygons with convex interiors will be called *convex polygons*. Another way of thinking of a convex polygon is as a polygon such that for each side, one of the two closed half-planes having that side as an edge contains each of the other sides of the polygon.

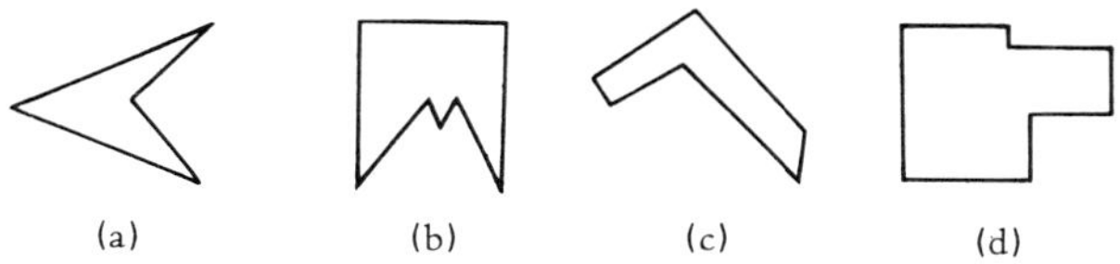

Figure 4.3.4

Example 1

See Figs. 4.3.5 and 4.3.6. Note neither of the closed half-planes whose edge contains $\overline{CD}$ contains each of the other sides of the quadrilateral. We can define the interior of a convex polygon.

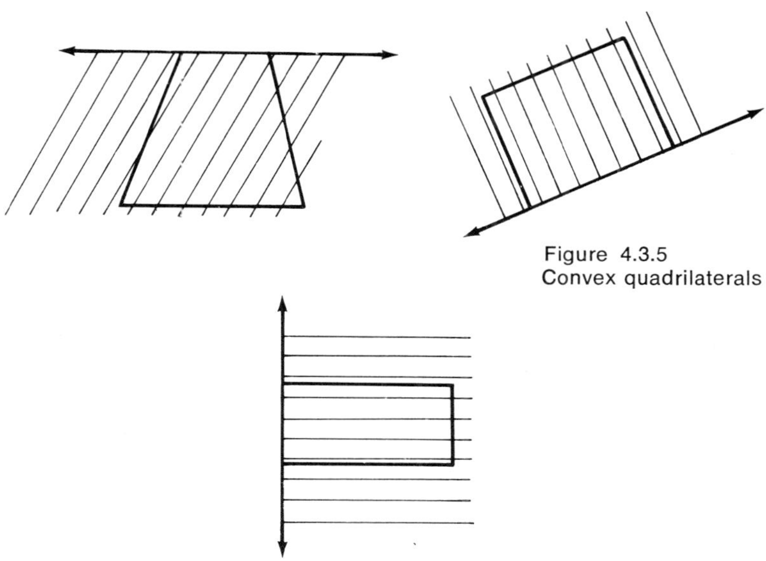

Figure 4.3.5
Convex quadrilaterals

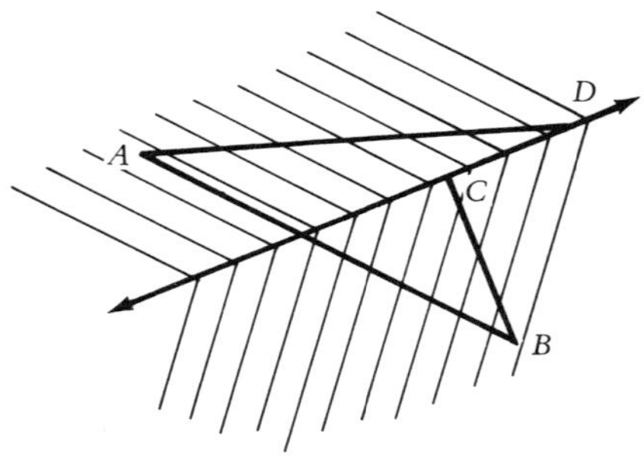

Figure 4.3.6
Concave quadrilateral

4.3.2 Definition

If $P_1P_2 \cdots P_n$ is a polygon which is convex, then its interior *is the intersection of the interiors of its angles.*

comment . . .

Students should be encouraged to find examples of various polygons in their everyday world. As a preparation for driver education, they should be able to identify each of the shapes in Fig. 4.3.7 and what information the shapes give.

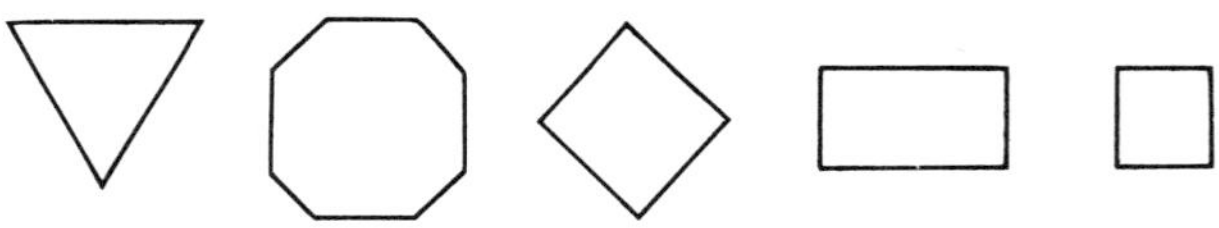

Figure 4.3.7

Experiment 3

Consider a polygon with four sides, such as quadrilateral $PQRS$ pictured in Fig. 4.3.8. Notice that quadrilateral is separated into two triangles by diagonal $\overline{PR}$. We also note that a quadrilateral has two diagonals. (We will be limiting our investigation to convex polygons.)

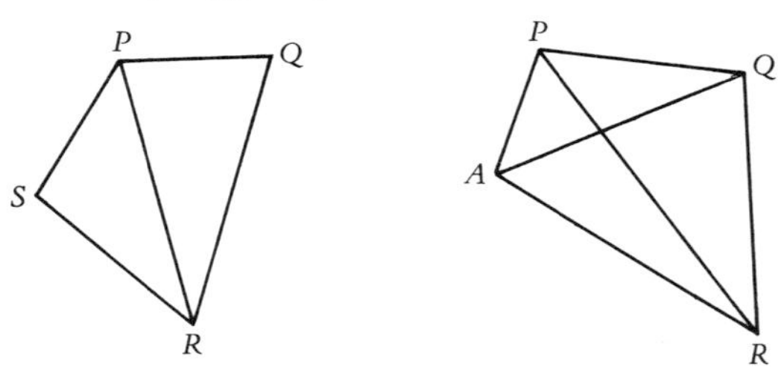

Figure 4.3.8

A polygon with five sides is a *pentagon* and can be separated into three triangles by drawing two of its diagonals from the same vertex as in Figure 4.3.9.

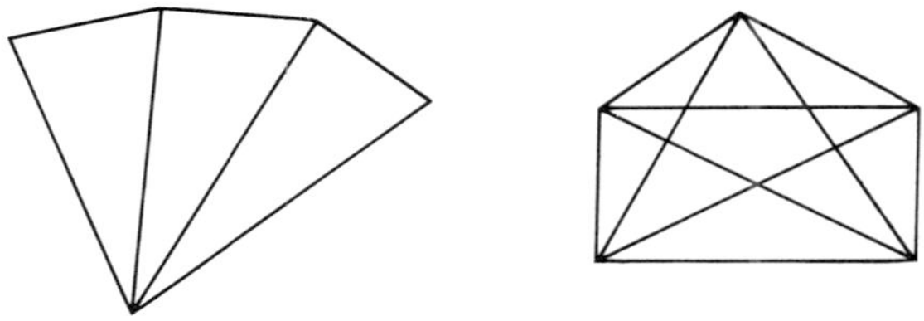

Figure 4.3.9

It is easy to check that a pentagon has five diagonals. In Fig. 4.3.10 we have separated a *hexagon* into four triangles.

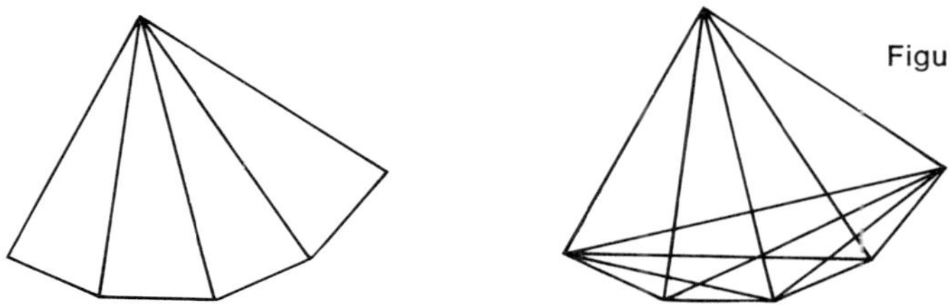

Figure 4.3.10

We find that a hexagon has nine diagonals. Complete Table 4.3.2 determining a formula giving the number of triangles a polygon can be separated into and the number of diagonals of the polygon.

Table 4.3.2

Number of sides	*Name of polygon*	*Number of triangles*	*Number of diagonals*
4	Quadrilateral	2	2
5	Pentagon	3	5
6	Hexagon	4	9
7	Heptagon		
8	Octagon		
9	Nonagon		
10	Decagon		
.	.		
.	.		
.	.		
n	n-gon		

One other important type of simple closed curve is that of a *circle*. The circle is not a polygon as it is not the union of line segments. It is defined as follows.

4.3.3 Definition

A circle *is a set of all points in a plane which are the same distance from a given point in the plane called the* center *of the circle.*

Note that the center of a circle is not on the circle but rather is in the interior of the circle. Each line segment from the center of the circle to a point on the circle is called a *radius* of the circle. In Fig. 4.3.11, $\overline{OA}$ is a radius of the circle. A segment having both endpoints on a circle such as $\overline{AB}$ is called a *chord* of the circle. A chord that contains the center of the circle such as $\overline{XY}$ is called a *diameter* of the circle. A line cointaining a chord of the circle is a *secant* of the circle. A line such as l in the plane of the circle which contains one and only one point of the

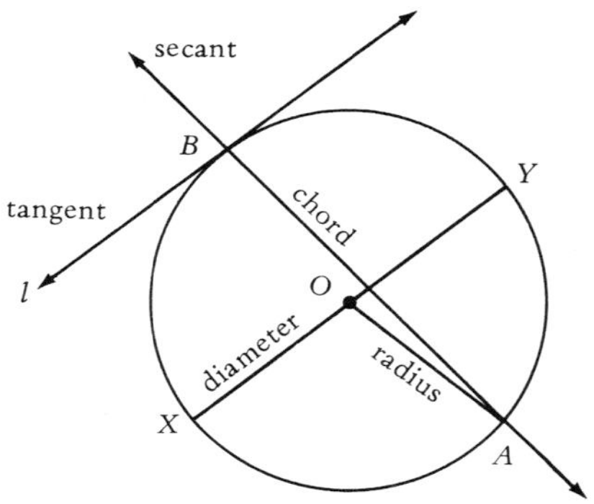

Figure 4.3.11

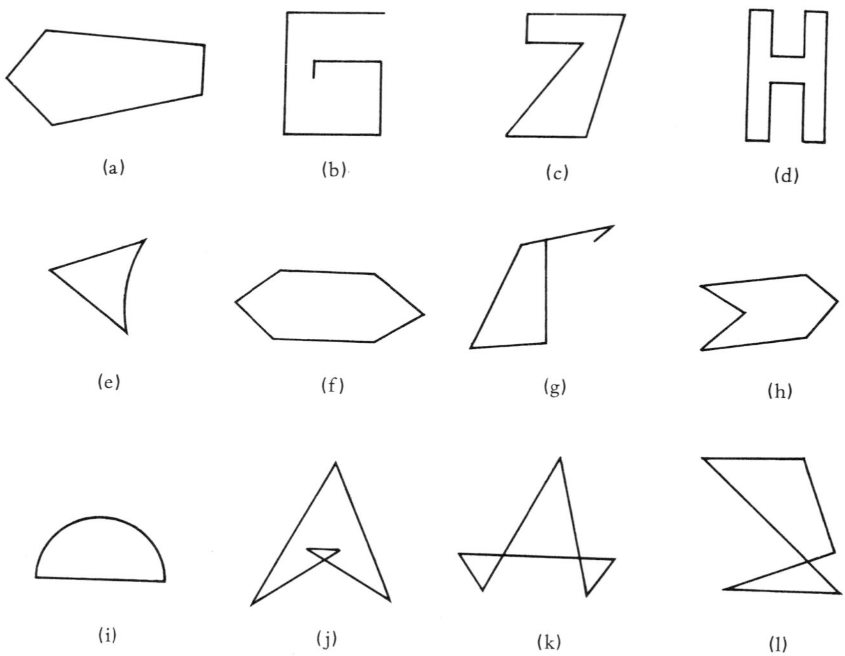

Figure 4.3.12

circle is said to be *tangent* to the circle. The point of intersection is called the *point of tangency*.

Exercise 4.3.1

1. Classify the curves in Fig. 4.3.12 as polygons or not polygons. Classify the polygons as convex or concave.
2. a. Draw convex polygons having 4, 5, 6, 7, and 8 sides.
   b. Draw all the diagonals in each of these convex polygons.
   c. Draw concave polygons having 4, 5, 6, 7, and 8 sides.
   d. Draw all the diagonals for each of these concave polygons.
   e. Does there appear to be a definition of a convex polygon dependent on the diagonals of the polygon?
3. Give arguments to support the following statements.
   a. Any two diagonals of a convex quadrilateral intersect at a point in its interior.
   b. If $ABCD$ is a quadrilateral, then A and B lie on the same side of $\overleftrightarrow{CD}$ or C and D lie on the same side of $\overleftrightarrow{AB}$. (Does it matter whether $ABCD$ is convex or concave?)

4.4 GEOMETRIC SOLIDS

Until now, we have restricted our study to figures formed in a plane. We now consider certain sets of points, not all of which lie in the same plane. The term *solid* is not widely used in geometry as a precise term, but it commonly appears as the name of certain special classes of *geometric configurations* (that is, any set of points in space). One way of thinking of a solid is as follows: First, consider the three-dimensional analogy of a circle, namely a *sphere*.

4.4.1 Definition

A sphere *is the set of all points in space that are the same distance from a given point called the* center *of the sphere (Fig. 4.4.1).*

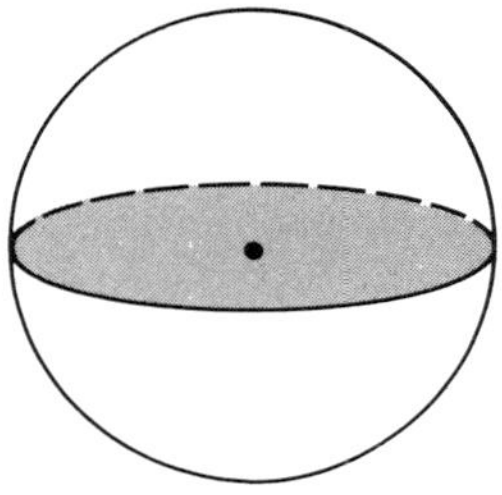

Figure 4.4.1

A *boundary point* of a *configuration* is a point such that every sphere having it as center contains points of the configuration and also points that do not belong to the configuration. A boundary point of a configuration need not belong to the configuration. The set of all boundary points of a configuration is called the *boundary of configuration.*

Example 1
A configuration will be called a *solid* if each nonboundary point of the configuration is the center of some sphere whose interior contains only points of the configuration (Fig. 4.4.2). We must also demand that such a configuration is not composed of two configurations which have no points in common. For example, the union of the two solids is not necessarily a solid.

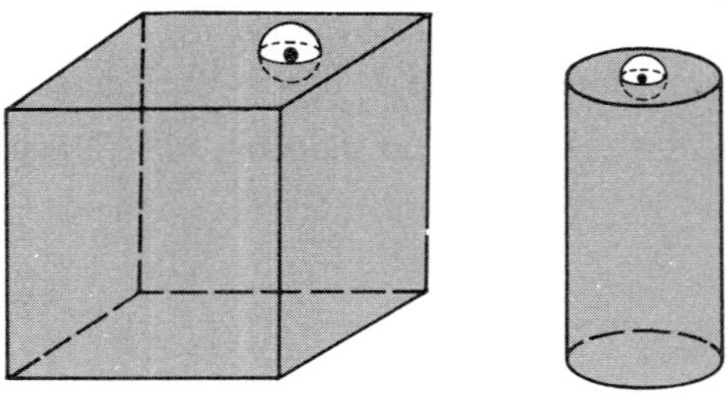

Figure 4.4.2

Example 2
It is easy to see, in Fig. 4.4.3, that the sphere and its interior is a solid. The classification of solids is an enormous problem. We will be interested in a very small part of this problem. A useful procedure in attacking the problem of classification of solids is to study their boundaries. The solids which we are interested in are those whose boundaries are *closed surfaces.* A sphere, a cylinder, and a torus are examples of closed surfaces. No precise definition of the concept of closed surface will be given except to say it is analogous to a closed curve so we will have to rely heavily on our intuition.

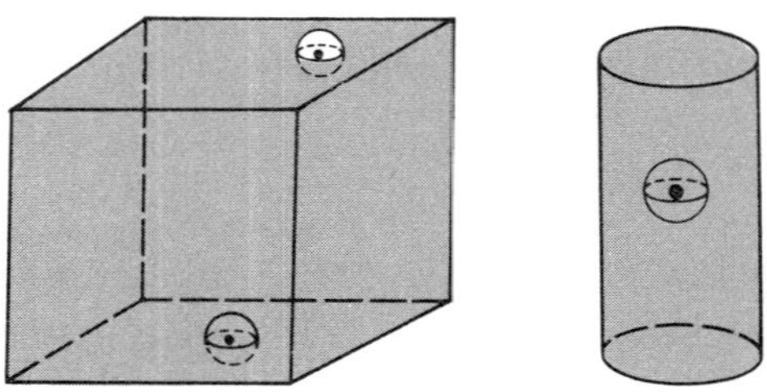

Figure 4.4.3

We will be particularly interested in a special class of solids: namely, *polyhedral solids* (that is, solids bounded by configurations formed by polyhedra). A *polyhedron* is a surface formed by four or more polygonal regions, situated in space such that space is separated into three disjoint subsets. Each polygonal region forming the polyhedron is called a *face* of the polyhedron.

comment . . .

This is not meant to be considered a formal definition of a polyhedron. The task of including everything we want to be considered a polyhedron and excluding everything we do not want to consider a polyhedron involves a very elaborate description. We will settle for identifying certain polyhedra.

If, for each face of a polyhedron, one of the two closed half-spaces that has that face as a boundary contains each of the other faces of the polyhedron, then the polyhedron is called a *convex polyhedron.* A *concave polyhedron* is a polyhedron that is not convex.

Example 3
See Figs. 4.4.4 and 4.4.5.

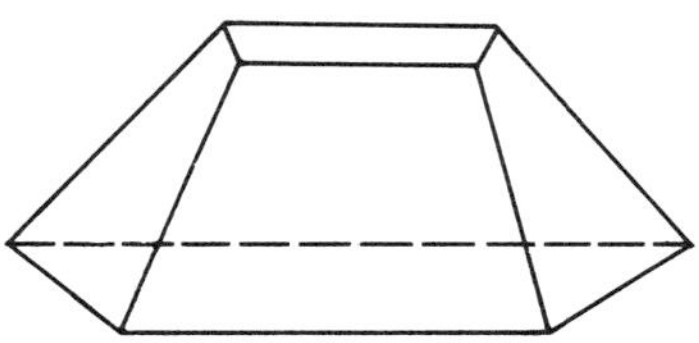

Figure 4.4.4
Convex polyhedron

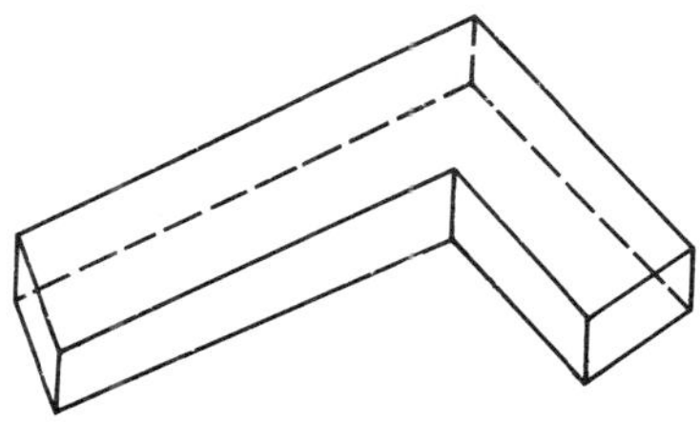

Figure 4.4.5
Concave polyhedron

comment . . .

The Greek word from which polyhedron is derived means literally "having many bases." (Plural: polyhedra.) Similarly, the word polygon is derived from a Greek word meaning "having many sides."

Experiment 4
Consider the polyhedron in Fig. 4.4.6. We note that the polyhedron has eight vertices. The intersection of the plane $\overleftrightarrow{ABC}$ with the polyhedron is a face of the polyhedron. Counting, we find that the polyhedron has six faces. The intersection of plane $\overleftrightarrow{ABC}$ with

plane $\overleftrightarrow{BCF}$ and the polyhedron is line segment $\overline{BC}$ which is called an *edge* of the polyhedron. The figure has twelve edges.

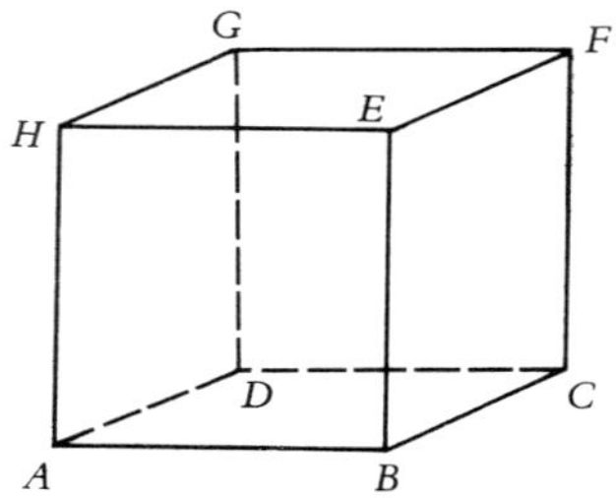

Figure 4.4.6

Consider each of the polyhedra in Fig. 4.4.7. Count the number of vertices (V), the number of faces (F), and the number of edges (E) for each of these polyhedra and complete Table 4.4.1.

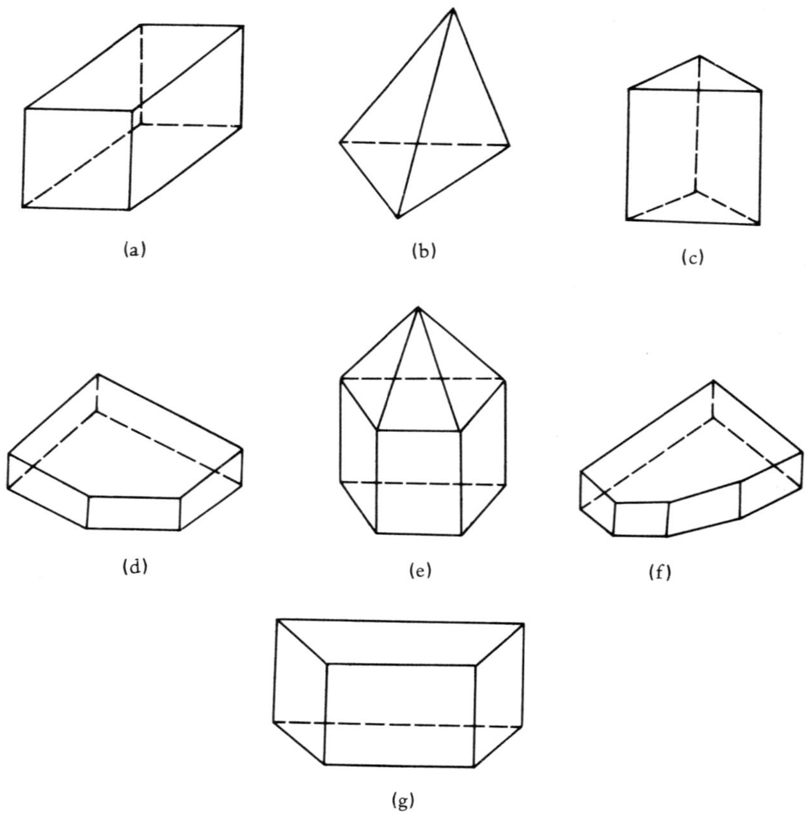

Figure 4.4.7

Table 4.4.1

Polyhedron	*F*	*V*	*E*
(a)	6	8	12
(b)	4	4	6
(c)	5	6	9
(d)			
(e)			
(f)			
(g)			

The mathematician Euler noticed a relationship between the vertices, faces, and edges of polyhedra. He was able to express this relationship in an equality which is called *Euler's formula.* That Euler's formula is true for all polyhedra can be proven, but it is beyond the scope of this book.

Try and make a conjecture concerning the relationship between the sum of the faces and the vertices of any polyhedron and the number of edges of the polyhedron. Check this conjecture by determining the number of faces, vertices, and edges of each of the polyhedra in Fig. 4.4.8.

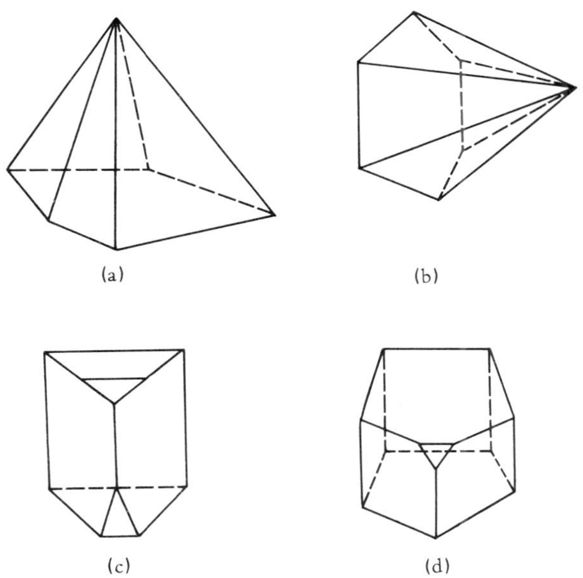

Figure 4.4.8

Exercise 4.4.1

A polyhedron with four faces is called a *tetrahedron*. Polyhedrons with five, six, seven, eight, and ten faces are called, respectively: *pentahedron, hexahedron, heptahedron, octahedron,* and *deca-*

hedron. Construct models of each of these polyhedra. Calculate the number of edges and the number of vertices of each polyhedra. Note there is more than one possibility for polyhedra with more than four faces. A *diagonal* of a polyhedron is a line segment connecting any two of its vertices not in the same face. How many diagonals does a tetrahedron have? A pentahedron? A hexahedron? A heptahedron? An octahedron? A decahedron?

4.5 PYRAMIDS

Just as a triangle is the simplest polygon, a tetrahedron is the simplest polyhedron.

4.5.1 Definition

A tetrahedron *is a polyhedron whose faces are triangular regions (Fig. 4.5.1).*

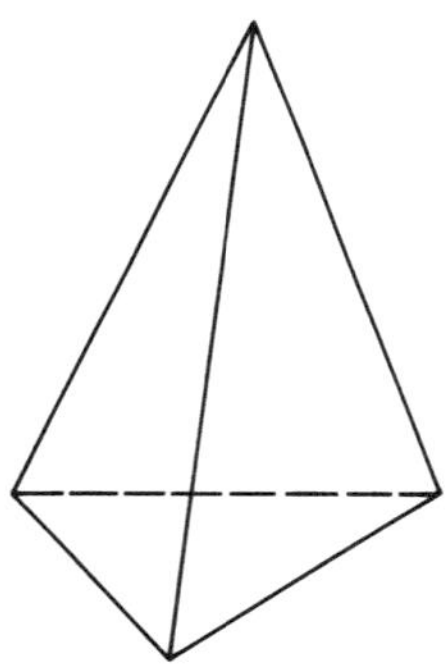

Figure 4.5.1

A tetrahedron is a special kind of pyramid which is itself a special type of polyhedron. Suppose we have some plane polygonal region and a point in space which is not in the same plane as the polygonal region. The figure formed by joining the vertices of the polygon to this point is called a pyramid.

4.5.2 Definition

A pyramid *is a polyhedron; one face called the* base *is some polygonal region while all other faces are triangular regions.*

Example 1

See Figs 4.5.2, 4.5.3, and 4.5.4. Note that a pyramid consists of one polygonal region and a number of triangular regions as its

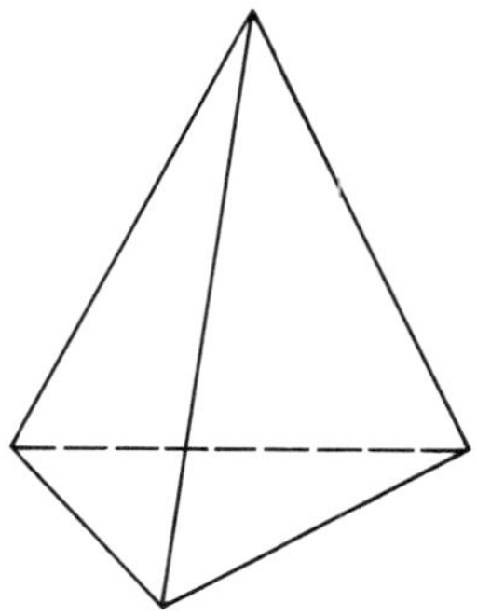

Figure 4.5.2
Triangular pyramid

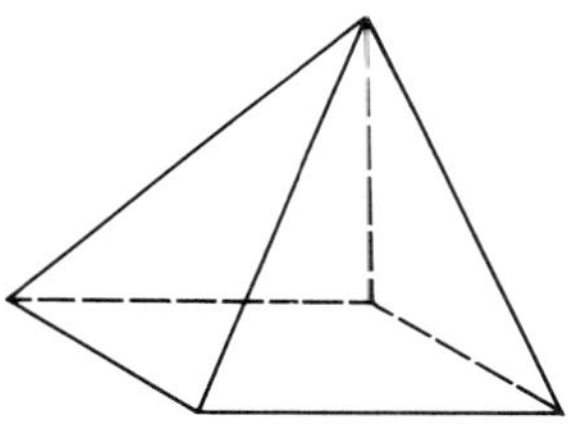

Figure 4.5.3
Quadrilateral pyramid

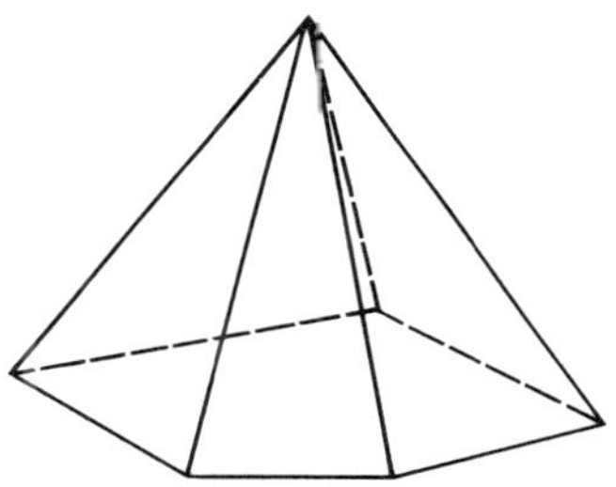

Figure 4.5.4
Pentagonal pyramid

faces. We refer to the polygonal region as the *base* of the pyramid and the triangular regions are called *lateral faces.* The common vertex of the triangular regions is called the *apex* of the pyramid. The edges of the pyramid that meet at the apex are called *lateral edges.*

As illustrated in the examples, pyramids are ordinarily named by their bases. This, of course, is not always the case as a triangular pyramid is more frequently called a tetrahedron. Of course, since all faces of a tetrahedron are triangular, any of its faces could be considered the base, any of its vertices could be considered its apex, and so on. Perhaps because of this, it deserves two names.

4.6 OTHER CLOSED SURFACES

Another important family of closed surfaces is that of *cones.* Consider a circle in a plane and some point not in the plane of the circle. Imagine line segments connecting every point of the circle to this point not in the plane of the circle. The resulting figure is called a *circular cone,* but because this is the only type of cone that we will study in this book, we will just refer to such a figure as a cone (Fig. 4.6.1). The circular region is called the *base* of the cone and the point P is called the *vertex* of the cone. The union of all the line segments from the vertex to points on the circle is called the *lateral surface* of the cone.

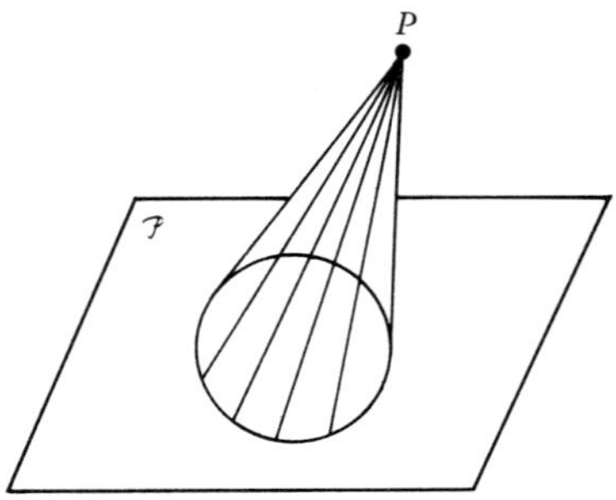

Figure 4.6.1

comment . . .

We can form a cone from any simple closed curve lying in a plane and some point not in the plane of the simple closed curve. The point not in the plane of the curve is joined by line segments to every point of the curve. For instance, Fig. 4.6.2 is a cone. In fact, pyramids can be considered special types of cones.

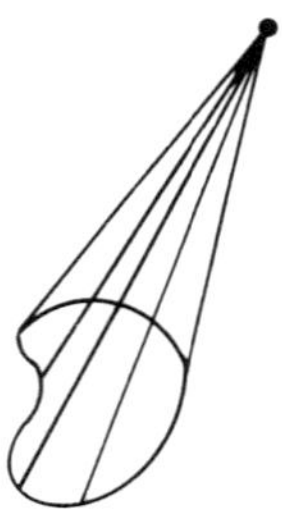

Figure 4.6.2

Another important class of closed surfaces is *cylinders*. Suppose we have two parallel planes and two circles, one in each plane. Let us suppose further that the radii of both circles are the same. Now connect the centers of the two circles with a line segment. Next, connect each point of one circle to a point in the other by means of a line segment that is parallel to the line segment connecting the centers of the two circles. The resulting figure is called a *circular cylinder*. Again, since this is the only type of cylinder we will be studying, we drop the adjective circular. The circular regions are called the *bases* of the cylinder and the union of all parallel lines connecting points of the two circles is called the lateral surface of the cylinder (Fig. 4.6.3).

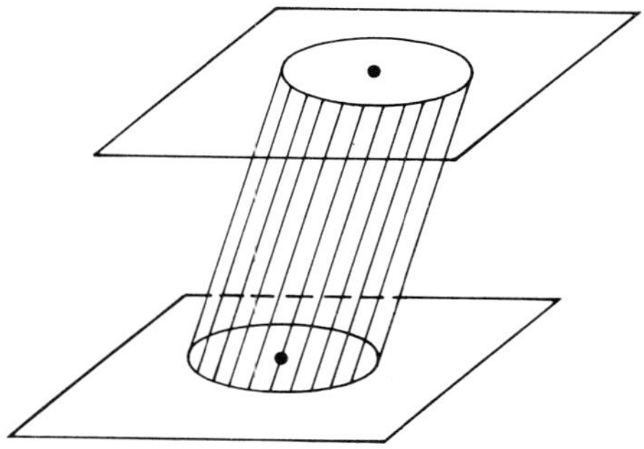

Figure 4.6.3

comment . . .

We can form a cylinder from any two simple congruent (that is, same size and shape) closed curves located in parallel planes. Join any two points of the curves by a line segment. Then connect pairs of points of the two curves by line segments parallel to the first line segment. For example, Fig. 4.6.4 is a cylinder.

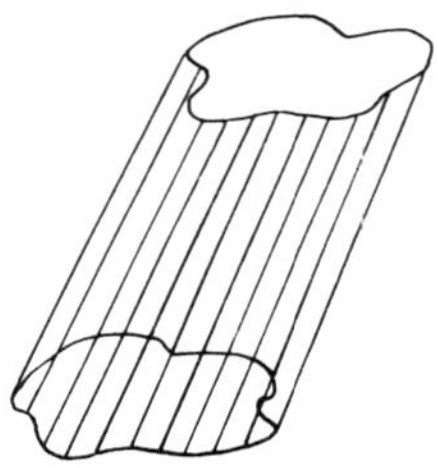

Figure 4.6.4

As we have already noted, a sphere is the set of all points in space that are the same distance from a given point of the space called the center of the sphere.

Any line segment such as $\overline{OA}$ connecting the center of the sphere to a point on the sphere is called a *radius* of the sphere. A line segment passing through the center of the sphere and having both endpoints on the sphere is called a *diametrical chord* or simply a *diameter* of the sphere. The endpoints of a diameter are called *antipodal points*. In Fig. 4.6.5, $\overline{CB}$ represents a diameter of the sphere.

In the picture we have drawn three planes that are parallel. The plane $\mathcal{S}$ has but one point in common with the sphere, namely P. We say that plane $\mathcal{S}$ is *tangent* to the sphere with point of *tangency* P. The plane $\mathcal{R}$ contains the center of the sphere. The intersection of a sphere with any plane that contains the center of the sphere is called a *great circle* Finally, plane $\mathcal{B}$ intersects the sphere in a circle but does not contain the center of the sphere. Such circles are called *small circles* of a sphere.

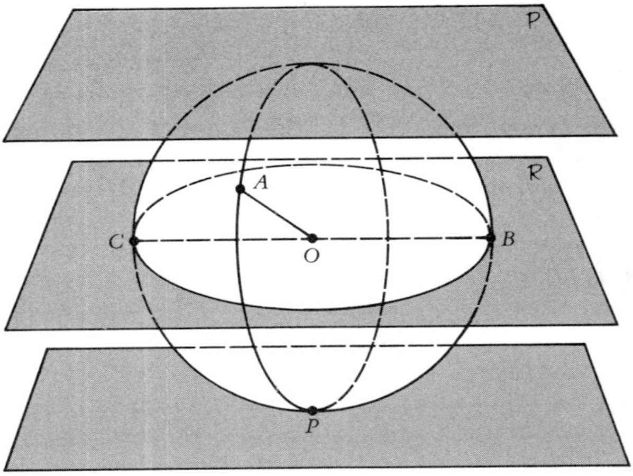

Figure 4.6.5

comment . . .

It is often said that the shortest distance between two points is a line segment. Although this may be true in a plane, on a sphere such as our earth, the shortest distance between two points is along the arc of a great circle which connects the two points.

There are other simple closed curves such as: squares, rectangles, parallelograms, trapezoids, rhombi; and other closed surfaces such as prisms, cubes, parallelepipeds which will be discussed after we have introduced the notion of measurement of lengths and angles.

in our opinion . . .

In this chapter, many important ideas were left undefined or only intuitively defined. This is partially due to the fact that we do not have all the necessary mathematical concepts available to us to make accurate definitions. As the student of this book progresses in his study of mathematics, both the need for better definitions and the prerequisite ideas will become more apparent. Part of the problem of mathematics is just this: Make good definitions, classify sets in a useful manner, and learn to describe mathematical observations.

Even with incomplete definitions or no definitions at all, a student can have an intelligent grasp of a mathematical idea. For instance, the notion of continuity is experienced by a child in water play. A puddle of water is not separate items but something continuous. This differs markedly from the pieces of a puzzle which can be put together to form a simple picture, yet is made up of individual pieces and is discontinuous. This is a naive notion of continuity but will contribute to a mathematically correct version eventually.

The child plays with blocks and notes the various structures he can build. Given a collection of three-dimensional shapes such as cubes, pyramids, spheres, cylinders, and so on, he may find that he can stack cubical shapes on top of each other in several ways, his cylinders can be built up on top of each other but not as easily as cubes, and that pyramids cannot be stacked very efficiently. In doing this, he has begun the classification problem.

There is an interesting child's toy which is formed by putting together two half-spheres. At various places on the surface, there are holes in the shape of squares, rectangles, stars, triangles, and so on. The child has several blocks whose faces are squares, rectangles, stars, triangles, and so on. By fitting these blocks properly in the holes in the sphere, they can be put inside the sphere. Children of ages two and up learn to place all their blocks in the sphere in a very short length of time. They have learned a great deal of geometry in the process.

In all these activities, the child has reached a level of abstraction on his own which, unfortunately, may not be improved in school. They do not possess correct definitions of geometric ideas, but this does not mean that they do not have definitions. An interesting experiment in an elementary school class is to ask students to answer such questions as: What is a triangle? What is a cube? What is a box? And so on. The wise teacher can determine from the students' answers much about the extent of their mathematical ability and their ability to abstract and classify.

Exercise 4.6.1

1. **4.6.1 Definition**

   The space geometric figure that is analogous to an angle is the union of two noncoplanar closed half-planes sharing the same boundary line and is called a dihedral angle *(Fig. 4.6.6).*

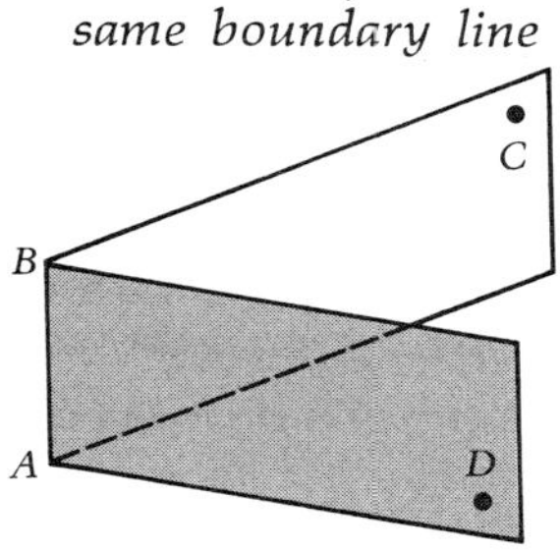

Figure 4.6.6

The half-planes are called the *faces* of the dihedral angle and the boundary line is called the *edge* of the angle. The standard notation for naming the dihedral angle pictured in Fig. 4.6.6 is $C - \overleftrightarrow{AB} - D$ where C is a point in one of the faces, $\overleftrightarrow{AB}$ is the edge, and D is a point in the other face.

Let $\mathcal{P}$ be the plane determined by C and $\overleftrightarrow{AB}$ and $\mathcal{R}$ be the plane determined by D and $\overleftrightarrow{AB}$. Find each of the following sets.

a. $\mathcal{P} \cap \mathcal{R}$ b. $\overleftrightarrow{AC} \cap \mathcal{P}$

c. $\overleftrightarrow{AC} \cap \mathcal{R}$ d. $\overleftrightarrow{DC} \cap \mathcal{R}$

e. $\overleftrightarrow{DC} \cap \mathcal{P}$ f. $\overleftrightarrow{AB} \cap \overleftrightarrow{DC}$

g. $\overleftrightarrow{BD} \cap \mathcal{R}$ h. $\overleftrightarrow{BD} \cap \mathcal{P}$

2. Let $X_1, X_2, \ldots, X_n$ be the vertices of a convex polygon and P a point not in the plane of the polygon. The union of all possible rays $\overrightarrow{PX_1}, \overrightarrow{PX_2}, \ldots, \overrightarrow{PX_n}$ is called a *polyhedral angle* (Fig. 4.6.7). Point P is called the *vertex* of the angle, the rays $\overrightarrow{PX_1}, \ldots, \overrightarrow{PX_n}$ are called the *edges* of the polyhedral angle. A plane angle formed by two successive edges is called a *face angle* of the polyhedral angle. The interiors of the face angle are called the *faces* of the polyhedral angle. We name the polyhedral angle by $P - X_1X_2 \cdots X_n$. A polyhedral angle is named depending on the number of faces it has. A polyhedral angle of three, four, five, six, and eight faces is, respectively, called *trihedral, tetrahedral, pentahedral, hexahedral,* and *octahedral.* Draw each of these angles. Give a physical model of each of these polyhedral angles. What is the least number of face angles that can make up a polyhedral angle?

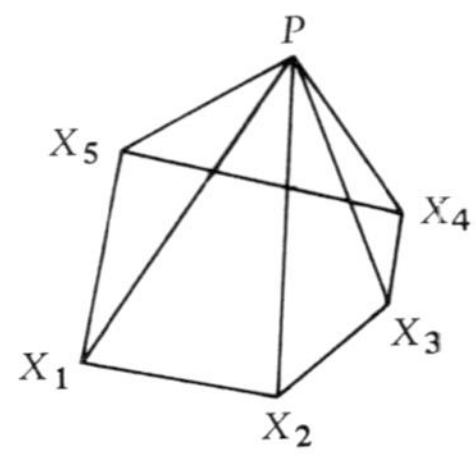

Figure 4.6.7

4.7 OTHER EXPERIMENTS

Experiment 5

Another way of describing polyhedra is in terms of *simplexes.* A *0-simplex* is a point, a *1-simplex* is a line segment, a *2-simplex* is a closed triangular region, a *3-simplex* is a solid tetrahedron.

The vertices of a polyhedron are 0-simplexes and the edges are 1-simplexes. The faces are plane regions called polygons, each of which is a union of 2-simplexes (see Experiment 3). Further, a solid polyhedron may be dissected into three simplexes. The study of polyhedra could be facilitated by regarding polyhedra as figures built up systematically out of simplexes.

Consider the polyhedron in Fig. 4.7.1. Describe a dissection of this polyhedron into three 3-simplexes.

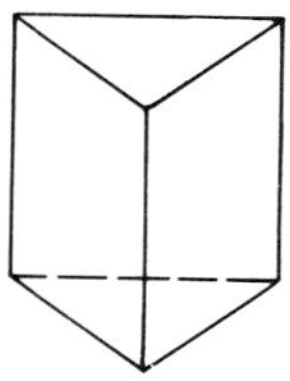

Figure 4.7.1

Experiment with other polyhedra, describing a dissection of each polyhedron. If π is a polyhedron and s_1 and s_2 are two 3-simplexes of π, how do s_1 and s_2 meet? That is, describe $s_1 \cap s_2$ in terms of simplexes.

By using simplexes we can also define the notion of the *dimension* of any subset ξ of space. Define $\rho(\xi) = \xi \cup \{X | P - X - Q \text{ for some } P, Q \in \xi\}$. Determine $\rho(\xi)$ if ξ is a singleton set, a doubleton set (that is, a set of two distinct points), a set consisting of three distinct noncollinear points, a set consisting of four distinct noncoplanar points. In each case, is $\rho(\xi)$ a convex set? If not, describe $\rho(\rho(\xi))$; is it convex? In each case, repeat application of ρ until you reach a convex set. The *dimension* of a set is the number of times the operation ρ must be applied to the set of vertices to get a convex set.

4.8 SUMMARY

1. A curve is a set of points forming a continuous path.
2. Any curve which is an allowable distortion of a circle is a simple closed curve.
3. A simple closed curve is convex if and only if the line segment joining any two points of its interior is itself a subset of the interior.
4. The union of the interior of a simple closed curve and the curve itself is called a region.
5. A plane polygonal path is a simple curve passing through n distinct points, no three consecutive points being collinear

and the curve formed by taking the union of the line segments connecting the n points in order.

6. A polygon is a plane polygonal path which is a closed curve.
7. All three-sided polygons are called triangles. A triangle is the union of three line segments such that each line segment intersects the other two line segments in exactly one point, an endpoint of the segment.
8. All four-sided polygons are called quadrilaterals.
9. Polygons whose interiors are convex sets are called convex polygons.
10. A diagonal of a polygon is a segment that joins two vertices and is not a side of the polygon.
11. A circle is a set of points in a plane, each of which is the same distance from a given point called the center of the circle. Any line segment from the center of the circle to a point on the circle is called a radius.
12. A sphere is the set of points in space, all of which are the same distance from a given point called the center of the sphere.
13. A polyhedron is formed by four or more parts of planes with boundaries that are polygons whose pair of edges are coincident, situated in space, such that space is separated into three disjoint subsets.
14. A tetrahedron is a polyhedron whose faces are triangular regions.
15. A pyramid is a polyhedron; one face called the base is some polygonal region while all other faces are triangular regions.

Exercise 4.8.1

Determine whether the following are true or false.

1. The angles of a triangle belong to the interior of the triangle.
2. Every pentagon is a convex polygon.
3. Points of the interior of a simple closed curve are not part of the curve.
4. The exterior of a triangle is a convex set.
5. A triangle has no diagonals.
6. All simple curves are closed curves.
7. If two simple closed curves intersect in exactly one point, the union of these two curves is a simple closed curve.
8. The surface of a ball is a closed surface which is not a polyhedron.

ANOTHER LOOK AT COORDINATE GEOMETRY

5

5.1 PROLOGUE

Whenever we identify a new set of points in space such as a line segment, a ray, a half-plane, a triangle, and so on, we could consider the related question: How can we describe the coordinates of this set of points? Whereas we shall be able to answer this question for some of these sets now, we shall have to set aside our consideration of others until we have developed some additional geometric ideas. In most cases, we will find that it is relatively easy to describe certain special cases of these subsets. For example, we will not find it difficult to describe the coordinates of points on lines which are parallel to one of the axes. However, for lines in general, we will have to delay our discussion to a later chapter. In our investigation, we will also want to consider the the coordinates of planes and half-planes. But since we know that the boundary of a half-plane is a line, we will also have to consider lines. Yet, consideration of lines reminds us that lines are determined by the intersection of planes. So our investigation leads us back to where we started. At some point, we will find that in order to break out of this circuitous investigation, it will be necessary to abandon the coordinate approach and return to some synthetic developments of prerequisite geometric notions.

in our opinion . . .

In the prologue of Chapter 2, we pointed out that until Descartes, algebra and geometry were treated as separate unrelated ideas. If the modern attitude rejects this idea, then it should be reflected in the way one develops algebraic and geometric ideas. To be sure, synthetic geometry can motivate algebraic concepts which can, in turn, lead to further excursions into synthetic geometric thoughts. Thus, algebra and geometry complement each other, uniting to form a blend of mathematical thought which possesses more conceptual power than either of its parts viewed separately. We believe that the devotion of a single section of this book to coordinate geometry destroys this attitude. Thus, we are steadfast in our belief that the only way one can become aware of the interdependence of algebraic and geometric ideas is to freely move from one to the other throughout the book.

5.2 THE COORDINATES OF LINES IN THREE-DIMENSIONAL SPACE

If we recognize that lines can be determined by the intersection of two planes, perhaps we should be able to describe the coordinates of a line in space by designating the coordinates of the planes which intersect to form it. Let us first consider the x-axis. What are two planes which intersect to form it? Look at Fig. 5.2.1.

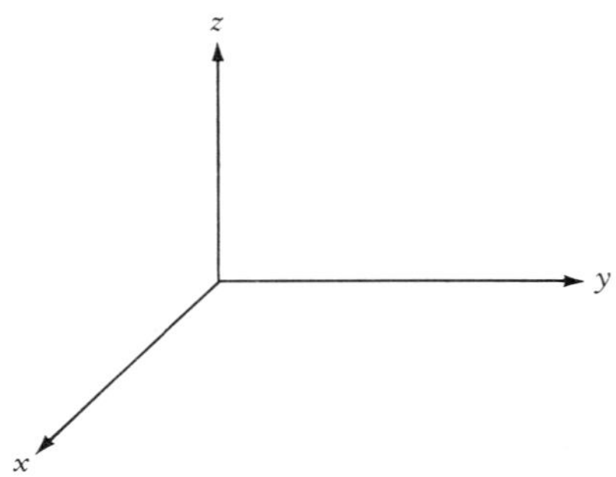

Figure 5.2.1

Can you see that the x-axis is formed by the intersection of the x-y plane with the x-z plane? Thus, we could say

$$\{P|P \in x\text{-axis}\} = \{P|P \in x\text{-}y \text{ plane}\} \cap \{P|P \in x\text{-}z \text{ plane}\}$$

Thus, we should be able to describe the coordinates of the x-axis by describing the coordinates in the x-y and x-z planes. To do this, we should consider the next experiment.

Experiment 1

a. Which of the following coordinates correspond to points in the x-y plane? $(0, 0, 0)$, $(2, 1, 1)$, $(0, 5, 2)$, $(-2, 1, 0)$, $(\sqrt{5}, -1, -4)$, $(8, 0, 1)$, $(3, -1, 0)$, $(-2, -5, 0)$, $(-1, 1, 5)$, $(0, 0, 8)$, $(-6, 7, 0)$, $(\pi, -2, 0)$, $(0, \sqrt{2}, \pi)$, $(-.1010010001\ldots, .32, 0)$.
b. Write down five more ordered triples which correspond to points on the x-y plane. What do you observe?

What is true about all coordinates in the x-y plane? From the previous experiment, you should have seen that the third member of the triple is zero and the first two members can be any real numbers. This description for the coordinates of the x-y plane can be denoted as $\{(x, y, z)|z = 0\}$. Likewise, you should be able to convince yourself that the x-z plane can be denoted as $\{(x, y, z)|y = 0\}$. Therefore, the statement

$$\{P|P \varepsilon x\text{-axis}\} = \{P|P \varepsilon x\text{-}y \text{ plane}\} \cap \{P|P \varepsilon x\text{-}z \text{ plane}\}$$

can be replaced by the statement

$$\{P|P \varepsilon x\text{-axis}\} \text{ is the graph of } \{(x, y, z)|z = 0\} \cap \{(x, y, z)|y = 0\}$$

That is, the x-axis consists of the intersection of the set of all points with the third member zero with the set of all points with the second member zero. Thus, the coordinates of such points in this intersection have zero for both the second and third members. Or, we could say

$$\{P|P \varepsilon x\text{-axis}\} \text{ is the graph of } \{(x, y, z)|z = 0\} \cap \{(x, y, z)|y = 0\} = \{(x, y, z)|y = 0, z = 0\}$$

Now, how might we describe only the positive half of the x-axis? What restrictions are now imposed on the *first* member of the ordered triples? Since we want the positive half, the x-coordinate must be positive. We could say that this half-line could be described as $\{(x, y, z)|x > 0, y = 0, z = 0\}$. To see if you understand this discussion, you should try some of the problems in the next exercise set.

Exercise 5.2.1

Describe the following sets using set-builder notation.

1. z-axis
2. The line in the y-z plane parallel to the y-axis but 3 units above it.
3. What is the name of the set of points which correspond to $\{(x, y, z)|z = 5, x > 0\}$?
4. Describe the coordinates of the ray which extend along the z-axis from the origin in a negative direction.
5. Describe the coordinates of the line segment which is a subset of the positive half of the x-axis with one endpoint 3 units from the origin and its other endpoint 7 units from the origin.
6. Describe the coordinates of the line segment which is a subset of the negative half of the y-axis, is $4\frac{1}{2}$ units long and the endpoint which is closest to the origin is 2 units from it.

The preceding discussion suggested that one way to describe the coordinates of lines and their subsets was by describing the coordinates of the planes whose intersection was the line. But we must be able to describe the coordinates of any plane in general. Eventually we hope to be able to do this, but sometimes we may be able to avoid this problem. If the line in question happens to be in the x-y plane, then we need not concern ourselves with deciding what second plane intersects with the x-y plane to form this line. If we knew how to describe the ordered pairs of points in the x-y plane which correspond to the given line, we would then be able to describe the ordered triple for that same line in space by adding to that description the additional requirement that places the points in the x-y plane. Therefore, we shall confine our consideration for a time to the two-dimensional case. Let us first consider all lines in the x-y plane which are parallel to the x- or y-axis.

The next experiment should help you investigate this idea.

Experiment 2

Step 1

a. Which of the following coordinates would correspond to points in the x-y plane which are on a line parallel to the x-axis and 5 units above it? (2, 3), (2, 5), (0, 0), (−2, 5), (0, 5), (2, −5), (8, 9), (10, 5), $(\sqrt{2}, 5)$, (5, 2), (−5, 8), (5, 5), (−5, 5), (−5, −5), $(-\pi/9, 0)$
b. Suggest four more ordered pairs which would be on this line.
c. What is true about coordinates of all the points on this line? Use set notation to describe these coordinates.

Step 2

a. Which of the following coordinates correspond to points in the x-y plane which are on a line parallel to the y-axis and 7 units to the left of it? (0, 0), (7, 2), (8, −7), (−7, 8), (5, −7), (−7, 5), (2, 3), (5, 10), (−7, 6), (−7, 0), (7, −7), (−7, −7), (−7, 7), $(\sqrt[3]{4}, -7)$, $(-7, -\sqrt[3]{4})$
b. Suggest four more ordered pairs which would be on this line.
c. What is true about the coordinates of all the points on this line? Use set notation to describe these coordinates.

From the preceding experiment, you should have seen that, in general, if we have a line in the x-y plane parallel to the x-axis and k units above it, the second member of all coordinates would be k, as is shown in Fig. 5.2.2. Thus, the coordinates for such a line would be $\{x, y)|y = k\}$. Likewise, a line parallel to the x-axis but k units below would be described as $\{(x, y)|y = -k\}$.

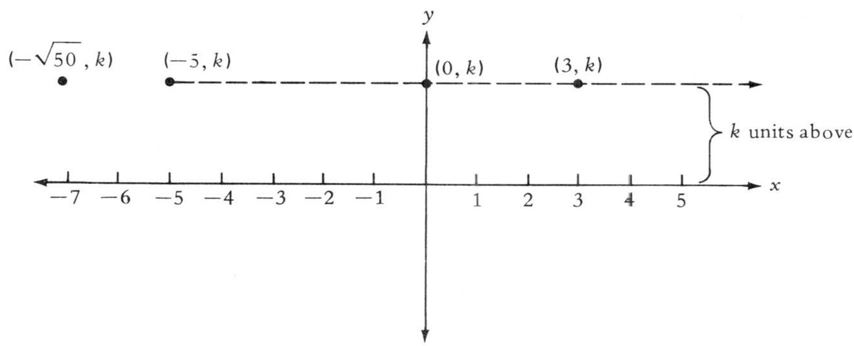

Figure 5.2.2

comment . . .

It is imperative that the reader note that this discussion is valid only as long as k is positive. Indeed, if k were a negative number, then $\{(x, y)|y = k\}$ could correspond to a line parallel to the x-axis but below it, and $\{(x, y)|y = -k\}$ would be the line parallel to it and above it. An analogous discussion is true for $\{(x, y)|x = k\}$.

Without too much difficulty, you should also be able to see that a line parallel to the y-axis and k units to its right or left is $\{(x, y)|x = k\}$ and $\{(x, y)|x = -k\}$, respectively.

As for determining other lines in the x-y plane, we shall have to wait for a later chapter. In some of your previous exercises you may recall that such sets as $\{(x, y)|x + y = 2\}$, $\{(x, y)|y = 2x + 5\}$ seemed to be straight lines. However, in order to be able to completely justify our conjectures, we need some additional knowledge about parallelism and similarity of triangles. For the time being, then, we shall assume that they are straight lines.

Example 1

Now let us consider the following example. How would you describe the coordinates of points on a line which is in the x-y plane (parallel to the x-axis and $\sqrt{5}$ units above it)?

If we know that in the x-y plane it is described as $\{(x, y)|y = \sqrt{5}\}$, what does the corresponding description $\{(x, y, z)|y = \sqrt{5}\}$ give us in 3-space? From our earlier work, we recall that it is a plane. Thus, to say that the set of points are in the x-y plane, we must add the requirement that $z = 0$. Thus, we get $\{(x, y, z)| y = \sqrt{5}, z = 0\}$.

Example 2

Suppose we assume that $\{(x, y)|x + y = 2\}$ is a line in the x-y plane. What would be a description of this corresponding line in 3-space? We would use the same basic description and add the requirement that it is in the x-y plane. Thus, this line would be $\{(x, y, z)|x + y = 2, z = 0\}$.

Exercise 5.2.2

1. Describe the coordinates in 3-space of points which are on a line in the x-y plane, parallel to the y-axis, and 8 units to its right.
2. Decribe the coordinates in 3-space of points which are on a line in the y-z plane which is parallel to the y-axis and $12/\pi$ units below it.
3. Describe the coordinates in 3-space of points which are on the line in the x-z plane which corresponds to the description $\{(x, z)|2x + 3z = 4\}$.
4. a. What name is given to the graph on the number line of $\{x|x \geqq 3\}$? What is $\{x|x > 3\}$?

   b. Use set-builder notation to describe the ray which is opposite to $\{x|x \leqq 6\}$.

   c. How would you use set-builder notation to describe the coordinates in 2-space which correspond to the rays in the x-y plane as indicated in Fig. 5.2.3?

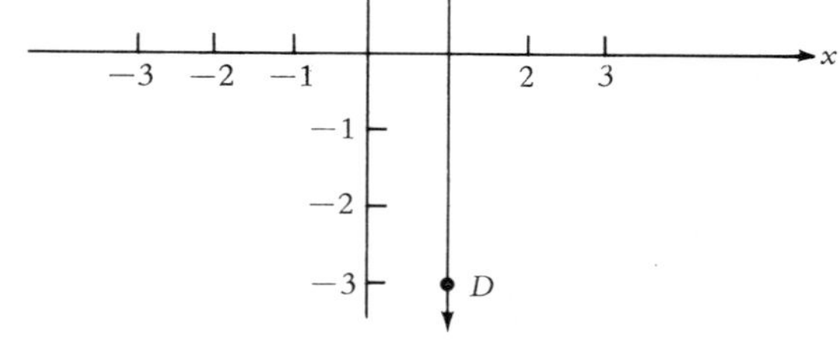

Figure 5.2.3

   d. Describe the same rays in part c if they were in the x-y plane in 3-space.

   e. Describe the same rays in part c if they were in a plane parallel to the x-y plane but 8 units above the x-y plane.

f. Suppose in Fig. 5.2.4 that $\overleftrightarrow{AB}$ can be described as $\{(x, y)|\ x - y = 2\}$. Use set-builder notation to describe the following rays: $\overrightarrow{AB}$, $\overrightarrow{CB}$, $\overrightarrow{AD}$, $\overrightarrow{CA}$, $\overrightarrow{BA}$.

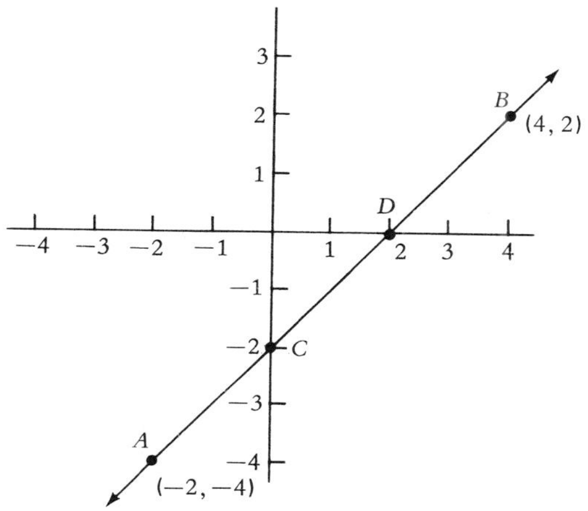

Figure 5.2.4

g. Describe the same rays in 3-space if they are in the x-y plane.
h. Describe the same rays in 3-space if they are in a plane parallel to the x-y plane and 15 units below it.

In the previous exercise, you learned how to describe the coordinates of all the points on a ray. The problem of describing the points of a line segment is quite similar since we are able to describe a line segment in terms of two rays. Consider Fig. 5.2.5.

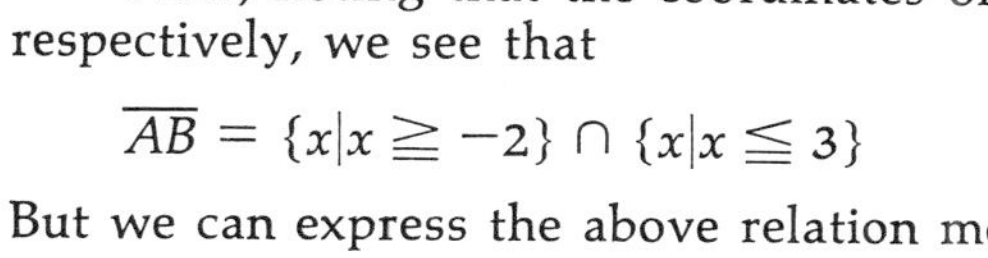

Figure 5.2.5
$AB = \overrightarrow{AB} \cap \overrightarrow{BA}$

Can you see that $\overline{AB} = \overrightarrow{AB} \cap \overrightarrow{BA}$?

Now, noting that the coordinates of A and B are -2 and 3, respectively, we see that

$$\overline{AB} = \{x|x \geqq -2\} \cap \{x|x \leqq 3\}$$

But we can express the above relation more conveniently as

$$\overline{AB} = \{x|x \geqq -2 \text{ and } x \leqq 3\} = \{x|-2 \leqq x \leqq 3\}$$

Now consider the next set of exercises.

Exercise 5.2.3

1. Use set-builder notation to describe the line segments $\overline{AB}$, $\overline{CD}$, $\overline{AC}$, $\overline{AD}$, $\overline{BC}$, $\overline{DB}$ as shown on the number line in Fig. 5.2.6.

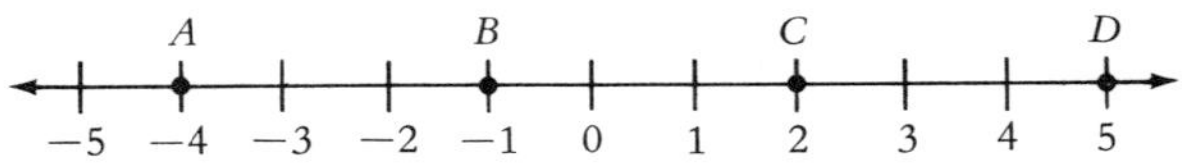

Figure 5.2.6

2. a. Describe the line segments $\overline{AB}$, $\overline{CD}$, $\overline{EF}$, and $\overline{GH}$ in the x-y plane in Fig. 5.2.7.

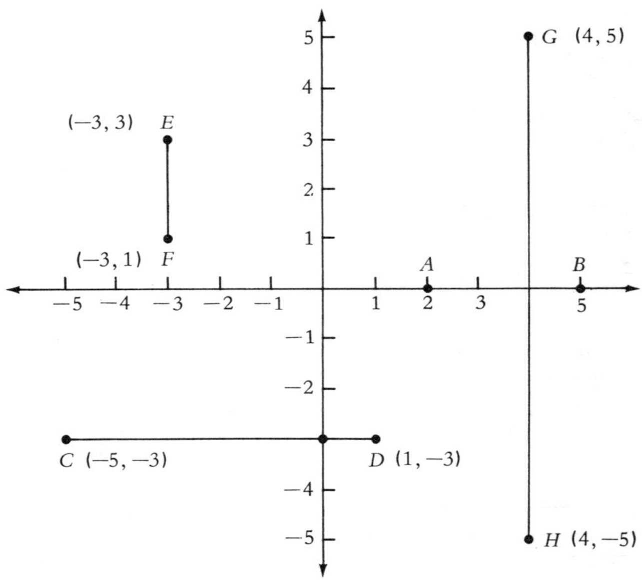

Figure 5.2.7

   b. Describe the same line segments in 3-space in the x-y plane.
   c. Describe the same line segments in 3-space in a plane parallel to the x-y plane but $\pi/4$ unit below it.

3. a. Assume that the line in the y-z plane in Fig. 5.2.8 can be described as $\{(y, z) | 2y + z = 10\}$. Describe segments $\overline{AB}$, $\overline{AC}$, $\overline{CB}$, and $\overline{AD}$.
   b. Describe the same line segments in the y-z plane in 3-space.
   c. Describe the same line segments in 3-space in a plane parallel to the y-z plane but 8 units in front of it.

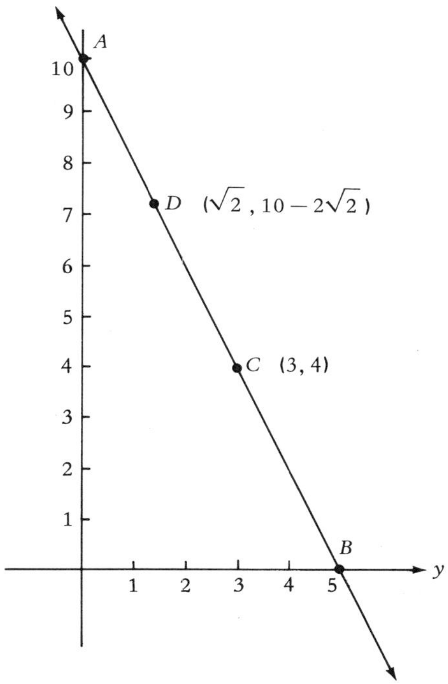

Figure 5.2.8

5.3 THE LENGTH OF LINE SEGMENTS

Now that we have developed notation for describing the coordinates of all the points in a line segment, we are able to discuss the length of these segments. Closely associated with the length of a segment is the notion of *absolute value*. When we discuss numbers on the number line, we have come to associate each number with two components: *direction* and *magnitude*. The direction is indicated by the *sign* of the number, either $+$ or $-$. The magnitude of the number is called the absolute value of the number and, in general, the magnitude or absolute value of any real number x is designated as $|x|$. Intuitively, we could think of the absolute value of a number as the distance of the graph of that number from the graph of zero. Thus, $|2| = 2$, $|-7| = 7$, $|0| = 0$.

Experiment 3

Step 1

When is the absolute value of a number equal to the number? That is, for what real numbers x will the equation $|x| = x$ be true?

Step 2

Suppose x is a number for which $|x| \neq x$; what does $|x|$ equal? If you took some time to think about the above ideas, you should now be ready to understand the following formal definition of absolute value.

5.3.1 Definition

$$\forall\, x \in R \begin{cases} \text{(a) } \textit{if } x \geqq 0, \textit{ then } |x| = x \\ \text{(b) } \textit{if } x < 0, \textit{ then } |x| = -x \end{cases}$$

Example 1
What is $|5|$?
Since $5 > 0$, we use part (a): $|x| = x$ *and get* $|5| = 5$.

Example 2
What is $|-7|$?
Since $-7 < 0$, we use part (b) and get $|x| = -x$. Thus, $|-7| =$ the additive inverse of -7 which is 7. Thus, $|-7| = -(-7) = 7$.

in our opinion . . .

The terminology that we choose in communicating mathematical ideas can either contribute to our understanding of a mathematical concept, or it can hinder it. An excellent example of this concerns this very definition for absolute value. There are three different ways that one can be encouraged to read the symbol $-x$: "minus x," "negative x," or "the additive inverse of x." We assert that the use of either of the first two interpretations can interfere with our comprehension of the meaning of the statement if $x < 0$, then $|x| = -x$. If we are in the habit of saying "the absolute value of x is minus x—or negative x," then we can easily become confused, believing that the absolute value is being defined as a negative number. We believe that interpreting the statement as "if x is negative, then the absolute value of x is the additive inverse of x" eliminates this possible source of confusion and helps clarify the definition of absolute value.

In our discussion of absolute value, we found it useful to think of $|x|$ as the distance from the graph of x to the graph of 0. Thus, $|x|$ becomes the length of the line segment whose endpoints are at 0 and x. What about lengths of segments which have neither of their endpoints at 0?

Experiment 4

Step 1

Consider the segment $\{x|2 \leqq x \leqq 8\}$. What is its length? That is, what is the distance from one endpoint to the other?

Step 2

Can you express this distance as a formula using only the numbers 2 and 8?

Step 3

Do the same for the segments $\{x|-5 \leqq x \leqq 14\}$ and $\{x|-7 \leqq x \leqq -3\}$.

Step 4

In general, what formula can you use for the length of the line segment $\{x|a \leqq x \leqq b\}$?

Step 5

How could you express the length of the line segments with endpoints p and q as coordinates, but you do not know which coordinate is the larger?

From the preceding problems, you should be able to see that, in general, the length of the segment $\{x|a \leqq x \leqq b\}$ can be expressed as $|b - a|$. A natural question might be: Why do we bother with the absolute value? The answer is: if a line segment is indicated by set notation $\{x|a \leqq x \leqq b\}$, we do not need to bother since we know that $b \geqq a$. Thus, we know that $b - a$ is not negative. However, suppose you were asked to find the length of the line segments whose endpoints have the coordinates u and v, but you did not know which point had the greater coordinate. Thus, you would not be sure whether the length of the segment was $u - v$ or $v - u$. However, if we say $|u - v|$ or $|v - u|$ we are correct for $|u - v| = |v - u|$. Thus, we are able to define the length of a line segment.

5.3.2 Definition

The length of a line segment with endpoints having coordinates a and b is $|a - b|$ or $|b - a|$.

Likewise, $|a - b|$ can be thought of as indicating the distance between a and b.

Example 3

Find the solution set to $|x - 4| = 3$.

We can interpret x as the coordinate of a point whose distance from the graph of 4 is 3 units. Thus, if we locate the graph of 4 on the number line of Fig. 5.3.1 we can now ask what points are a distance of 3 units from the graph of 4. We see there are two such points. Thus, x could be either the coordinate $+1$ or $+7$. Therefore, the solution set is $\{+1, +7\}$.

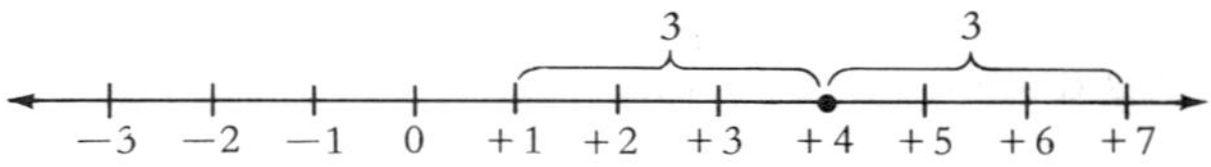

Figure 5.3.1

Example 4

Find the solution set to $|x + 2| = \sqrt{3}$.

Since $|x + 2| = |x - (-2)|$ we can interpret x as being the coordinate of a point which is $\sqrt{3}$ units from -2. Thus, on the number line of Fig. 5.3.2 we can see that x can correspond to two numbers, $-2 + \sqrt{3}$ and $-2 - \sqrt{3}$. Thus, the solution set is $\{-2 - \sqrt{3}, -2 + \sqrt{3}\}$.

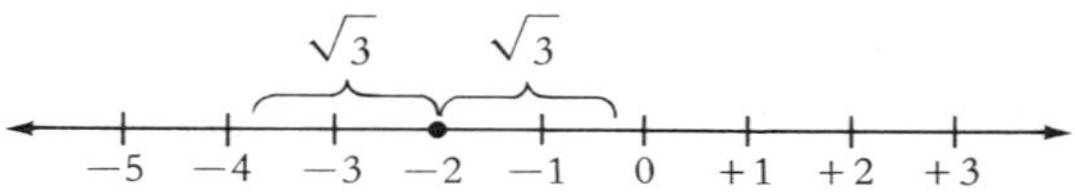

Figure 5.3.2

Example 5

Find the solution set to $|2x - 5| = 4$.

Here we can interpret $2x$ as the coordinate of a point which is 4 units from the graph of 5. An examination of our number line, Fig. 5.3.3, shows that the graph of $2x$ could correspond to either $+9$ or $+1$. Since $2x$ is either $+9$ or $+1$, x is either $+4.5$ or $+.5$. Thus, the solution set is $\{+.5, +4.5\}$.

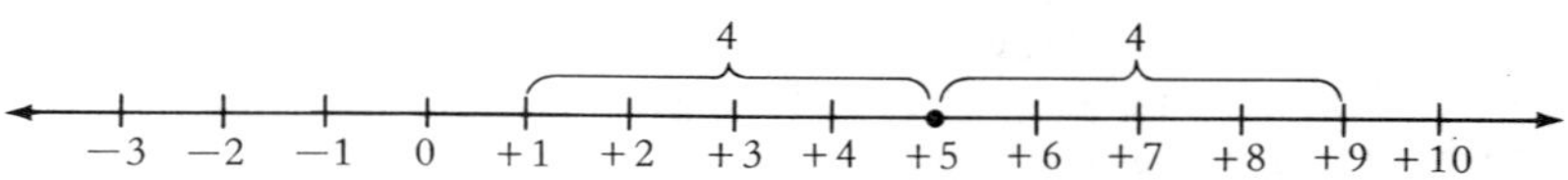

Figure 5.3.3

Exercise 5.3.1

1. Find the length of the following line segments:
   a. $\{x | 5 \leqq x \leqq 7\}$ b. $\{x | -3 \leqq x \leqq 4\}$
   c. $\{x | -13 \leqq x \leqq -3\}$ d. $\{x | r \leqq x \leqq 5\}$
   e. $\{x | k \geqq x \geqq j\}$
2. Express the length of line segments whose endpoints are:
   a. p and q b. p and $-q$
   c. $-p$ and q d. $-p$ and $-q$
3. Use the above interpretation to find the solution set of each of the following:

a. $|x - 2| = 6$ b. $|3 - x| = 7$
c. $|x + 4| = 10$ d. $|2x - 6| = 8$
e. $|2x + 6| = 8$ f. $|3x + 7| = -6$

4. Find the solution set for each equation.
   a. $|x| = 2$ b. $|x| + 3 = \sqrt{5}$
   c. $2|x| + 7 = 8$ d. $3|x| + 3 = 2$
5. Answer each of the following questions.
   a. If $x > 0$, $|x| = ?$ b. If $x < 0$, $|x| = ?$
   c. If $x > 0$, $|-x| = ?$ d. If $x < 0$, $|-x| = ?$
   e. If $-x > 0$, $|x| = ?$ f. If $-x < 0$, $|x| = ?$
   g. If $-x > 0$, $|-x| = ?$ h. If $-x < 0$, $|-x| = ?$
   i. Is it ever possible for $|x| = x$ and $|x| = -x$ at the same time? Explain.

Experiment 5

Step 1

Evaluate the following expressions where $a = 2$, $b = 3$, $c = -4$, and $d = -5$.

a. $|a + b|$ and $|a| + |b|$ b. $|a + c|$ and $|a| + |c|$
c. $|c + d|$ and $|c| + |d|$ d. $|a + d|$ and $|a| + |d|$
e. $|a \cdot b|$ and $|a| \cdot |b|$ f. $|a \cdot c|$ and $|a| \cdot |c|$
g. $|c \cdot d|$ and $|c| \cdot |d|$ h. $|a \cdot d|$ and $|a| \cdot |d|$
i. $|a|^2$ and a^2 j. $|b|^2$ and b^2
k. $|c|^2$ and c^2 l. $|d|^2$ and d^2

Step 2

Make a conjecture about the relationship between $|a + b|$ and $|a| + |b|$, $\forall a, b \in R$.

Step 3

Make a conjecture about the relationship between $|a \cdot b|$ and $|a| \cdot |b|$, $\forall a, b \in R$.

Step 4

Make a conjecture about $|x|^2$ and x^2, $\forall x \in R$.

Thus far, we have been concerned with the length of line segments on a given line. Let us extend this idea to line segments which are subsets of lines in a given plane.

Experiment 6

Step 1

Determine the lengths of the line segments joining the points (2, 5) and (7, 5); (3, −7) and (−4, −7); (.5, π) and ($\sqrt{2}$, π); (x, y) and (u, y).

Step 2

Determine the lengths of the line segments whose endpoints have the following coordinates: (8, 3) and (8, 10); (−9, −2) and (−9, 8); (.21, $\sqrt[11]{12}$) and (.21, 6); (x, y) and (x, v).

Step 3

Determine the lengths of the following segments:

a. $\{(x, y) | -4 \leqq x \leqq 10, y = 20\}$
b. $\{(x, y) | -8 \leqq y \leqq -3, x = -7\}$
c. $\{(x, y) | a \leqq x \leqq b, y = k\}$
d. $\{(x, y) | c \leqq y \leqq d, x = p\}$

Step 4

You have just considered the problem of determining the lengths of certain special line segments in the plane. Can you state what is so special about these?

Step 5

Can you make a definition for determining the lengths of these kinds of line segments?

In the previous experiment, we considered a problem of determining the lengths of line segments where the segment is parallel either to the x- or y-axis.

In the first case, if the segment is parallel to the x-axis, the second members of the endpoints would be the same. Thus, if (x, y) is one endpoint, the second endpoint would have to be (a, y). In this case, the length of the segment would be $|x - a|$ (Fig. 5.3.4). Likewise, for line segments parallel to the y-axis (Fig. 5.3.5), the length would be $|y - b|$.

Figure 5.3.4

Thus, so far we are able to define the length of a segment with endpoints (x_1, y_1) and (x_2, y_2) in the following way. If d is the distance between (x_1, y_1) and (x_2, y_2), then

$$d = |x_1 - x_2| \text{ if } y_1 = y_2$$
$$d = |y_1 - y_2| \text{ if } x_1 = x_2$$

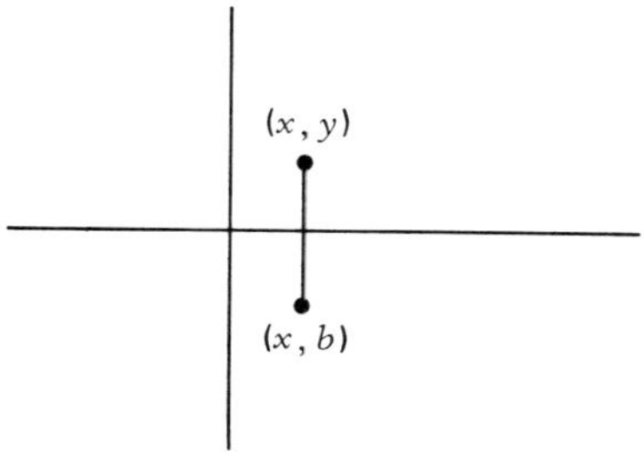

Figure 5.3.5

Experiment 7

Step 1

What case has not yet been taken care of?

Step 2

Plot the points (1, 2) and (4, 6) on a graph. Draw a line segment between the two points.

a. Does this segment fall into either one of the categories mentioned above? Can you introduce a third point (a, b) so that the segment joining (1, 2) and (a, b) falls into one of the above cases and the segment joining (4, 6) and (a, b) falls into the other case?
b. Determine the lengths of these segments. How can you use these two lengths to help you determine the length of the segment joining (1, 2) and (4, 6)?

Obviously, we have yet to determine the length of a segment which is not parallel to either axis. That is, we must next consider the segment with endpoints (x_1, y_1) and (x_2, y_2) where $x_1 \neq x_2$ and $y_1 \neq y_2$ (Fig. 5.3.6).

Figure 5.3.6

Since we know how to determine the lengths of segments which are parallel to one of the axes, let us introduce such segments into our problem. Suppose we draw a line through (x_1, y_1) parallel to the x-axis and one through (x_2, y_2) parallel to the y-axis (Fig. 5.3.7). What would the coordinates of the point of intersection be?

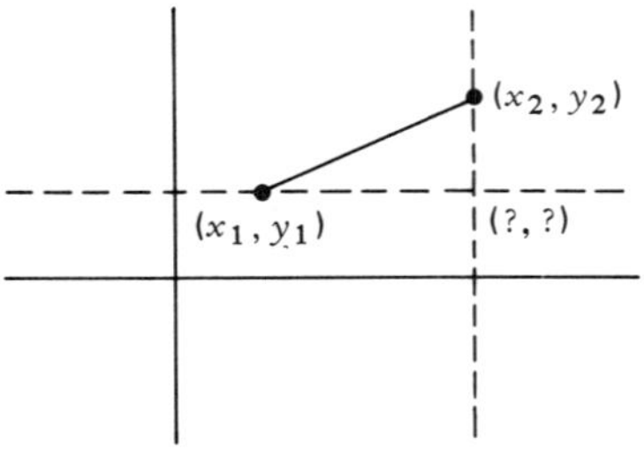

Figure 5.3.7

Be sure you are convinced that it has coordinates (x_2, y_1). Now look at Fig. 5.3.8. What is the length of the horizontal segment? From our previous discussion, it must be $|x_1 - x_2|$. Likewise, the length of the vertical segment would be $|y_1 - y_2|$. Now, we have a right triangle with the right angle having (x_2, y_1) as its vertex and the segment joining (x_1, y_1) and (x_2, y_2) as its hypotenuse. Thus, we can use the Pythagorean Theorem to determine d. So, $d^2 = |x_1 - x_2|^2 + |y_1 - y_2|^2$, or, since from an earlier experiment, we know $|x|^2 = x^2$, we can write $d^2 = (x_1 - x_2)^2 + (y_1 - y_2)^2$ or $d = \sqrt{(x_1 - x_2)^2 + (y_1 - y_2)^2}$.

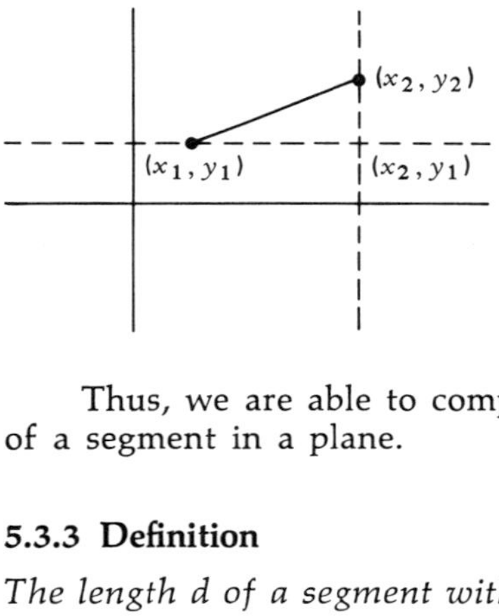

Figure 5.3.8

Thus, we are able to complete our definition of the length of a segment in a plane.

5.3.3 Definition

The length d of a segment with coordinates (x_1, y_1) *and* (x_2, y_2) *is*

$$d = |x_1 - x_2| \text{ if } y_1 = y_2$$
$$d = |y_1 - y_2| \text{ if } x_1 = x_2$$
$$d = \sqrt{(x_1 - x_2)^2 + (y_1 - y_2)^2} \text{ if } x_1 \neq x_2 \text{ and } y_1 \neq y_2$$

Actually, this last formula can work for the first two cases as well, as you will be asked to show in an exercise.

Example 6
Determine the length of the segment joining $(-2, 4)$ and $(-8, -5)$. If we arbitrarily set $(-2, 4) = (x_1, y_1)$ and $(-8, -5) = (x_2, y_2)$

then, since $x_1 \neq x_2$ and $y_1 \neq y_2$,

$$\begin{aligned} d &= \sqrt{(x_1 - x_2)^2 + (y_1 - y_2)^2} \\ &= \sqrt{((-2) - (-8))^2 + (4 - (-5))^2} \\ &= \sqrt{(6)^2 + (9)^2} \\ &= \sqrt{36 + 81} \\ &= \sqrt{117} \end{aligned}$$

Example 7

Determine the length of the segment joining (8, 5) and (8, $12/\pi$). Suppose (8, 5) = (x_1, y_1) and (8, $12/\pi$) = (x_2, y_2). Then, since

$$\begin{aligned} x_1 &= x_2 \\ d &= |y_1 - y_2| \\ &= |5 - 12/\pi| \end{aligned}$$

Exercise 5.3.2

1. Determine the lengths of the following segments with endpoints as indicated.
   a. (2, 1) and (3, 5)
   b. (5, 9) and (13, 9)
   c. (−2, 7) and (3, −9)
   d. (−8, −2) and (−1, −3)
   e. (.7, 1.6) and (.7, −3.5)
   f. (.3, 4) and (−.1, 5)
   g. $(\sqrt{7}, -5)$ and (−7, 9)
   h. $(\pi, 3)$ and $(2\pi, 10)$
   i. $(6, \pi)$ and (12, 22/7)
   j. $(6, \pi)$ and (12, 3.14)
   k. $(6, \pi)$ and $(12, \pi)$
2. Determine the lengths of the following segments:
   a. $\{(x, y)|x + y = 10, 5 \leqq x \leqq 11\}$
   b. $\{(x, y)|2x + y = 0, -3 \leqq y \leqq 4\}$
3. a. Show that for a segment joining (x_1, y_1) and (x_2, y_2) where $y_1 = y_2$, $\sqrt{(x_1 - x_2)^2 + (y_1 - y_2)^2} = |x_1 - x_2|$.
   b. Also show that for any segment joining (x_1, y_1) and (x_2, y_2) with $x_1 = x_2$, $\sqrt{(x_1 - x_2)^2 + (y_1 - y_2)^2} = |y_1 - y_2|$.
4. See if you can develop a formula for the distance between (x_1, y_1, z_1) and (x_2, y_2, z_2).

5.4 AN APPLICATION OF THE DISTANCE FORMULA

Experiment 8

A teacher brought a box of different colored marbles into class. She asked one student, John, to come up and take out of the box all the marbles, and only the marbles, that were red. After John had done this, the teacher asked various other students to check and see if John had followed her directions. Several students responded by looking into the box and they confirmed that John had indeed followed directions because the box did not contain any red marbles.

1. Do you believe the students' check was adequate?
2. If John had taken *all* the red marbles but also two black marbles, would he have taken *all* the red marbles? Would he have taken *only* the red marbles? What could the students have done to see if John had taken *only* the red marbles?

In mathematics, it is essential that one learns to distinguish between *all* and *only*, or between *if* and *only if*, or between *necessary* and *sufficient*. Actually, the distinction between *all* and *only* which was emphasized in the last episode is very similar to the differences between *if* and *only if* and *necessary* and *sufficient*. If you are able to grasp the full implication of this discussion, you should have little difficulty with *if* and *only if* and with *necessary* and *sufficient* when they are discussed in a later section. The main point here is to realize that if one imposes conditions described by *all* the elements and *only* those elements which satisfy certain conditions, this is describing two distinct requirements. Recall that the students in Experiment 8 should not only have looked in the box to see if John had taken *all* the red marbles, but they also should have looked in John's hand to see if he took *only* the red marbles. Look at Figs. 5.4.1 and 5.4.2, then do the following set of exercises.

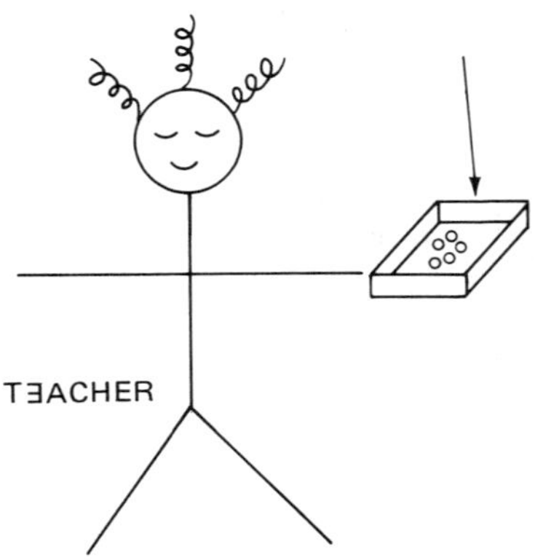

Figure 5.4.1
Look in the box to see if John took *all* the red marbles.

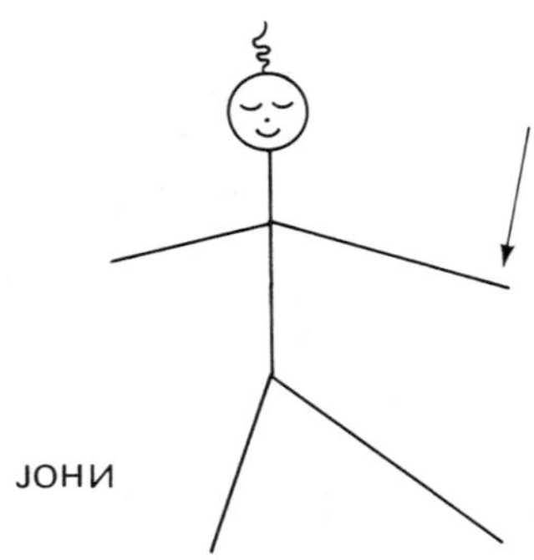

Figure 5.4.2
Look in John's hand to see if John took *only* red marbles.

Exercise 5.4.1

1. Let $A = \{1, 2, 3, 5, 8\}$ and $B = \{1, 2, 3, 5\}$. Answer the following questions.
   a. Does A contain all the points that are in B?

b. Does A contain only the points that are in B?
c. Does A contain all the points and only the points in B?

2. Answer the same questions for $A = \{5, 9, 6, 4, 1\}$ and $B = \{10, 9, 8, 7, 6, 5, 4, 3, 2, 1\}$.
3. Answer the same questions for $A = \{2, 3, 4, 5, 6, 11\}$ and $B = \{11, 5, 4, 2, 3, 6\}$.

The preceding set of exercises should have shown that if we wish to determine if set A contains *all* the elements and *only* the elements that are in set B, we must first determine if set A contains *all* the elements of set B by seeing if every element in B is in A, and then we must find out if A contains *only* the elements of B by checking if every element in set A is also in B. Thus, we see that the statement, *Set A contains all the elements and only the elements of set B*, is equivalent to saying, *Every element in set B is in set A and every element in set A is in set B*. Now suppose that two sets A and B did satisfy both conditions listed above. Then we would say set A equals set B (written as $A = B$). Thus, if we are ever asked to demonstrate that two sets A and B are equal, we must first show that set A contains all the elements in set B and then we must show that set A contains only those elements that are in set B. This is made precise in the following definition.

5.4.1 Definition

Two sets A and B are equal if and only if every element in A is in B and every element in B is in A.

Example 10
Show that the graph of $\{(x, y)|x^2 + y^2 = 4\} = \{$the set of all points on a circle with radius 2 and center at (0, 0)$\}$.

Suppose we let $A =$ the graph of $\{(x, y)|x^2 + y^2 = 4\}$ and $B = \{$the set of all points on a circle with radius 2 and center at (0, 0)$\}$. To show that $A = B$ we have two things to do.

1. Show that A has *all* the points which are in B.
2. Show that A has *only* those points which are in B.

To see the distinction between these two parts, let us consider them separately. Statement 1 suggests that we have to be sure that when we graph the relation $\{(x, y)|x^2 + y^2 = 4\}$ we will cover *all* the points of the circle with radius 2 and center at (0, 0). We know this circle is the one shown in Fig. 5.4.3.

Now, if we proceed to graph $\{(x, y)|x^2 + y^2 = 4\}$ we see almost immediately that the graph contains such points as (2, 0), (−2, 0), (0, −2), and (0, 2). Furthermore, by observing that

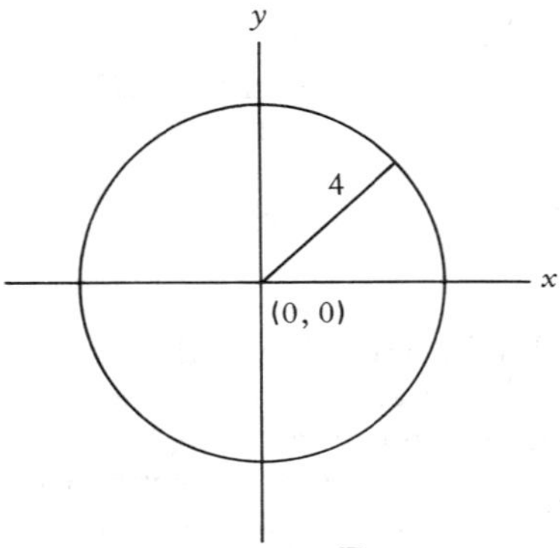

Figure 5.4.3

$$1^2 + (\sqrt{3})^2 = 1 + 3 = 4$$
$$1^2 + (-\sqrt{3})^2 = 1 + 3 = 4$$
$$(-1)^2 + (\sqrt{3})^2 = 1 + 3 = 4$$
$$(-1)^2 + (-\sqrt{3})^2 = 1 + 3 = 4$$

we know that $(1, \sqrt{3})$, $(1, -\sqrt{3})$, $(-1, \sqrt{3})$, and $(-1, -\sqrt{3})$ are also on the graph. Likewise, we could see that $(\sqrt{3}, 1)$, $(\sqrt{3}, -1)$, $(-\sqrt{3}, 1)$ and $(-\sqrt{3}, -1)$ are on the graph. We could continue this investigation, but let us see what the graph is so far (Fig. 5.4.4).

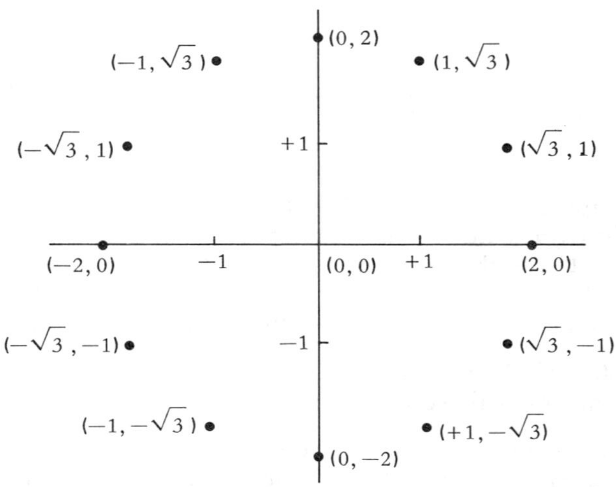

Figure 5.4.4

Now, indeed, these points appear to be on the circle. But the question is, if we graph this relation, will we get *all* the points on this circle? That is, how do we know that maybe after graphing the entire relation we won't end up with a circle with a few gaps such as in Fig. 5.4.5? That is, we must prove that

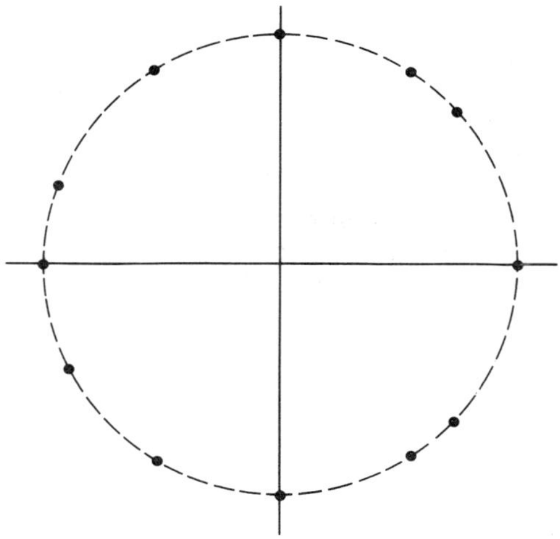

Figure 5.4.5

the graph of $\{(x, y)|x^2 + y^2 = 4\}$ contains *all* the points of the circle with center at (0, 0) and radius of 2. In order to show that the graph contains all the points of the circle, we can show that if a point is on the circle, then it has to be on the graph. Suppose P is any point of the circle; then it has a pair of coordinates, say (u, v). Now, because (u, v) is on the circle, its distance from the center of the circle is 2. The center has coordinates (0, 0) (Fig. 5.4.6). Thus, applying the distance formula, we see that

$$2 = \sqrt{(x_1 - x_2)^2 + (y_1 - y_2)^2} = \sqrt{(u - 0)^2 + (v - 0)^2}$$
$$= \sqrt{u^2 + v^2}$$

Since this distance is 2, we get $2 = \sqrt{u^2 + v^2}$, or, squaring both sides, we get $u^2 + v^2 = 4$. Thus, P is a point whose coordinates satisfy the equation $x^2 + y^2 = 4$. Therefore, P must be on the graph. In other words, if a point is on the circle, it must be on the graph of $\{(x, y)|x^2 + y^2 = 4\}$.

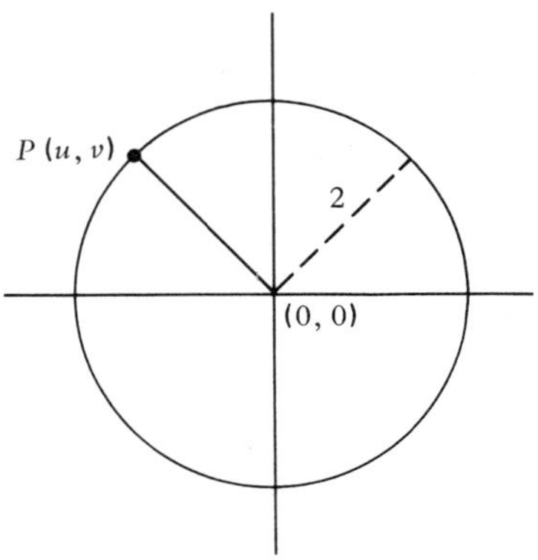

Figure 5.4.6

Now that we have argued that the graph contains *all* the points of the circle, we must show that the graph contains *only* those points which are on the circle. What does this mean? We have just shown that the graph of $\{(x, y)|x^2 + y^2 = 4\}$ contains all the points of the circle, but how do we know it does not contain some other points as well? Maybe our graph really looks like Fig. 5.4.7. That is, although the graph contains *all* the points of the circle, maybe it does not contain *only* the points of the circle. Thus, we must also show that the graph of $\{(x, y)|x^2 + y^2 = 4\}$ contains *only* the points of the circle with center at (0, 0) and radius of 2.

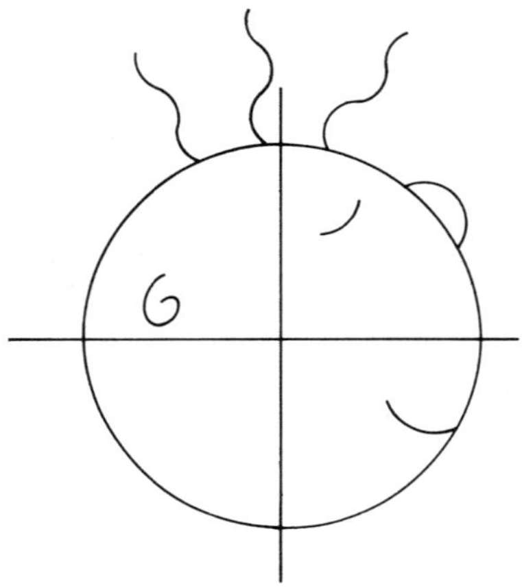

Figure 5.4.7

To do this, we can suppose that Q is any point on the graph and show that it is on the circle. Suppose Q has the coordinates (s, t). Then, since it is on the graph we know it satisfies the equation $x^2 + y^2 = 4$. That is, $s^2 + t^2 = 4$. Now to show that it is on the circle, we must show that its distance from (0, 0) is 2 units. Let us determine the distance between (s, t) and (0, 0). By the distance formula, we get $d = \sqrt{(s - 0)^2 + (t - 0)^2} = \sqrt{s^2 + t^2}$ or $d^2 = s^2 + t^2$. But since (s, t) is on the graph, we know that $s^2 + t^2 = 4$. Thus, $d^2 = 4$ and $d = 2$. That is, the distance from (s, t) to (0, 0) is 2 units. By the definition of a circle, (s, t) is on a circle whose center is at (0, 0) and whose radius is 2. Thus, (s, t) is on the circle. That is, if a point is on the graph, then it must be on the circle. Thus, we have shown that the graph of $\{(x, y)|x^2 + y^2 = 4\}$ equals the circle with center of (0, 0) and radius of 2.

Exercise 5.4.2

1. Use the same kind of argument to show that the graph of $\{(x, y)|(x - 1)^2 + (y - 2)^2 = 25\}$ is the circle of radius 5 with center at (1, 2).

2. If you want to demonstrate that the graph of $\{(x, y)|y = 2x + 1\}$ is the line with slope of 2 which goes through the point (0, 1), what two ideas must you show?
3. Does $\{(x, y)|y = x^2\}$ contain all the ordered pairs of $\{(x, y)|x = \sqrt{y}\}$? Does $\{(x, y)|y = x^2\}$ contain only the ordered pairs of $\{(x, y)|x = \sqrt{y}\}$? Does $\{(x, y)|y = x^2\}$ equal $\{(x, y)|x = \sqrt{y}\}$? Explain why or why not. Draw the graph of each relation.

5.5 SUMMARY

1. To describe the coordinates for a particular subset of a line, first determine what conditions make it a subset of the line, then determine what condition places the line in a plane, then determine what conditions place the plane in space.
2. All ordered pairs with a fixed first member correspond to a line parallel to the y-axis and those with a fixed second member correspond to a line parallel to the x-axis.
3. All ordered triples with a fixed first member correspond to a plane parallel to the y-z plane; those with a fixed second member correspond to a plane parallel to the x-z plane; and those with a fixed third member correspond to a plane parallel to the x-y plane.
4. $|a - b|$ can be interpreted as representing the length of a line segment with endpoints of a and b.
5. For all real numbers a and b, $|a + b| \leqq |a| + |b|$, $|a \cdot b| = |a| \cdot |b|$ and $|x|^2 = x^2$.
6. The distance d between two points (x_1, y_1) and (x_2, y_2) is $d = \sqrt{(x_1 - x_2)^2 + (y_1 - y_2)^2}$.
7. To determine if set A equals set B we must show that A contains all the elements of B and that A contains only the points of B.

INTRODUCTION TO MEASUREMENT

6

6.1 PROLOGUE

Measurement is so much a part of our daily life that we often take it for granted. We are awakened by the clamor of the alarm clock, only the first of several times during the day when we will check the time. We step on the scales to determine if our new diet has had any effect. On the way to work, we stop at a service station to buy ten gallons of gas and a quart of oil. Listening to the car radio, we learn that the temperature is 78°, the barometric pressure is 29.92, humidity is 90%, and that we had a half-inch of rain. We remember that on the way home we must stop at the store and buy a pound of butter, a gallon of milk, and a bushel of apples.

These are only examples of measurement that we encounter every day. Not as obvious, but just as important in our lives, is the measurement that is an integral part of almost all commerce, manufacturing, and the building trades. How could we buy and sell land, build homes and bridges, or manufacture airplanes and automobiles without measurement? In addition, scientists and mathematicians depend on measures to help them make discoveries and, most recently, there is a growing concern over increasing pollution levels which threaten our very existence.

Measurement may be defined, in general, as the process of quantifying the properties of a set by a comparison with some standard unit. An important question is what properties are measurable. The process of discovering that a property is measurable and setting up a procedure for measuring is one that rests entirely upon experimental inquiry. The British physicist and philosopher of science, Norman Campbell, puts it in this way:

> Whenever a new branch of physics is opened up, the first step is always to find some process for measuring the new properties that are investigated; and it is not until this problem has been solved that any great progress can be made along the branch. Its solution demands the discovery of new laws.

Along the same lines, Lord Kelvin has said:

> I often say that when you can measure what you are speaking about and express it in numbers you know something about it, but when you cannot measure it, when you cannot express it in numbers, your knowledge is of a meagre kind.

In this chapter we will investigate the measurements of line segments and angles. In doing so, we will try to determine an answer to the question of what geometric properties are measurable and how we may develop processes of measurement. In the United States, length is measured chiefly in inches, feet, yards, and miles. These are units of length or linear units. Early man measured lengths long before there were such units. One of the earliest types of linear measurement was body measures. In ancient times, four body measures were used. One was the "cubit," which was the distance from the point of the elbow to the end of the middle finger. Another was the "span," which was the distance from the end of the thumb to the end of the little finger when the hand was spread out. The third was the "palm," which was the breadth of the four fingers held close together. The fourth body measure was the "digit," which was the breadth of the first finger or the middle finger (Fig. 6.1.1).

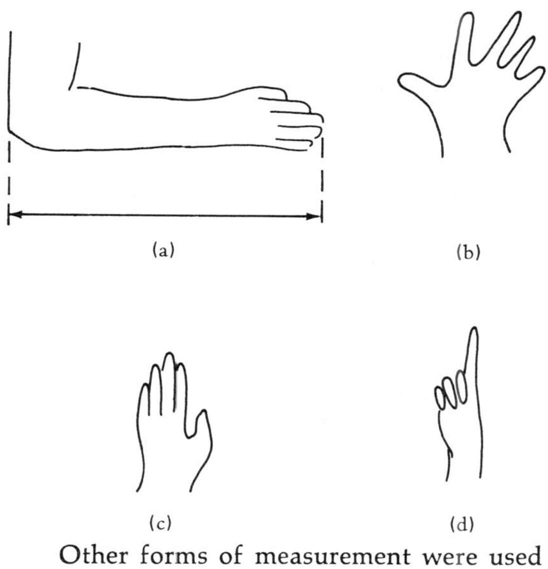

Figure 6.1.1
(a) Cubit; (b) Span; (c) Palm; (d) Digit

Other forms of measurement were used to answer the question, "How far away is it?" The technique was to tell the questioner how long it would take to get there. The American Indian might say that some place was two suns away, meaning that it would take two days to reach there. Even today, ask a resident of Connecticut how far away a certain town is and he is more likely to tell you how many minutes it will take to reach it by car rather than how many miles away it is.

Counting steps was another method of measuring distances in earlier times (Fig. 6.1.2). Even today, we often "step off" distances.

Figure 6.1.2

The early Babylonians, Egyptians, and Hebrews relied heavily on the cubit for measuring lengths. They had realized the need for a standard cubit which may have been the length of the forearm of the "ruler" of the country. Thus, today one of our instruments for measuring length is called a "ruler."

We do not know exactly when the "foot," which gets its name from the human foot, became a standard measure. The Greeks used it and the Romans borrowed it from them, making it the chief measure of length for short distances. Although the Romans divided a foot into 12 in., these measurements are not the same lengths used today. The Roman "inch" was the breadth of the thumb. The "mile" was also introduced by the Romans, deriving from *mille passum,* meaning a thousand paces. The Roman pace was a double step about 5 ft long.

The Romans had rods showing exactly how long a foot should be. During the Middle Ages every city and town in Europe had its own measuring rods. It was a long time before standards were set for entire countries. In England, Henry I declared that a "yard" was to be the distance from the tip of his nose to the end of his thumb. Edward II abolished all body measurements and declared that an "inch" would be the length of three good grains of barley laid end to end (Fig. 6.1.3).

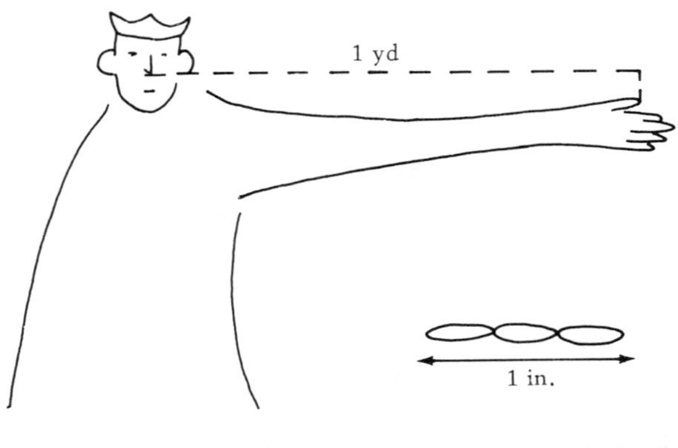

Figure 6.1.3

At the time of Napoleon and the French Revolution, a new set of measures was introduced. The chief measure of length was the "meter," which was one ten-millionth of the distance from the equator to the North Pole measured along a meridian passing through Paris. The whole system of weights and measures got its name from the meter and became the *metric system.*

In the United States, the standard measuring rod is a meter rod made mostly of plantinum. The standard yard is 3600/3937 of this meter. This rod is always kept at the same temperature to prevent it from expanding with heat or contracting when cooled. The United States, along with such countries as Trinidad, Malawi, Ghana, Sierre Leone, Tonga, Gambia, Liberia, Nauru, Burma, Jamaica, and Southern Yemen, has not yet converted to the metric system, although at the time of this writing Congress has passed legislation to change to the metric system within ten years.

Although such a change will prove to be very costly to industry, it will prove very advantageous in the long run. It also may prove to destroy some of our cherished cliches. The sportscaster will excitedly announce, "Notre Dame has the ball, first and 9.144," and the famous cry "Gentlemen, start your engines," will ring out at the Indianapolis 804.5. (Fig. 6.1.4).

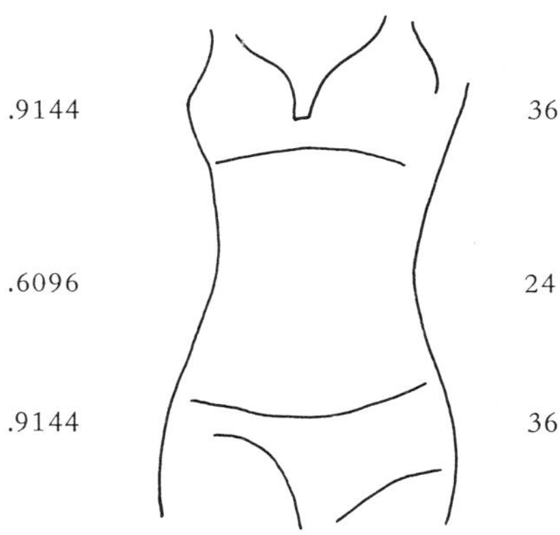

Figure 6.1.4

Experiment 1

In this experiment only body measures will be used. First of all, using your own finger, hand, and arm, determine relationships between the body measurements: cubit, span, palm, and digit. (You will probably need the assistance of a partner.) Measure your neck, wrist, leg, and ankle. Which body measurement will you use to do this? Using your personal table of conversions, translate these measurements to the other body measurements. How many "wrist measurements" will go around your neck? (Estimate first.) How many "neck measurements" will go around your waist? (Estimate first.)

Next, measure the length of your desk and your book using (a) cubits, (b) spans, (c) palms, (d) digits. Make a rough estimate before you measure. Which do you find gives you the best measurement? Which took the longest to do?

Using the length of your span, your pace, and your cubit, or the length of your foot, measure: (a) the room, (b) the door, (c) the diagonals of the room, (d) a window. Think about the best units to be used before you measure and estimate the measurement first. Compare your results with others.

Start a discussion of these activities. Discuss the importance of estimation. Which unit of measurement seems the most generally useful and accurate? What are the disadvantages of body measurements? Are there any circumstances where body measurements are appropriate and useful?

Experiment 2

In this experiment, rulers, yard sticks, and tape measures will be used. Estimate the lengths and heights in inches, feet, and yards of the following: (a) the door of the classroom, (b) the length of your stride or pace, (c) your teacher's height, (d) the length and breadth of your book, (e) the dimensions of the room (including height). Make these measurements with appropriate measuring devices. Compare with your estimates.

Try and find something in the room that you feel is approximately 1 yd long, something that is approximately 1 ft long, something that is approximately 1 in. long. When you have estimated, measure and check to the nearest unit.

Estimate the length of your foot; measure it to the nearest unit. How do the results compare? Draw a line (without using a ruler) which you think is about 1 in. long. Now measure it and see how far off you are. Now draw lines (again, without using a ruler) which you think are: 2 in. long, 3 in. long, 6 in. long. Now check these with a ruler and note how much too long or too short you are.

Next, draw several curved or zig-zag lines (polygonal paths) which you think are approximately 6 in. long. How are you going to check these? Find as many round objects as you can and see if you can measure the distance around them to the nearest unit. Make an estimate first. How well did you do this?

Start a discussion of these activities. Discuss your estimation ability. Are inches, feet, and yards improvements over body measurements? If so, in what way? What difficulties did you encounter? How certain are you of the accuracy of your measurements?

Experiment 3

As we have seen, measurement ordinarily involves a comparison of two or more objects. If we say that the height of a person is 72 in., this information would be meaningless if it were not for the fact that we are comparing this information to a standard unit of measurement, the inch. The particular standard unit of measurement used is arbitrary but constant. A person who has height of 72 in. is no shorter or no taller if we say his height is 6 ft or 2 yd. Still, we need standard units of measurement if we are to be able to compare two different objects to be measured.

In determining the measure of a line segment such as $\overline{AB}$ in Fig. 6.1.5, we may first choose another segment as an arbitrary unit of measure. We then determine the number of these segments—called unit segments—contained in $\overline{AB}$.

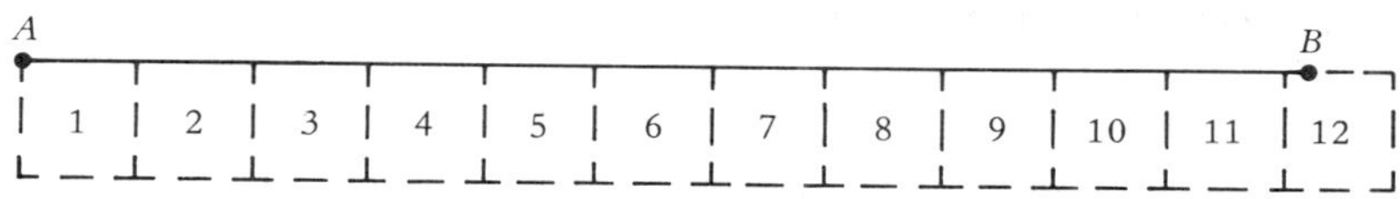

Figure 6.1.5

Notice that 11 of the units are included within $\overline{AB}$ and the end of the twelfth unit extends beyond point *B*. Because 12-unit segments are required to include $\overline{AB}$, we say that 12 is the *outer measure* of $\overline{AB}$. The number 11 is the *inner measure* of $\overline{AB}$ because 11-unit segments are completely included in $\overline{AB}$. (What should we say if a segment such as $\overline{AB}$ contains an exact number of the unit segments?)

Use the 3-unit segments represented in Fig. 6.1.6 by *x*, *y*, and *z* to measure each of the segments $\overline{AB}$, $\overline{DC}$, $\overline{EF}$, and $\overline{GH}$. For each unit, give the inner and outer measures. Which of the two measures is to your mind more accurate in each case? For each of the units *x*, *y*, and *z*, what do you feel will be the *greatest possible error* in measurement to the unit used? In each case, how does this greatest possible error compare with the total measure? Which unit seems to measure the segments most accurately?

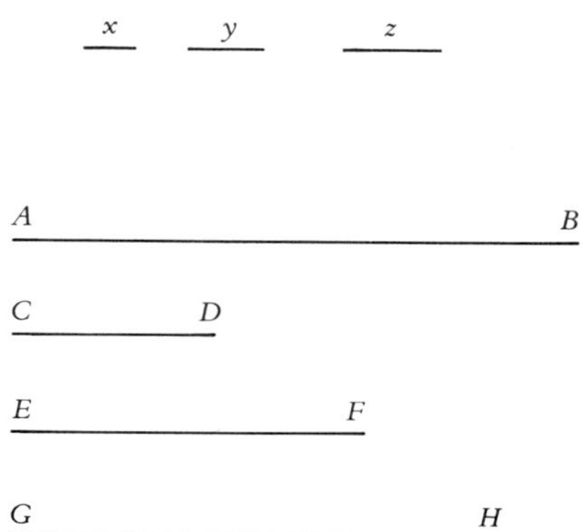

Figure 6.1.6

in our opinion . . .

Measurement is not a collection of formulas, although formulas are useful. Understanding of measurement is not dependent nor is it indicated by the ability to recite comparison tables, although knowledge of their existence is important. Students at all levels should learn of the nature of measurement, its uses and limitations. They should become aware of those features which are common to all measurement. They should learn how to determine which properties are and which properties are not measurable. They should be able to interpret the significance of particular measurements. They should develop an openness to new measurements and new uses of measurement. All this should be acquired in addition to the ability to measure.

6.2 MEASURING LENGTHS

Our experiments and comments are intended to illustrate the fundamentals of measurement which are:

1. Measurement involves comparing a unit with the object or item being measured.
2. For convenience and consistency in communication, standard units have been established.
3. As the result of the comparison, measurement produces a number called a measure.
4. The measurement will be stated in terms of a measure and a unit; for example, a length of 6 ft.

To facilitate measurement of lengths, we have devised certain instruments, such as the ruler (Fig. 6.2.1). The ruler has marks on it to indicate the unit of measurement and various multiples of the unit. In Fig. 6.2.1 we have indicated a part of the ruler with the 1, 2, 3, and 4 representing, respectively, the *unit length* (1 in.), twice the unit length (2 in.), three times the unit length (3 in.) and four times the unit length (4 in.). In addition, we have indicated fractional parts of the unit lengths. The ¼, ½, ¾

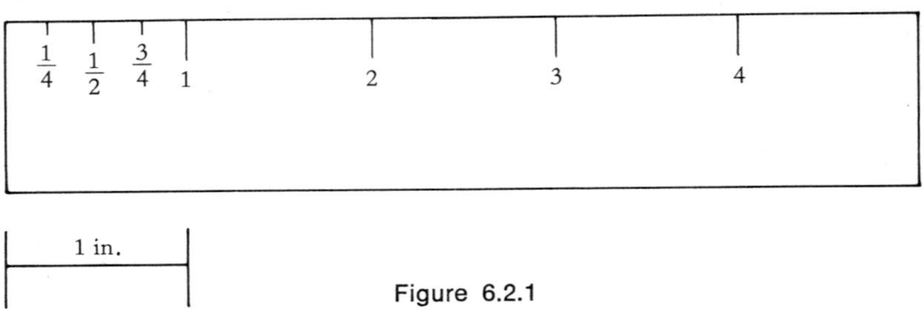

Figure 6.2.1

marks represent, respectively, ¼ of the unit length (¼ in.), ½ of the unit length (½ in.), and ¾ of the unit length (¾ in.).

To use a ruler to measure the length of an object, say the width of a desk top, we place the edge of the ruler at one edge of the desk top. We then look across the ruler and find the part which crosses the opposite side of the desk top from which we have started. Suppose this point is between 14 and 15 on the scale of the ruler. Then 14 is the inner measure of the desk and 15 is the outer measure of the desk in terms of inches. As we examine the scale more closely, perhaps we note that the edge of the desk intersects the ruler at a point between 14½ and 14¾. Thus, in terms of our new unit of measurement, in this case ¼ in., the inner measure is 14½ and the outer measure is 14¾. If the ruler's scale is marked any finer, we can gain a better approximation. If not, we must approximate the width of the desk top as being something between 14½ and 14¾ in.

When we are determining the measure of the segment to the nearest unit, we choose either the inner or the outer measure of a segment as the measure of the segment. The *greatest possible error* in a measure is equal to ½ the unit of measure. The more *precise* of two measures is the measure with the lesser greatest possible error. For the desk top, our two measurements suggest that to the nearest inch, the width is 15 in. with a greatest possible error of ½ in. If we take 14½ in. and then measure to the nearest ¼ in., the greatest possible error is ⅛ in.

The *relative error* of a measure is the amount of error per unit which is determined by comparing the greatest possible error with the total measure. The more *accurate* of two measures is the measure that has the lesser relative error. If the inch is used as the unit of measurement, the relative error in measuring the desk top would be $\frac{1}{2}/15 = \frac{1}{30}$ in. In the second measure, the relative error would be $\frac{1}{8}/14\frac{1}{2} = \frac{1}{116}$ in.

An important fact about the measurement of real objects is that every measurement is an approximation. This is not to say that an object does not have an exact length (that is, a number representing this length), it is just that human limitations make the process of measuring inaccurate.

Exercise 6.2.1

1. Using a ruler, measure each of the line segments in Fig. 6.2.2 to the nearest inch, the nearest ½ in., the nearest ¼ in. In each case, what is the greatest possible error? What is the relative error?

Figure 6.2.2

2. The linear measurement instrument of the metric system is known as a meter stick. The advantage of the meter stick is that every measure can be represented by a decimal numeral. The meter stick is subdivided into ten congruent units, each called a *decimeter*. The decimeter is divided into ten congruent units, each called a *centimeter*. The centimeter is divided into ten congruent units, each called a *millimeter*.

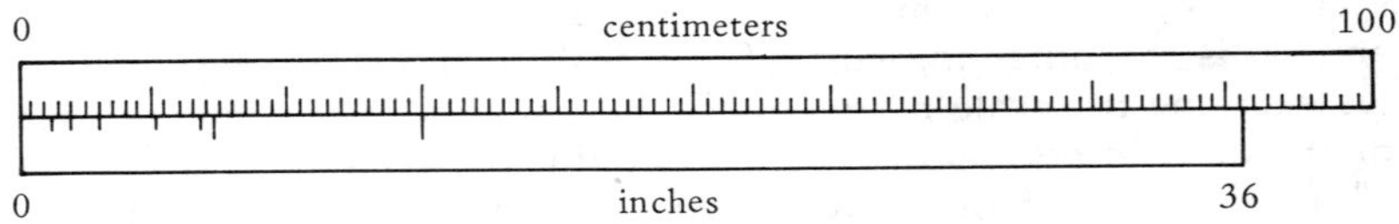

Figure 6.2.3

a. In Fig. 6.2.3 we have compared a meter stick to a yard stick. Using a meter stick and a yard stick, determine a relationship between a meter and an inch which is accurate to the nearest hundredth.

b. We have the following relationships:

.1 m (meter) = 1 dm (decimeter)
.01 m = 1 cm (centimter)
.001 m = 1 mm (millimeter)

Deduce a relationship between decimeter and centimeter, between centimeter and millimeter.

c. A measurement of 3.145 m can be read as "3 m, 1 dm, 4 cm, 5 mm." In a like manner, interpret the following measurements:
1. 5.003 m
2. 6.723 m
3. .458 m

Translate the following measurements to decimal notation:
4. 7 m, 3 dm, 5 cm
5. 2 m, 6 cm, 2 mm

d. This system of units can be extended to the left of the decimal point similar to the decimal system of numeration.

10 m = 1 dekameter (dkm)
100 m = 1 hectometer (hm)
1000 m = 1 kilometer (km)

Determine a relationship between dekameter and hectometer, between hectometer and kilometer. Note that changing from one unit to another in the metric system amounts to only a repositioning of a decimal point if decimal numerals are used.

1000	100	10	1	1/10	1/100	1/1000
km	hm	dkm	m	dm	cm	mm

Complete the following:

1. 6.25 km = ____ hm
2. 62.5 hm = ____ dkm
3. 625 dkm = ____ m
4. .63 m = ____ dm
5. 6.3 dm = ____ cm
6. 63 cm = ____ mm
7. 6705 m = ____ km
8. 6705 m = ____ hm
9. 39 dm = ____ m
10. 76 cm = ____ m
11. 748 m = ____ dm
12. 748 m = ____ cm
13. .84 km = ____ m
14. 76 km = ____ m
15. 6708 m = ____ km
16. 7.9 cm = ____ mm
17. 9.13 km = ____ m
18. 80,000 mm = ____ dkm
19. 1 dkm = ____ mm
20. 1 hm = ____ dm

e. Deduce a relationship between each of the following:
   1. a centimeter and an inch
   2. an inch and a centimeter
   3. a millimeter and an inch
   4. an inch and a millimeter
   5. a kilometer and a mile
   6. a mile and a kilometer

f. Formulate rules to express
   1. a number of centimeters as a number of millimeters
   2. a number of meters as a number of millimeters
   3. a number of millimeters as a number of centimeters

3. Using a ruler divided into centimeters and millimeters, measure each of the line segments of Fig. 6.2.2 to the nearest centimeter, to the nearest millimeter. In each case, what is the greatest possible error? What is the relative error?
4. Compile a complete list of the metric system of units for length, area, and volume and the necessary conversion tables between the metric system and the English system. Which of these units would you introduce in elementary school? Why? Develop a lesson plan for a suitable grade level of your choice to teach these metric concepts.

6.3 THE GENERAL PROPERTIES OF LINEAR MEASURE

So far we have seen that there are several well-known units of linear measure. Furthermore, we have seen that our choice of a standard length as a unit is quite arbitrary. In Chapter 2, we saw that we could coordinatize points on a line, in a plane, or in three-dimensional space by using coordinate axes which were divided into unit segments. We then saw that if P and Q are two points in the "space," we could then determine the length of $\overline{PQ}$ if we knew the coordinates of P and Q.

For instance, suppose on the real number line point P is assigned the real coordinate x and point Q has the coordinate y.

Then the distance between P and Q is given by $|x - y|$. This real number $|x - y|$ is the length of the segment $\overline{PQ}$ (that is, $PQ = |x - y|$).

In the case of points lying in a plane, let us say that point P has coordinates (x_1, y_1) and point Q has coordinates (x_2, y_2). Then the distance between P and Q is given by $\sqrt{(x_1 - x_2)^2 + (y_1 - y_2)^2}$. This number is then the length of $\overline{PQ}$. It can be shown that in three-dimensional space if P has coordinates (x_1, y_1, z_1) and Q has coordinates (x_2, y_2, z_2), the distance between P and Q is $\sqrt{(x_1 - x_2)^2 + (y_1 - y_2)^2 + (z_1 - z_2)^2}$. This number is the length of $\overline{PQ}$.

All of these distance formulas share common properties. First of all, each of these distance expressions can be considered as a function

$$d: \mathcal{S} \times \mathcal{S} \to \mathbb{R}$$

where $\mathcal{S}$ stands for the space of all points under construction. This particular function satisfies each of the following properties.

1. $d(P, Q) = 0$ if and only if $P = Q$ for all $P, Q \varepsilon \mathcal{S}$
2. $d(P, Q) \geqq 0$ for all $P, Q \varepsilon \mathcal{S}$
3. $d(P, Q) = d(Q, P)$ for all $P, Q \varepsilon \mathcal{S}$
4. $d(P, R) \leqq d(P, Q) + d(Q, R)$ for all $P, Q, R \varepsilon \mathcal{S}$

Property 4 says in essence that "the shortest distance between two points is the line segment joining them." We call it the *triangle inequality*. If points P, Q, R are collinear such that Q is between P and R, the statement becomes an equality. Otherwise, it is an inequality.

We can easily see that these conditions are satisfied in the case where $d(P, Q) = |x - y|$ with x the coordinate of P and y the coordinate of Q.

1. $|x - y| = 0$ if and only if $x = y$
2. $|x - y| \geqq 0$ for all $x, y \varepsilon \mathbb{R}$ by definition of the absolute value function
3. $|x - y| = |y - x|$ as was shown earlier
4. Suppose the coordinates of P, Q, and R are x, y, and z, respectively. Then $d(P, R) = |x - z|$. Now $|x - z| = |(x - y) + (y - z)|$. From an earlier experiment, we saw that $|a + b| \leqq |a| + |b|$. $|(x - y) + (y - z)| \leqq |x - y| + |y - z|$. Thus, we have $d(P, R) \leqq d(P, Q) + d(R, Q)$.

Exercise 6.3.1

Verify that each of the measure properties is true if P and Q are points in a plane with coordinates of $P(x_1, y_1)$ and $Q\ (x_2, y_2)$ and with $d(P, Q) = \sqrt{(x_1 - x_2)^2 + (y_1 - y_2)^2}$ by substituting several values for x_1, x_2, y_1, and y_2.

6.4 EQUAL, EQUIVALENT, AND CONGRUENT

In the second paragraph of the Declaration of Independence, there is the statement, "We hold these truths to be self-evident, that all men are created equal. . . ." Because Thomas Jefferson was one of our most brilliant presidents, it should be obvious that in using the word equal, he was not expressing the opinion that all men are exactly the same. Instead, he proceeded to describe in just what manner all men are created equal: ". . . that they are endowed by their Creator with certain inalienable rights, that among these are life, liberty, and the pursuit of happiness." Thus, our equality is by definition the treatment we are entitled to receive under law.

In mathematics, when we say that two things or two ideas are equal, we mean that they are identically the same. But this does not mean that the representations of a single idea are necessarily the same. For instance, if we speak of the Empire State or if we speak of New York State, we are referring to the same thing. The names are not identical but they do represent the same state. We can write the idea of the number seven in many ways, all of which differ somewhat in appearance but still refer to the same idea—for instance, "$2 + 5$," "$3 + 4$," "$8 - 1$," "$14 \div 2$," "VII," and so on. As another example, a container of milk holds the same amount whether we describe it as holding 1 qt or 2 pt. Similarly, 1 ft = 12 in. and 1000 m = km. Numerous elementary problems in mathematics amount to showing that two representations of ideas actually refer to the same idea. For instance, a student in first year algebra learns that $3(x + y) = 3x + 3y$ and $x^2 + 5x + 6 = (x + 3)(x + 2)$, the inference being that the "names" on either side of the equality sign represent the same idea.

Mathematics does have a concept which is similar to the concept of equality as used by Jefferson; namely, the concept of equivalence. In this case, a class of perhaps distinct ideas are classified by definition depending on whether or not they share a common property. For instance, suppose we consider the set of integers I. We will say that integers a and b are equivalent if their difference is divisible by 2. Thus, for instance, 7 and 11 are equivalent because $11 - 7$ (or $7 - 11$ for that matter) is divisible by 2. Other examples of integers that are equivalent in this way are 4 and 10, 6 and -4, -7 and -13, 4004 and 2, and so on. (Can you find other examples? Can you find examples of integers that are not equivalent in this sense?)

6.4.1 Definition

More generally, suppose we have a set X of objects that we wish to classify in accordance with some property. We then define two elements a and b of X as related if they share this property. Symbolically, we write aRb if and only if a and b have this property. We now have a set X with a relation R defined on it. If this relation satisfies the following three properties, we say that R is an equivalence relation.

1. *Reflexive:* For each $a \in X$, aRa.
2. *Symmetric:* For each $a, b \in X$, if aRb, then bRa.
3. *Transitive:* For each $a, b, c \in X$, if aRb and bRc, then aRc.

comment . . .

In Chapter 2, we defined a relation R from set A to B as any subset of ordered pairs from $A \times B$. This does not contradict the definition given here. The relation R on set X can be considered as a subset of ordered pairs of $X \times X$, where aRb can be interpreted as: aRb if and only if $(a, b) \in R$. Thus, the equivalence properties can be rewritten as:

1. *Reflexive:* For each $a \in X$, $(a, a) \in R$.
2. *Symmetric:* For each $a, b \in X$, if $(a, b) \in R$, then $(b, a) \in R$.
3. *Transitive:* For each $a, b, c \in X$, if $(a, b) \in R$ and $(b, c) \in R$, then $(a, c) \in R$.

The reader should keep in mind both ways of interpreting a relation, that is, as a property or as a set of ordered pairs. Sometimes one interpretation will be more useful than the other one, depending on the circumstances.

Example 1

The prototype of an equivalence relation is the relationship of *equality*. Given a set X with $a, b \in X$, we say that a is related to b if $a = b$. This relation of equality on a set X obviously satisfies the above three properties.

Reflexive: because every element is equal to itself.
Symmetric: because if $a = b$, then $b = a$.
Transitive: because if $a = b$ and $b = c$, then $a = c$.

Example 2

Let X be the set of integers. Define the relation on X as follows: aRb if $a - b$ is divisible by 2. This relation on X is an equivalence relation.

Reflexive: Each integer a is related to itself because $a - a = 0$ is divisible by 2.

Symmetric: Suppose aRb; this means that $a - b$ is divisible by 2, but then so is $b - a$ divisible by 2 since $b - a = -(a - b)$. Thus, bRa.

Transitive: Suppose aRb and bRc. This means that $a - b$ is divisible by 2 and $b - c$ is divisible by 2. The sum of two even numbers is even; hence, $(a - b) + (b - c) = (a - c)$ is divisible by 2. Thus, aRc.

Example 3

Let X be the set of all closed curves in a plane. We will say that curve C is related to curve C' if they both separate the plane into the same number of regions. By this relation all simple closed curves are related, but a curve in the shape of a figure eight is not related to a simple closed curve. This particular region is easily seen to be an equivalence relation.

Reflexive: Every closed curve is certainly related to itself.

Symmetric: If curve C is related to curve C', surely C' is related to C; that is, they separate the plane into the same number of regions.

Transitive: If curve C separates a plane into the same number of regions as C' and C' separates a plane into the same number of regions as C'', then C separates a plane into the same number of regions as C''.

Exercise 6.4.1

Let X be the set of integers. Define aRb if $a - b$ is divisible by 5. Show that R is an equivalence relation on X.

One type of equivalence of particular interest in the study of geometry is that of *congruence*. Intuitively, congruent figures are those which have the same size and shape. Another way of saying this is that two figures are congruent if one of the figures can be moved without changing its size or shape so as to coincide with the second figure.

The idea of congruence is found in manufacturing in the principle of interchangeable parts. If a car's muffler needs to be replaced or if the picture tube burns out in a television set, it is not necessary to have the replacement parts specially made for the particular automobile or particular television set. The manufacturers have certain standards for their products and all parts of a particular piece of equipment can be replaced by thousands of parts which are, for all practical purposes, exactly the same. We might say that, in a sense, interchangeable parts are congruent.

We will not pursue the idea of interchangeable parts but will consider what it means to say that two geometric figures are congruent. All line segments have the same shape, but they may differ in size. Size in this case is given in terms of length.

6.4.2 Definition

If we let X be the set of all line segments, two line segments $\overline{AB}$ *and* $\overline{CD}$ *will be said to be* congruent, *written* $\overline{AB} \cong \overline{CD}$, *if and only if the length of AB equals the length of CD, written* $AB = CD$.

comment . . .

Explain the difference between saying $AB = CD$ and $\overline{AB} = \overline{CD}$.

It is easy to show that this relation of congruence of line segments is an equivalence relation.

1. *Reflexive:* For any line segment $\overline{AB}$, $\overline{AB} \cong \overline{AB}$.
2. *Symmetric:* For any two line segments $\overline{AB}$ and $\overline{CD}$, if $\overline{AB} \cong \overline{CD}$, then $\overline{CD} \cong \overline{AB}$.
3. *Transitive:* For any three line segments $\overline{AB}$, $\overline{CD}$, and $\overline{EF}$, if $\overline{AB} \cong \overline{CD}$ and $\overline{CD} \cong \overline{EF}$, then $\overline{AB} \cong \overline{EF}$.

The equivalence relation of congruence partitions the set of all line segments into equivalence classes. The equivalence class of $\overline{AB}$ is the set of all line segments which are congruent to $\overline{AB}$. If $\overline{AB}$ and $\overline{CD}$ are not congruent, they belong to different equivalence classes. Every segment in the equivalence class of $\overline{AB}$ has the same length as $\overline{AB}$.

Exercise 6.4.2

1. Let A, B, C be three points on line l. Let x, y, z be real numbers such that x, y, z are, respectively, the coordinates of points A, B, and C. Give an argument to support the contention that if $A - B - C$, then $AC = AB + BC$.
2. Explain how it is possible for points A, B, C to be collinear such that $d(A, C) < d(A, B) + d(B, C)$.
3. What can be said about points P and Q if $d(P, Q) = 0$?
4. Let R be the midpoint of $\overline{PQ}$. Describe this in terms of the distance function and betweenness.
5. Explain why the following statement should be true: If $\overline{AB} \cong \overline{PQ}$ and $\overline{CD} \cong \overline{PQ}$, then $\overline{AB} \cong \overline{CD}$.
6. Explain why the following statement should be true: If $A - B - C$ and $P - Q - R$, $\overline{AB} \cong \overline{PQ}$, $\overline{BC} \cong \overline{QR}$, then $\overline{AC} \cong \overline{PR}$.
7. Explain why the following statement should be true: For any

segments $\overline{AB}$ and $\overline{PQ}$, exactly one of the following is true: $\overline{AB} \cong \overline{PQ}$, $AB < PQ$, or $PQ < AB$.

6.5 ANGLES AND ANGLE MEASUREMENT

We will now consider the problem of determining the size of an angle such as $\angle ABC$ (Fig. 6.5.1). We cannot use the idea of linear measurement to determine the size of $\angle ABC$ since the sides of angles are rays, all of which have infinite extent and so in this respect they are indistinguishable. Furthermore, we cannot measure the size of the angle by determining the area of the interior of an angle, for once again this is infinite for all angles.

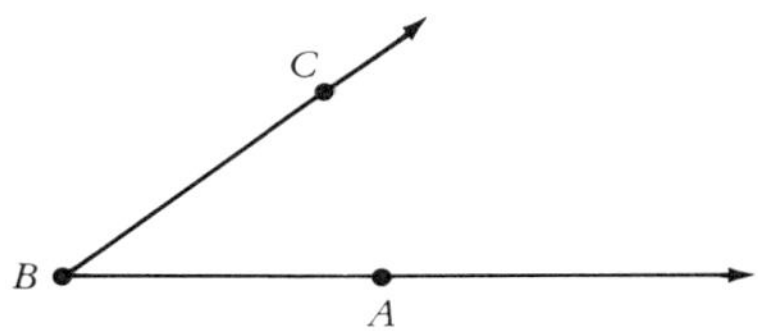

Figure 6.5.1

Let us recall our actual goal in measuring angles. To each angle we wish to assign a unique number. We wish to make this assignment using the following idea of order. Intuitively, in Fig. 6.5.2, $\angle ABC$ is larger than both $\angle ABD$ and $\angle DBC$. In assigning numbers to each angle, we wish to incorporate this idea. In addition, we want the sum of the numbers assigned to $\angle ABD$ and $\angle DBC$ to equal the number assigned to $\angle ABC$.

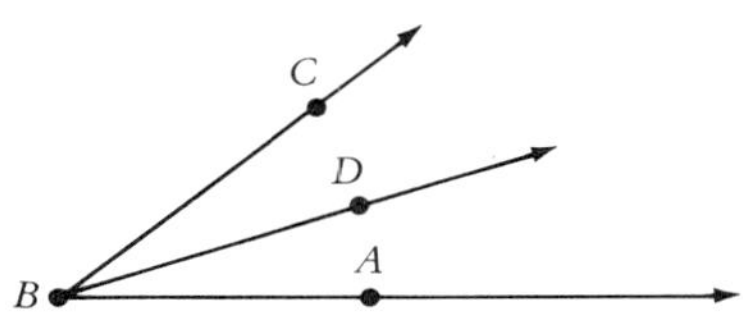

Figure 6.5.2

comment . . .

For a property to be measurable in other than an artificial manner, this additive condition must be present We might go so far as to say that the measurable properties are those which are changed by the combination of similar objects.

Our procedure is as follows. Choose a basic angle with which to compare $\angle ABC$. This then will be our arbitrary unit of mea-

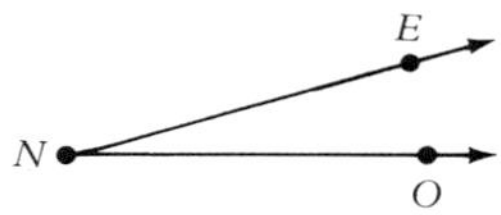

Figure 6.5.3

surement. Suppose our basic angle is $\angle ONE$ (Fig. 6.5.3). Figure 6.5.4 illustrates how $\angle ONE$ is used to measure $\angle ABC$. We have "copied" $\angle ONE$ on $\angle ABC$ three times by first matching point N with point B, ray $\overrightarrow{NO}$ with ray $\overrightarrow{BA}$, and point E in the interior of $\angle ABC$. In this way, ray $\overrightarrow{NE}$ coincides with ray $\overrightarrow{BX}$. Next, match point N with point B, and ray $\overrightarrow{NO}$ with ray $\overrightarrow{BX}$, point E in the interior of $\angle CBX$. Then ray $\overrightarrow{NE}$ coincides with ray $\overrightarrow{BY}$

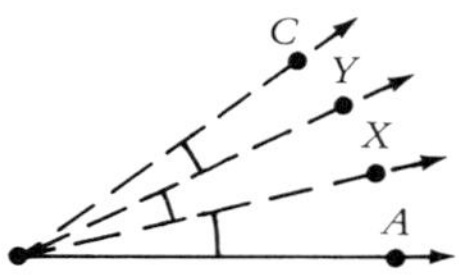

Figure 6.5.4

In this procedure, rays $\overrightarrow{BX}$ and $\overrightarrow{BY}$ have been chosen so that

1. $\angle ABX$, $\angle XBY$, and $\angle YBC$ are each the same size as $\angle ONE$.
2. The union of the interior of the three angles and the two half-lines BX and BY is the interior of $\angle ABC$.

If we assign the number one to $\angle ONE$, then to satisfy the conditions mentioned above we should assign the number three to $\angle ABC$. Of course, our choice for illustration is a special case in that $\angle ONE$ fit exactly three times. Such a situation would not occur for every angle. Ordinarily, we would be faced with an approximation problem similar to the one of linear measurement.

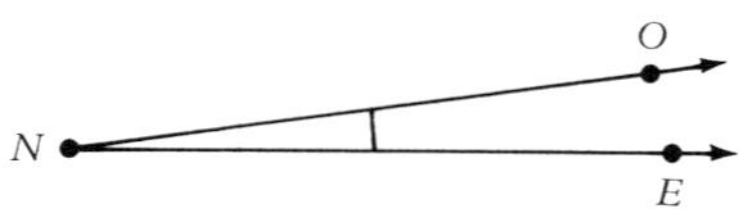

Figure 6.5.5
Our angle of 1° is $\angle ONE$

There is no system of measurement in which $\angle ONE$ is the *unit angle* (Fig. 6.5.5). The most usual measurement system of angles is the one that chooses a unit angle so that fitting together 180 of them forms a straight angle. In this system, the size of the base angle is named one *degree* (in symbols, 1°) and the measure of a straight angle is 180° (read, 180 degrees).

The best way to think of angle measure is as a function m° whose domain is the set of all angles $\mathcal{A}$ and the range is the set of real numbers $\mathbb{R}$.

$$m^\circ : \mathcal{A} \to \mathbb{R}$$

To represent the measure of $\angle ABC$, we will write $m^\circ \angle ABC$. For instance, if $\angle PQN$ is a straight angle, $m^\circ \angle PQN = 180$. We do not write $m^\circ \angle PQN = 180^\circ$ because the values of the function m° are simply real numbers. In labeling figures, we will use the degree sign to indicate that certain letters or numbers are meant to be the degree measure of angles.

In Fig. 6.5.6, angles are classified according to their measure.

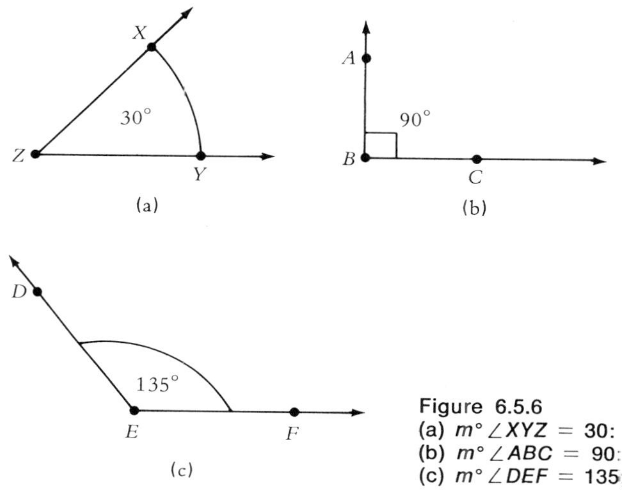

Figure 6.5.6
(a) $m^\circ \angle XYZ = 30$: acute angle;
(b) $m^\circ \angle ABC = 90$: right angle;
(c) $m^\circ \angle DEF = 135$; obtuse angle

6.5.1 Definition

An angle whose measure in degrees is between 0 and 90 is called an acute *angle. An angle of 90° is a* right *angle. An angle whose measure in degrees is between 90 and 180 is an* obtuse *angle (Fig. 6.5.6).*

In measuring line segments, we used an instrument called a ruler; to measure angles we use an instrument called a *protractor.* Figure 6.5.7 illustrates a protractor.

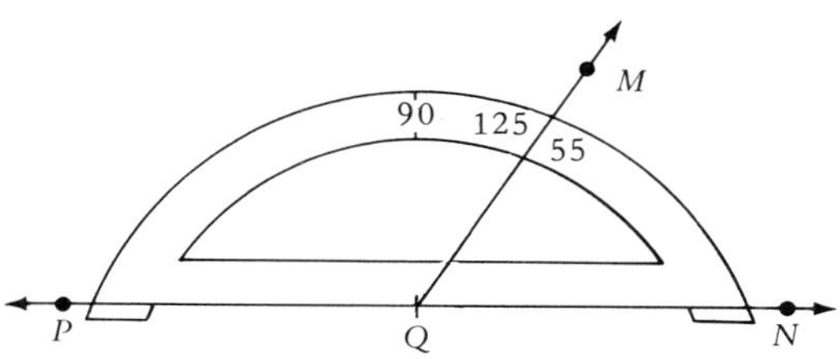

Figure 6.5.7

There are two scales on the protractor; the top one marked from left to right about the arc from 0 to 180, the lower scale marked from right to left about the arc from 0 to 180. Each of the marks on the two scales of the protractor suggest short segments of the ray $\overrightarrow{QM}$ that divide the straight angle $\angle PQN$ into 180 angles of the same size, each measuring 1°.

To measure a given angle:

1. Choose one of the scales of the protractor.
2. Place the protractor so that its center point Q coincides with the vertex of the angle and the zero point of the chosen scale lies on one of the sides of the angle.
3. Read the scale at the point where the other side of the angle crosses the scale. This reading gives the measure of the angle in degrees.

In Fig. 6.5.7, if we read the inner scale, the measure in degrees of $\angle NQM$ is 55; that is, $m^\circ \angle NQM = 55$. Reading the outer scale we see that $m^\circ \angle MQP = 125$. Note that the sum of the measures of $\angle NQM$ and $\angle MQP$ is 180.

6.5.2 Definition

Two angles are supplementary angles *if the sum of their measures in degrees is 180. Each angle is called the* supplement *of the other angle. Two angles are* complementary angles *if the sum of their measures is 90. Each angle is called the* complement *of the other angle.*

We may summarize the properties of angle measurement as follows:

1. For every angle $\angle ABC$, $m^\circ \angle ABC$ is between 0 and 180 inclusively.
2. Let $\overrightarrow{PQ}$ be a ray on the edge of the half-plane H. For every number x between 0 and 180, there is exactly one ray $\overrightarrow{PM}$ with M in H, such that $m^\circ \angle MPQ = x$ (Fig. 6.5.8).

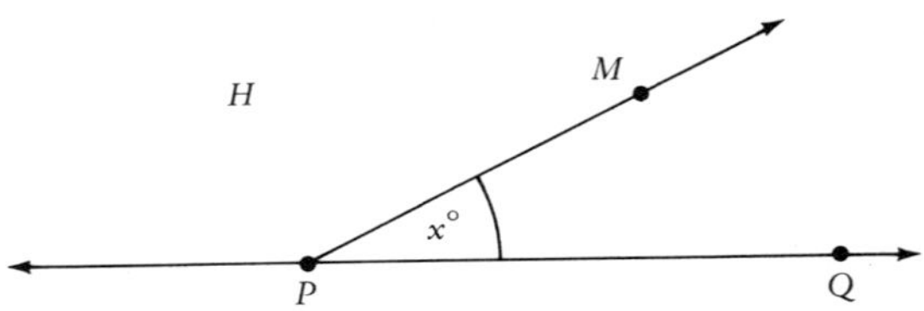

Figure 6.5.8

3. If D is the interior of $\angle ABC$, then $m^\circ \angle ABC = m^\circ \angle ABD + m^\circ \angle DBC$ (Fig. 6.5.9).

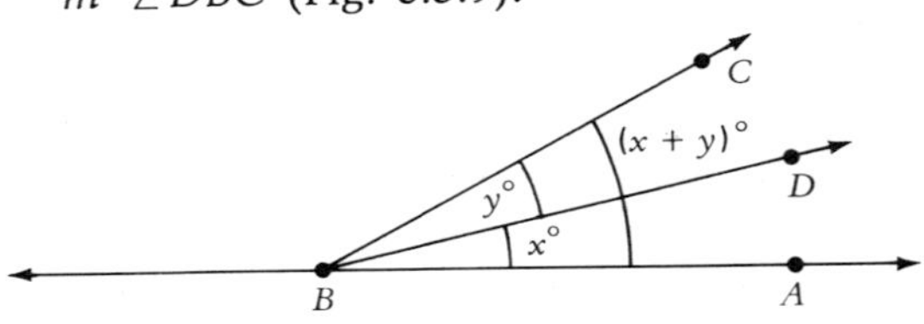

Figure 6.5.9

4. Two angles $\angle DAB$ and $\angle DAC$ form a *linear pair* if $\overrightarrow{AB}$ and $\overrightarrow{AC}$ are opposite rays and $\overrightarrow{AD}$ is any third ray. If two angles form a linear pair such as in Fig. 6.5.10, then they are *supplementary;* that is, $x + y = 180$.

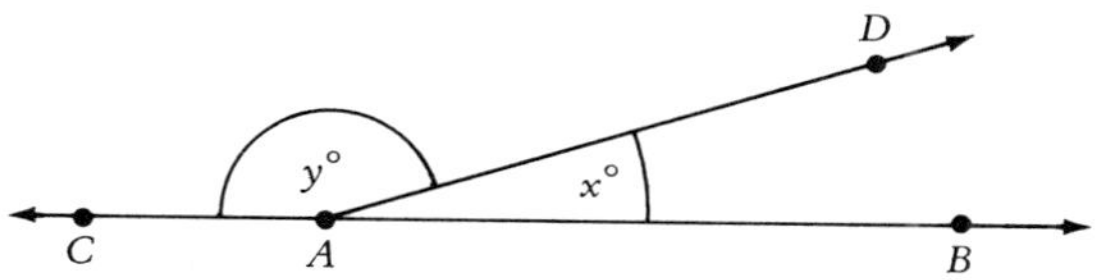

Figure 6.5.10

6.5.3. Definition

We say that two angles $\angle ABC$ and $\angle PQR$ are congruent *(written $\angle ABC \cong \angle PQR$) if and only if $m^\circ \angle ABC = m^\circ \angle PQR$. It is easy to see that angle congruence is an equivalence relation on the set of all angles. That is:*

1. *Each angle is congruent to itself.*
2. *If $\angle ABC \cong \angle PQR$, then $\angle PQR \cong \angle ABC$.*
3. *If $\angle ABC \cong \angle PQR$ and $\angle PQR \cong \angle XYZ$, then $\angle ABC \cong \angle XYZ$.*

Exercise 6.5.1

1. Classify the angles of Fig. 6.5.11 as acute, right, or obtuse. Estimate the measure in degrees and measure the angles with a protractor.

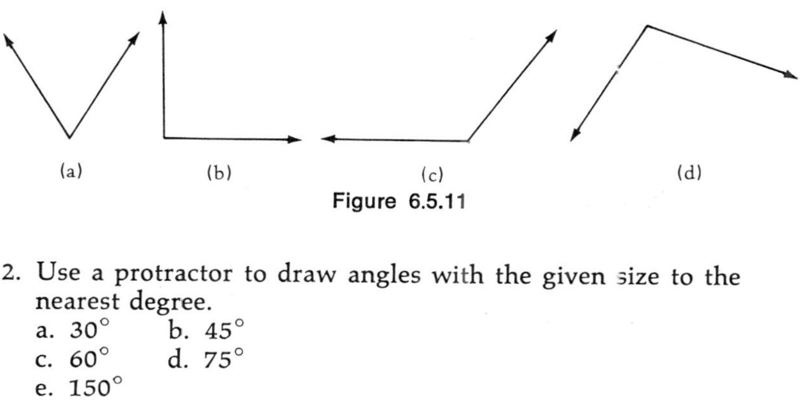

Figure 6.5.11

2. Use a protractor to draw angles with the given size to the nearest degree.
   a. 30° b. 45°
   c. 60° d. 75°
   e. 150°
3. Using a protractor, find the measure in degrees of each angle shown in Fig. 6.5.12.
4. If $\angle PQR \cong \angle ABC$ and $\angle PQR \cong \angle DEF$, what relationship exists between $\angle ABC$ and $\angle DEF$?

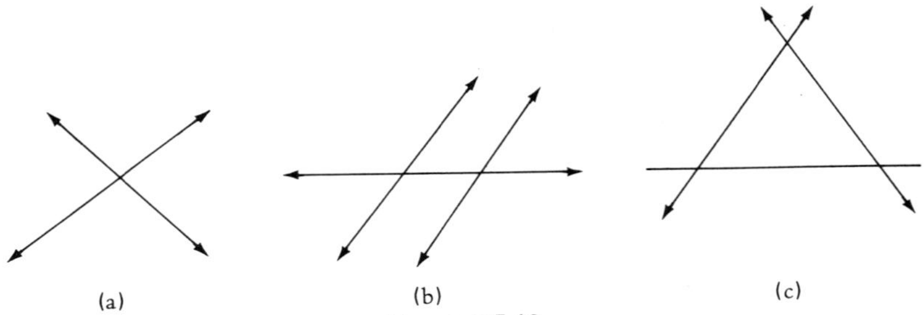

Figure 6.5.12

5. Give an argument to support each of the following statements.
   a. All right angles are congruent.
   b. If two angles are congruent and supplementary, then each is a right angle.
   c. If two nonzero angles are complementary, then each is acute.
   d. Supplements of congruent angles are congruent.
   e. Complements of congruent angles are congruent.
6. a. If $\angle ABC$ and $\angle CBD$ are supplementary and if $m^\circ \angle ABC = m^\circ \angle CBD$, find $m^\circ \angle ABC$ and $m^\circ \angle CBD$.
   b. If $\angle ABC$ and $\angle CBD$ are complementary and if $m^\circ \angle ABC = 2m^\circ \angle CBD$, find $m^\circ \angle ABC$ and $m^\circ \angle CBD$.
7. Explain why the following statement should be true: For any pair of angles, $\angle A$ and $\angle B$, exactly one of the following conditions hold: $\angle A \cong \angle B$, $m^\circ \angle A < m^\circ \angle B$, $m^\circ \angle B < m^\circ \angle A$.
8. Explain why the following statements should be true.
   a. If $m^\circ \angle A < m^\circ \angle B$ and $\angle A \cong \angle C$, then $m^\circ \angle C < m^\circ \angle B$.
   b. If $m^\circ \angle A < m^\circ \angle B$ and $\angle B \cong \angle C$, then $m^\circ \angle A < m^\circ \angle C$.
   c. If $m^\circ \angle A < m^\circ \angle B$ and $m^\circ \angle B < m^\circ \angle C$, then $m^\circ \angle A < m^\circ \angle C$.

6.6 SOME CONSEQUENCES OF ANGLE MEASUREMENT

Two angles form a *vertical pair* if their sides form pairs of opposite rays, as in Fig. 6.6.1. Here $\angle RPQ$ and $\angle SPM$ form a vertical pair. Similarly, $\angle RPS$ and $\angle QPM$ form a vertical pair.

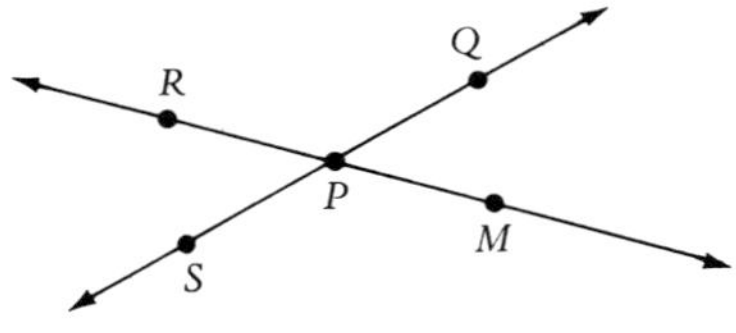

Figure 6.6.1

Another way of describing the vertical pair $\angle RPQ$ and $\angle SPM$ is as follows:

6.6.1 Definition

If $R - P - M$ and $S - P - Q$ *and the lines* $\overleftrightarrow{RP}$ and $\overleftrightarrow{PQ}$ *are distinct, then* $\angle RPS$ *and* $\angle QPM$ *form a* vertical pair.

In Fig. 6.6.1, we could use a protractor to measure the four angles. If we do so, we discover that $m^\circ \angle RPQ = m^\circ \angle SPM$ and $m^\circ \angle RPS = m^\circ \angle QPM$. In other words, $\angle RPQ \cong \angle SPM$ and $\angle RPS \cong \angle QPM$. This situation is true in general. That is, if $\angle RPS$ and $\angle QPM$ form a vertical pair, then they are congruent. This can be seen as follows: Let $m^\circ \angle RPQ = x$; then since $R - P - M$, it follows that $\overrightarrow{PR}$ and $\overrightarrow{PM}$ are opposite rays. Therefore, $\angle RPQ$ and $\angle QPM$ form a linear pair. This means these angles are supplementary and it follows that

$$m^\circ \angle QPM = 180 - x$$

On the other hand, since $S - P - Q$, it follows that $\overrightarrow{PS}$ and $\overrightarrow{PQ}$ are opposite rays and so $\angle RPS$ and $\angle RPQ$ form a linear pair. Thus, $\angle RPS$ and $\angle RPQ$ are supplementary. This means that

$$m^\circ \angle RPS = 180 - x$$

Therefore,

$$m^\circ \angle QPM = m^\circ \angle RPS$$

from which it follows that

$$\angle QPM \cong \angle RPS$$

Similarly, we could show that $\angle RPQ \cong \angle SPM$. To summarize, note the following.

6.6.2 Theorem

If two lines intersect in a point, the two angles of the vertical pair are congruent.

Suppose now that we have two lines l and m, not necessarily parallel, which are intersected by a transversal t as in Fig. 6.6.2.

comment . . .

A *transversal* is a line that intersects two or more given lines in distinct points.

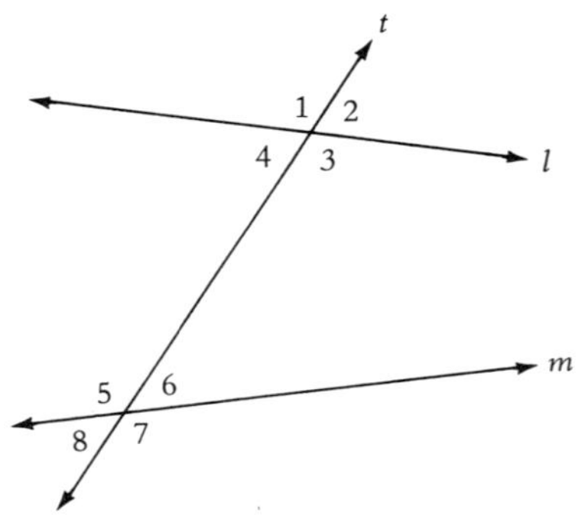

Figure 6.6.2

We are using the numbers 1 to 8 to denote the eight angles formed by the lines l and m and the transversal t. In pairs these angles are given special names, some of which we have already discussed. For instance, $\angle 1$ and $\angle 3$ are vertical angles and $\angle 2$ and $\angle 6$ are *corresponding angles*. The angles labeled 1, 2, 7, 8 will be referred to as *exterior angles,* while those labeled 3, 4, 5, 6 will be referred to as *interior angles*. We call $\angle 4$ and $\angle 6$ *alternate interior angles* while $\angle 1$ and $\angle 7$ are referred to as *alternate exterior angles*. Note that $\angle 3$ and $\angle 5$ form another pair of alternate interior angles and that $\angle 2$ and $\angle 8$ form another pair of alternate exterior angles.

If the lines l and m are parallel, it will happen that each pair of corresponding angles are congruent. More precisely, we will accept the following statement as a postulate.

6.6.3 Postulate

Two or more lines intersected by a transversal are parallel if and only if each pair of corresponding angles are congruent.

Now suppose we have two lines l and m intersected by a transversal t as in Fig. 6.6.3. Let us suppose that $l \parallel m$. Note that $\angle 2$ and $\angle 3$ are vertical angles and as such must be congruent. On the other hand, $\angle 1$ and $\angle 2$ are corresponding angles and since $l \parallel m$, they must be congruent. Therefore, it follows that $\angle 1$ and $\angle 3$ are congruent.

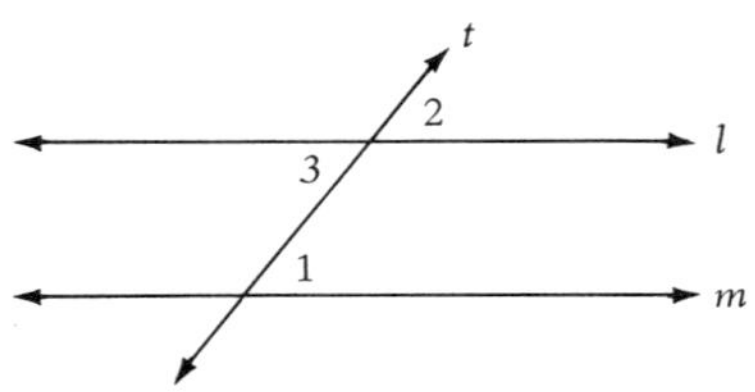

Figure 6.6.3

We have thus discovered the following.

6.6.4 Theorem

If two parallel lines are cut (intersected) by a transversal, then alternate interior angles are congruent.

What can we say about alternate exterior angles?

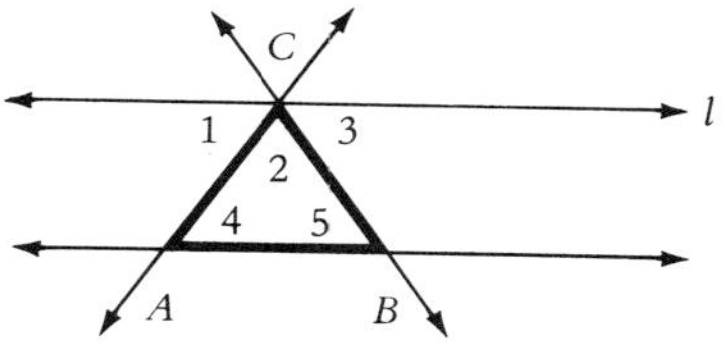

Figure 6.6.4

Suppose next that we have the triangle ABC as shown in Fig. 6.6.4 such that the line l passing through C is parallel to the line $\overleftrightarrow{AB}$. Since the line containing side AC can be thought of as a transversal cutting the two parallel lines l and $\overleftrightarrow{AB}$, we have that $\angle 1 \cong \angle 4$. Similarly, since the line containing side $\overline{BC}$ can be considered as a transversal cutting the two parallel lines l and $\overleftrightarrow{AB}$, it follows that the alternate interior angles, $\angle 3$ and $\angle 5$, are congruent. Now, $\angle 1$ and the angle formed by combining $\angle 2$ and $\angle 3$ form a linear pair and as such are supplementary. Thus, it follows that

$$m^\circ \angle 1 + m^\circ(\angle 2 + \angle 3) = 180$$

Note that $m^\circ(\angle 2 + \angle 3) = m^\circ \angle 2 + m^\circ \angle 3$. Thus,

$$m^\circ \angle 1 + m^\circ \angle 2 + m^\circ \angle 3 = 180 \qquad (6.1)$$

Since $\angle 1 \cong \angle 4$, it follows that $m^\circ \angle 1 = m^\circ \angle 4$. Similarly, since $\angle 3 \cong \angle 5$, we have that $m^\circ \angle 3 = m^\circ \angle 5$. Substituting in Eq. (6.1) $m^\circ \angle 3$ for $m^\circ \angle 5$ and $m^\circ \angle 4$ for $m^\circ \angle 1$, we have

$$m^\circ \angle 4 + m^\circ \angle 2 + m^\circ \angle 5 = 180$$

Thus, we can conclude the following.

6.6.5 Theorem

The sum of the measures, in degrees, of the angles of any triangle is 180.

Exercise 6.6.1

1. Refer to Fig. 6.6.5 to answer the following questions.
   a. Name two pairs of corresponding angles.
   b. Name two pairs of interior angles.
   c. Name two pairs of alternate exterior angles.
   d. Name two pairs of alternate interior angles.

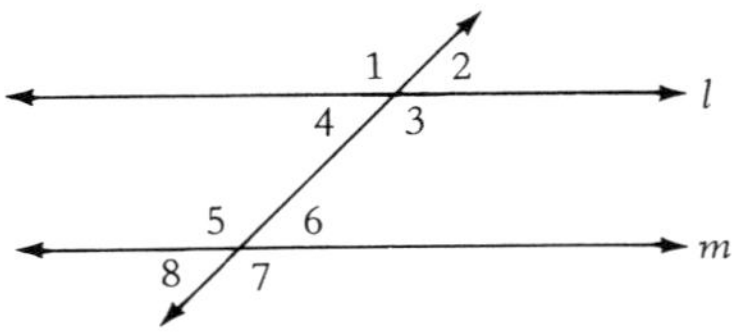

Figure 6.6.5

2. In Fig. 6.6.6, suppose lines l and m are parallel. If $m^\circ \angle 2 = 50$, what are all the other angles?

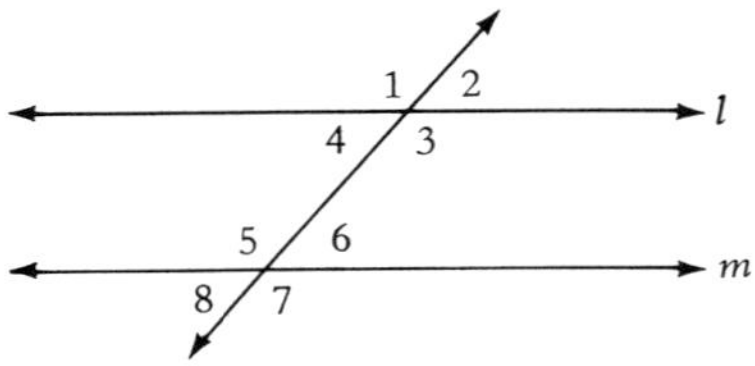

Figure 6.6.6

3. Explain why the following statements are true:
   a. If a transversal intersects two distinct coplanar lines and is perpendicular to each of them, then the lines are parallel.
   b. If a transversal intersects two distinct coplanar lines and if two consecutive interior angles determined by the transversal are supplementary, then the lines are parallel.

6.7 ANOTHER LOOK AT POLYGONS

Now that we have segment measurement and angle measurement, polygons can be classified both by the length of their sides and by the measure of their angles. For instance, see the following definitions.

6.7.1 Definition

A scalene *triangle is a triangle that has no two sides congruent. An* isosceles *triangle is a triangle with two or more congruent*

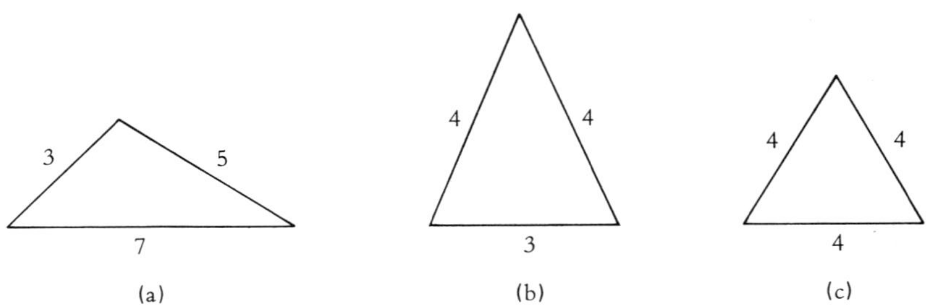

Figure 6.7.1
(a) Scalene triangle; (b) Isosceles triangle; (c) Equilateral triangle

sides, and an equilateral triangle *is a triangle that has three congruent sides (Fig. 6.7.1).*

6.7.2 Definition

In considering the angles of triangles, we say that an acute triangle *is one with three acute angles. A* right triangle *is a triangle with a right angle and an* obtuse triangle *is one with an obtuse angle (Fig. 6.7.2).*

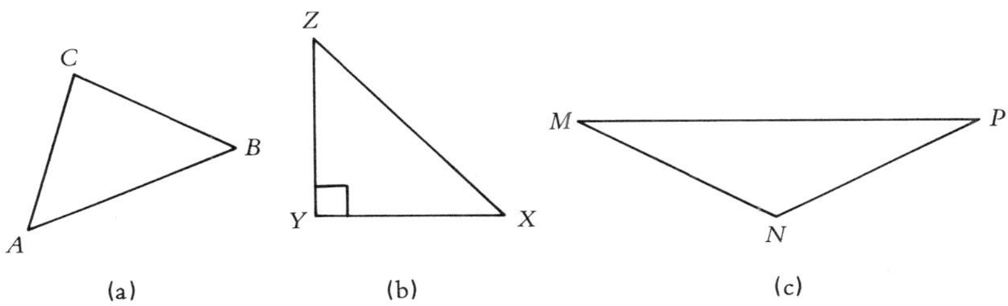

Figure 6.7.2
(a) Acute triangle; (b) Right triangle; (c) Obtuse triangle

In Fig. 6.7.3 are quadrilaterals, respectively called trapezoid, parallelogram, rhombus, rectangle, and square.

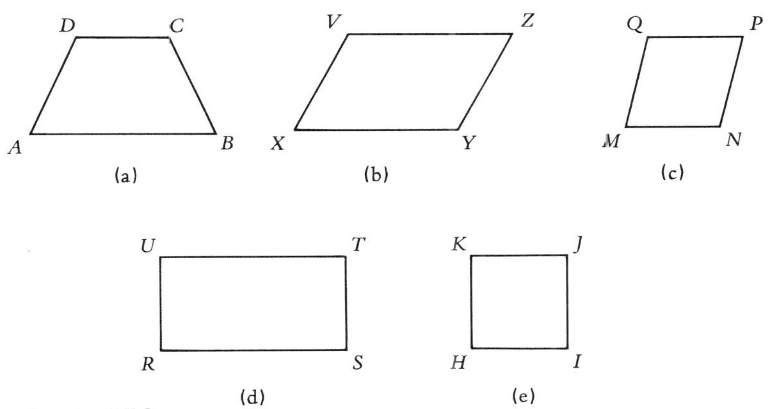

Figure 6.7.3
(a) Trapezoid; (b) Parallelogram; (c) Rhombus; (d) Rectangle; (e) Square

6.7.3 Definition

A trapezoid *is a quadrilateral with one and only one pair of opposite sides parallel. The parallel sides are the* bases *and the other two sides are the* legs *of the trapezoid. A* parallelogram *is a quadrilateral in which both pairs of opposite sides are parallel.*

A parallelogram in which all sides are congruent is called a rhombus. *A quadrilateral with all angles being right angles is called a* rectangle. *A rectangle with all sides congruent is called a* square.

Exercise 6.7.1
Let Q be the set of all quadrilaterals; T, the set of all trapezoids; P, the set of all parallelograms; Rh, the set of all rhombi; Re, the set of all rectangles; and S, the set of all squares. Determine all the possible set containment relations that exist among these sets.

Experiment 4

In each of the following, make the indicated conjectures and try to support the conjectures using drawings, rulers, protractors, and a compass. Give plausible arguments to support your conjectures.

1. Try to discover a relationship between the opposite sides of a parallelogram.
2. Try to discover a relationship between the opposite angles of a parallelogram.
3. Try to discover a relationship between a pair of consecutive angles of a parallelogram.
4. Try to discover a relationship between the diagonals of a parallelogram.
5. Try to discover a relationship between the opposite sides of a rectangle.
6. Try to discover a relationship between the diagonals of a rectangle.

6.7.4 Definition

A regular *polygon is a polygon having congruent sides and congruent angles. A regular triangle is an* equilateral *triangle and a regular quadrilateral is a square.*

Experiment 5

We have seen that the sum of the measure of the angles of a triangle is 180. But what is true of a polygon of 4 sides, of 6 sides, of n sides? Try to develop a generalization that will enable us to determine the sum of the measures of the angles of a polygon whenever we know how many sides the polygon has. As a start, let us consider any quadrilateral $ABCD$. Remember that the interior of the quadrilateral can be divided into regions by diagonal $\overline{AC}$ (Fig. 6.7.4). Diagonal $\overline{AC}$ separates $\angle BAD$ into two

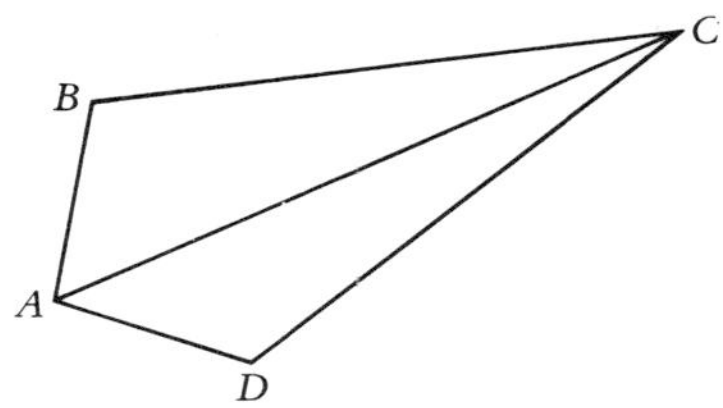

Figure 6.7.4

angles, one in each of the triangles ABC and ADC. It also separates $\angle BCD$ into two angles, one in each triangle. Because of our procedure for measuring angles,

$$m^\circ \angle BAD = m^\circ \angle BAC + m^\circ \angle CAD$$
$$m^\circ \angle BCD = m^\circ \angle BCA + m^\circ \angle ACD$$

We know that the sum of the measures of the angles of each triangle is 180, so the sum of the measures of the angles of the quadrilateral is 2×180 or 360.

Now try to find a formula which will give the sum of the measures of the angles of a convex n-gon. Can you find a formula for the sum of the measures of the interior angles of a concave polygon of n-sides?

Experiment 6

Any equilateral triangle is also equiangular so that each angle of an equilateral triangle has measure 60. Each angle of a square is a right angle. Try to find a formula that expresses the measure of each angle of a regular n-gon.

Experiment 7

Make conjectures in each of the following problems and try to support the conjectures using drawings, rulers, protractors, and a compass. Give plausible arguments to support your conjectures.

1. How shall we define a line that is perpendicular to a plane?
2. What should be the definition of the shortest distance from a point P outside a plane to the plane?
3. What should be the definition of the distance between two parallel planes?

6.7.5 Definition

If two polygonal regions having the same size and shape are placed in parallel planes in such a way that their corresponding sides are coplanar, and if their corresponding vertices are connected by line segments, the figure formed is called a prism.

We classify prisms according to the kind of polygonal regions that form their bases.

6.7.6 Definition

A rectangular *prism is one whose bases are rectangular regions; if the polygonal regions are pentagons, it is called a* pentagonal *prism, and so on. If the bases are bounded by parallelograms, the prism is called a* parallelepiped.

In Fig. 6.7.5, we show a triangular prism.

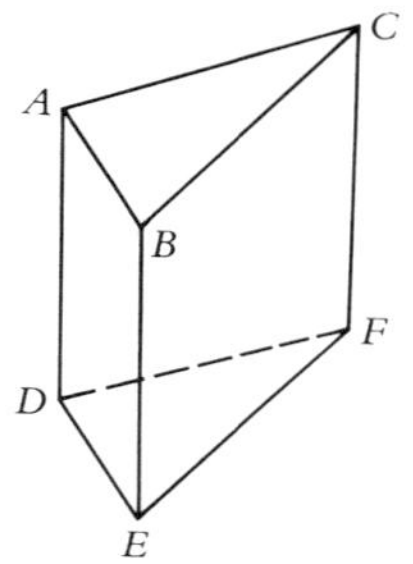

Figure 6.7.5

6.7.7 Definition

The two triangular regions are called the bases *of the prism. The other faces of the prism are called the* lateral faces *of the prism and the segments* $\overline{AD}$, $\overline{BE}$, *and* $\overline{CF}$ *are called the* lateral edges.

6.7.8 Definition

A right prism *is any prism in which the lateral edges are perpendicular to the planes of the bases.*

In such a prism the lateral faces are, therefore, rectangular regions. For instance, a *cube* is a right prism. A *regular prism* is a prism with bases that are regular polygonal regions.

Example 1

See Fig. 6.7.6.

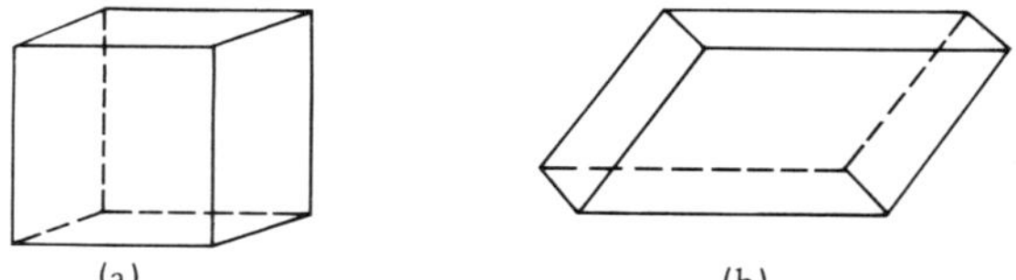

Figure 6.7.6
(a) Cube; (b) Parallelepiped

Experiment 8

If a prism has n lateral faces, how many faces in all does it have? How many vertices? How many edges? Does Euler's formula hold for a prism with n lateral faces?

Exercise 6.7.2

In Fig. 6.7.7 patterns for a cube and a parallelepiped are given. To make a model, copy on construction paper. Cut on the solid lines and fold on the dotted lines. Paste each tab to the inside of the adjoining face.

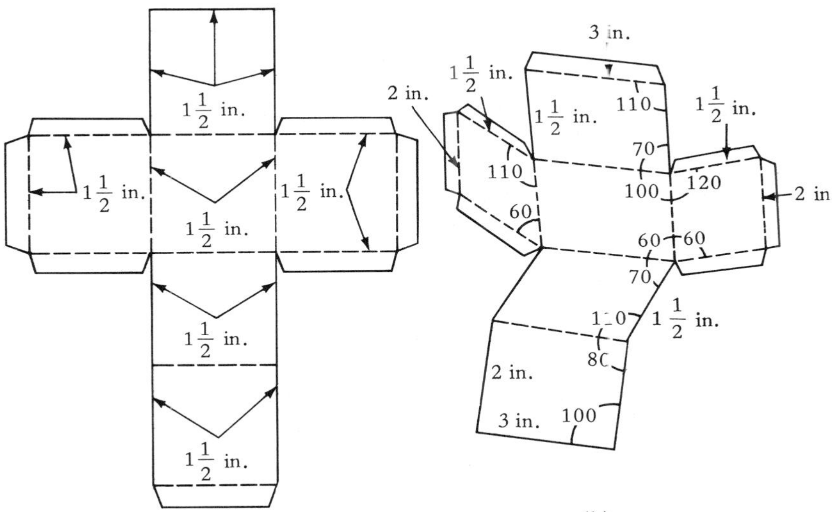

Figure 6.7.7
(a) Cube; (b) Parallelepiped

6.8 SUMMARY

1. The fundamentals of measurement are:
   a. Measurement involves comparing a unit with the object to be measured.
   b. Standard units have been established.
   c. Measurement produces a number called a measure.
   d. Measurement is stated in terms of a measure and a unit.
2. Every measurement is an approximation. The greatest possible error is one-half the unit of measurement.
3. Line segments $\overline{PQ}$ and $\overline{XY}$ are said to be congruent, $\overline{PQ} \cong \overline{XY}$, if they have the same length.

4. Congruence of line segments is an equivalence relation.
5. Angle measure is a function m° where the domain is the set of all angles and where the range is in the real numbers.
6. An acute angle has measure between 0 and 90. A right angle has measure 90. An obtuse angle has measure between 90 and 180. A straight angle has measure 180.
7. Two angles are supplementary angles if the sum of their measures is 180. Two angles are complementary angles if the sum of their measures is 90.
8. Two lines intersected by a transversal are parallel if and only if the corresponding angles formed by the transversal and the lines are congruent.
9. Vertical angles are congruent.
10. If two parallel lines are intersected by a transversal, their alternate interior angles are congruent.
11. The sum of the measures of the angles of any triangle is 180.
12. A scalene triangle has no two sides congruent. An isosceles triangle has two or more sides congruent. An equilateral triangle has three congruent sides.
13. A trapezoid is a quadrilateral with one and only one pair of opposite sides parallel. A parallelogram is a quadrilateral in which both pairs of opposite sides are parallel. A rhombus is a parallelogram with all sides congruent. A rectangle is a parallelogram with four right angles. A square is a rectangle with all sides congruent.
14. A regular polygon is a polygon having congruent sides and congruent angles.
15. A prism has lateral faces which are parallelograms.

Exercise 6.8.1

1. If the length of $\overline{PQ}$ is $3\frac{3}{13}$ in., which of the following statements are false? Explain why.
   a. $\overline{PQ}$ is a 3-in. segment.
   b. The length of $\overline{PQ}$ to the nearest inch is 3 in.
   c. The length of $\overline{PQ}$ is $3\frac{1}{2}$ in. to the nearest quarter inch.
2. Each given inequality below describes a measurement. What unit of measurement is being used in each case? What is the greatest possible error?
   a. $5\frac{1}{2}$ in. $\leqq d(A, B) \leqq 6\frac{1}{2}$ in.
   b. $3\frac{5}{8}$ in. $\leqq d(A, B) \leqq 3\frac{7}{8}$ in.
   c. 5.58 in. $\leqq d(A, B) \leqq$ 5.95 in.
3. Complete the following statements:
   a. All angles that are congruent to a given angle will have the same ______.
   b. All angles that have the same size are ______.
   c. A straight angle measures ______ degrees.

d. Vertical angles are ______ and therefore have the same ______.

e. The greatest possible error of an angle measurement made with a protractor scaled in degrees is ______.

4. Explain how to draw a triangle with one angle measuring 30°, another measuring 90°, and with the common side of these angles measuring 2 in.
5. Use a protractor and a ruler to draw the figure described.
   a. a square with sides of length 2 in.
   b. a rectangle with adjacent sides measuring 2 in. and 3 in.
   c. a parallelogram $ABCD$ with adjacent sides measuring 2 in. and 4 in., $m^\circ \angle ABC = 60$ and $m^\circ \angle DAB = 120$
   d. a rhombus with sides of length 4 in. and consecutive angles measuring 45° and 135°

PERIMETER AND INTRODUCTION TO AREA

7

7.1 PROLOGUE

We have seen how we can measure any one-dimensional object using a suitable linear measure. What about two-dimensional figures? Is it possible to apply linear measurement techniques to these as well? Just as in the one-dimensional case, we shall have to confine ourselves to segments. What types of plane segments are there? Of course, there are the common ones such as triangles, squares, parallelograms, *n*-gons, and circles, but you should also keep in mind that our techniques should also be applicable to such irregular shapes as are shown in Fig. 7.1.1.

Figure 7.1.1

So, now we come to an important question: Can we apply linear measure to two-dimensional geometric figures in some meaningful way?

Exercise 7.1.1

1. What properties do all the geometrical figures which are given above as examples of two-dimensional regions possess?
2. Do any of these properties concern themselves with one-dimensional figures?
3. What term is commonly used to designate the measure of the boundary of a two-dimensional region?

7.2 PERIMETER

If you spent some time considering the above exercise set, you should by now have come to realize that one way to apply linear measure to two-dimensional regions is to measure the length of the boundary which determines the segment. This is, of course, commonly referred to as *perimeter*. Does it seem sensible to suggest that we can adequately measure two-dimensional regions by measuring their perimeters? Perhaps the following set of exercises may clarify the issue for you.

Exercise 7.2.1

1. Determine the perimeter of the following shapes:
   a. a triangle with sides of 3 in., 5 in., and 7 in.
   b. a square with a side of 27 in.
   c. a parallelogram with a pair of adjacent sides being 3 ft and 10 ft, respectively
   d. a regular hexagon with one side of 52 in.
   e. a regular n-gon with one side s units
2. Figure out a way to estimate the perimeter of Fig. 7.2.1.

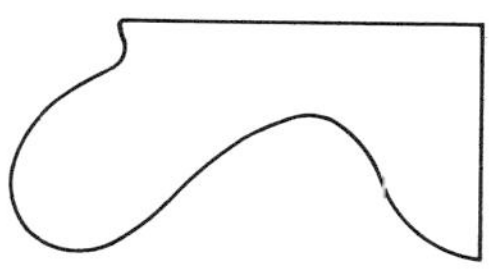

Figure 7.2.1

3. Use your ruler to determine the perimeter of the triangle, hexagon, 12-gon, and 24-gon in Fig. 7.2.2.
   a. triangle ________
   b. hexagon ________
   c. dodecagon (12-gon) ________
   d. 24-gon ________

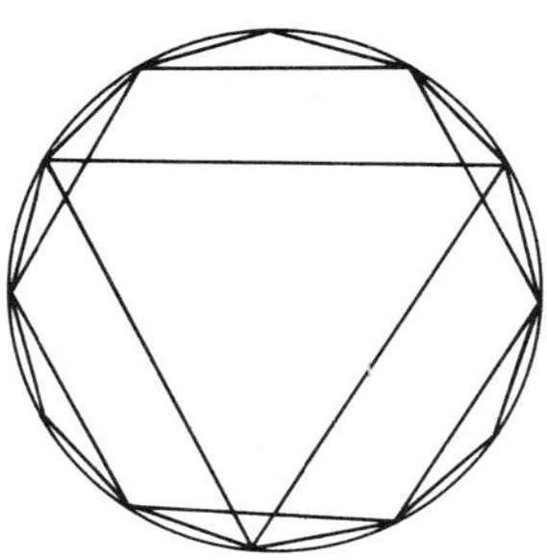

Figure 7.2.2

4. Use your calculations to estimate the perimeter of the circle. What term is normally used to refer to the perimeter of a circle?

5. Write algebraic expressions for the perimeter of the following figures:
   a. a scalene triangle with sides x, y, and z
   b. an isosceles triangle with one leg a and a base b
   c. an equilateral triangle with one side r
   d. a quadrilateral with sides p, q, r, and s
   e. a parallelogram with a pair of adjacent sides, a and b
   f. a square with one side t
   g. a rectangle with one side l and the other side three times as long
   h. a regular octagon with one side l.
   i. a rectangle with one side r and the other side half as long
6. Sketch a rectangle which would have a perimeter equal to the perimeter of the triangle in Fig. 7.2.3. What are the dimensions of your rectangle? Is this a unique answer?

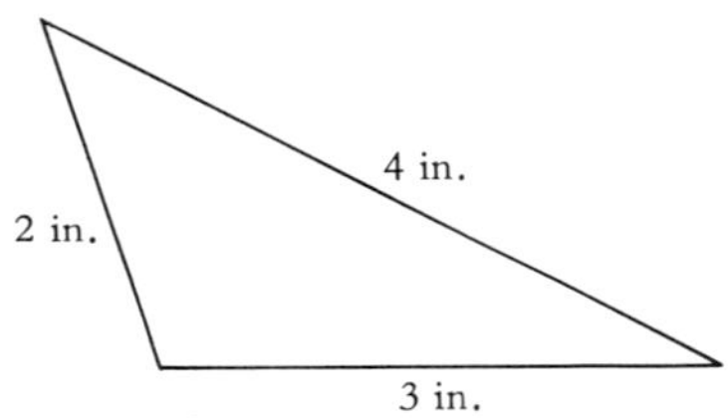

Figure 7.2.3

7. Sketch two different rectangles each of which has a perimeter of 10 in.
8. The following newspaper ad appeared in a local newspaper:

Gigantic Land Sale!

Prime farmland at a low, low price. Now available for a limited time only a limited number of lots with up to 900 ft of boundary for only 10¢ a linear foot. Every lot is guaranteed to be prime farm land with a minimum of 12 in. of lush, rich top soil, or your money will be cheerfully refunded. Lots are going fast.

Just last week, Mr. J. H. Brown of Hartford, Connecticut, purchased a 200 ft × 250 ft plot—that's more than an acre of land —for less than a hundred dollars! These lots are going fast.

Don't miss this sensational offer. Send your check for \$90 to Fertile Farms, Inc.

Would you recommend a friend to take advantage of this offer? Explain why or why not.

Experiment 1

Find two different-sized flat, circular objects such as a plate, cake tin, and so on. Make the following measures, making all measurements to the nearest hundredth of an inch. (Actually, you should make three measurements and then average them.)

	1st estimate	*2nd estimate*	*3rd estimate*	*Average (in tenths)*
Diameter	______	______	______	______
Circumference	______	______	______	______

(Measure it, don't use a formula.)

Use your average of the three estimates as the actual measurement. Now calculate the ratio of the circumference to the diameter (to the nearest tenth). Repeat the experiment on the second circle. What do you observe? What would you expect to get if you tried the experiment on a third circle? Do you know what the ratio of the circumference of a circle to its diameter is usually called? On the basis of your calculations, estimate what you think the circumference of a plate will be whose diameter is 25.27 cm. Estimate what you think the diameter of a plate will be if the circumference is 50.54 cm.

in our opinion . . .

The preceding experiment outlines an essential activity which every student should perform as an initial introduction to the real number pi. The student who can respond to the question, "What is pi?" with an answer of, "It is the ratio of the circumference of any circle to its diameter," demonstrates greater understanding than the student who can rattle off that pi is 3.14159. . . .

7.3 MEASURING THE INTERIOR OF A TWO-DIMENSIONAL OBJECT

If you took time to ponder Exercise Set 7.2.1, you should by now be aware that the application of a linear measure to the boundary of a two-dimensional shape is not adequate because it is possible to have shapes which have the same perimeter but which seem to differ substantially from one another in important ways; namely, in shape and in the size of their interiors. Thus, we need some way to measure their interiors. Linear measure does not seem adequate for this task. Consider for the moment trying to ask the question, "How many inches are there in the interior of a square which is 1 in. on a side?" (See Fig. 7.3.1.) We have difficulty envisioning what the question even means. Perhaps we should think of a series of closely packed line seg-

Figure 7.3.1

ments 1 in. long filling the square (Fig. 7.3.2). But something is wrong with that idea. Remember the printed line is but an aid to our concept of a line segment, which is to have no thickness. Thus, we are asking, "How many inch-long segments of no thickness can be packed inside the square?" The answer to such a question has no practical value. The fundamental problem here is that we have been attempting to measure a two-dimensional region with a unit which does not have a two-dimensional region.

Figure 7.3.2

When we considered the problem of measuring the length of a segment, we saw that we could in reality use any object which possessed length as a unit. Now that we wish to measure a two-dimensional region, we must use as a unit of measure any object which possesses a two-dimensional region.

Just as we have called the measure of the interior of a line segment its length, so shall we call the measure of the interior of a plane segment its *area*.

Exercise 7.3.1

1. Which of the following objects would most likely be considered as units of length and which as units of area?
   a. a match stick — b. a penny
   c. a tablecloth — d. a clothesline
   e. a three-cornered diaper — f. a washer
   g. a football field — h. a ruler
   i. a piece of graph paper — j. a human hair
   k. a piece of thread — l. the tip of a pin
2. The divergent thinker is liable to object to the above question with the remark that each object above could be used either as a unit of length or as a unit of area. Explain why.

Experiment 2 (for the very patient only!)

Take a solid rubber ball (one that is expendable) and cut it in half. The plane segment made by the knife should look like a circle. The outside of each piece is a half of a sphere (*hemisphere*).

Both the circle and the hemisphere have area. Which area is greater?

Take a long piece of string and fasten an end to a pin in the center of the circle. Slowly and carefully, coil the string around the pin, gradually covering the area of the circle (Fig. 7.3.3). (You may have to fasten the string from time to time with glue or tape.) Now do the same thing to the hemisphere by placing a pin in its center and coiling a string around it until it covers the hemisphere (Fig. 7.3.4). Now compare the lengths of the string used on the circle with that used to cover the hemisphere. What do you perceive? What seems to be the relationship between the area of the circle and the area of the hemisphere?

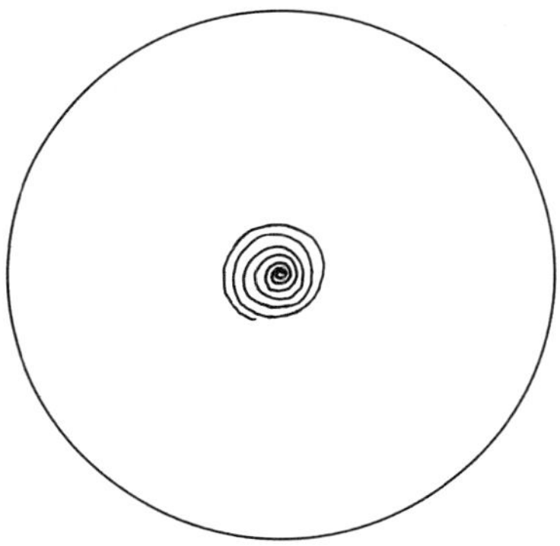

Figure 7.3.3

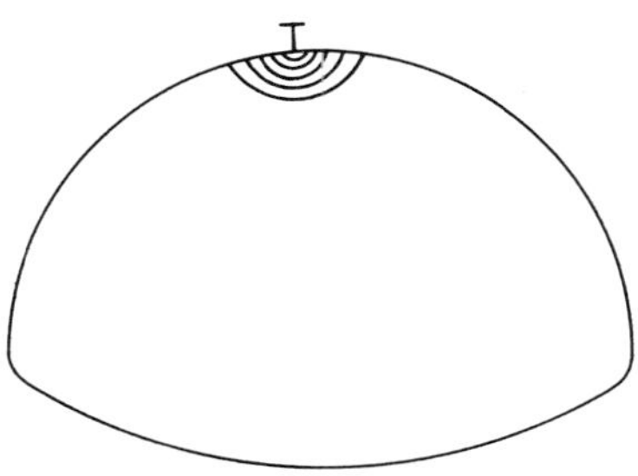

Figure 7.3.4

Doesn't this experiment violate our previous argument that you cannot use length to measure area? Explain.

7.4 SOME SUITABLE UNITS OF AREA

Theoretically, any shape which posseses area can be used as a unit of area. Practically speaking, some shapes lend themselves to this task more readily than others. Consider using a circular region as a unit (such as a penny). Thus, we could ask, "How many pennies will it take to cover a desk top?" Or, perhaps one might consider buying land by the plate (or for larger areas by the platter). But, unfortunately, it is not difficult to see that a circular unit presents some insurmountable difficulties, as the following experiment will suggest.

Experiment 3

1. Determine the area of a standard 8½ in. x 11 in. sheet of paper in some circular unit (such as pennies).
2. Determine the area of a plate in the same circular unit.

If you actually attempted the above experiment, you would have immediately seen that when you try to cover the sheet of paper or plate there are many spaces that are not covered. Circles simply do not fit together nicely to cover regions. What kinds of shapes will fit together nicely and compactly? Of course, we can see immediately that squares, rectangles, and some triangles have this property. In fact, there are many different shapes that we could think of which would have the desired property. However, even though we could have selected as a standard area unit any one of a variety of shapes, the square has emerged as the one most commonly used. An interesting problem arises from considering other polygons as our area unit. Experiment 4 considers the use of the triangle.

Experiment 4

1. If we were completely ignorant of area and did not know that squares were commonly used as units of area, we might quite logically suggest a triangle, for isn't the triangle the "building block" of all other polygons? So, consider for a moment using an equilateral triangle of 1 in. on a side as our unit of area. We might conceivably call it a "triangular inch." Use it to measure the area of a rectangle which is 6 in. by 4 in. Draw lines to show how the units would fit together.
2. Find the area in triangular inches of equilateral triangles with a side of 2 in., 3 in., 5 in., and ½ in. Can you make a conjecture? What will be the area of an equilateral triangle with a side of S units?
3. Find a formula for the area of a rhombus with a side of S units and with the measure of one pair of opposite angles 60° each.
4. Find the area of a parallelogram with the measure of one pair of opposite angles 60° each and with one pair of opposite sides S units each and the other pair $2S$, $3S$, $4S$, nS.
5. Find the area of an isosceles trapezoid with acute base angles 60°, the legs S units, and the longer base $2S$, $3S$, $4S$, nS.

6. Estimate the area of a square whose side is 4 in.
7. Estimate the area of an isosceles triangle with legs of 2 in. and base of 1 in.

Perhaps the previous experiment suggested that the triangular inch works very nicely for equilateral triangles and other polygons which are constructed by equilateral triangles, but not so nicely for other polygons such as nonequilateral triangles, squares, rectangles, parallelograms, and so on.

However, the same may be said about using the square as a unit. It works well for other squares and polygons composed of squares (rectangles), but not so nicely for other polygons. It comes down to a problem of which type of shape do we wish to measure more frequently. Perhaps one reason why the square was selected is because man frequently sells land in rectangular lots rather than in triangular ones. Also, perhaps the square was selected because it is somewhat easier to count, since when squares are packed in a region they line into rows and columns of equal numbers of squares.

Thus, the square becomes our *unit of area measure*. If we use a square 1 in. on a side, we call it 1 sq in., or 1 m on a side is called 1 sq m, and so on.

Experiment 5

1. Explain the difference between 2 sq. in. and a 2-in. square.
2. Use Fig. 7.4.1 to help you develop convincing formulas between square units and triangular units. ρ is defined to be the measure of the altitude of a triangular inch.
   a. What is the area of the rectangle in square inches?
   b. What is the area of the rectangle in triangular inches?

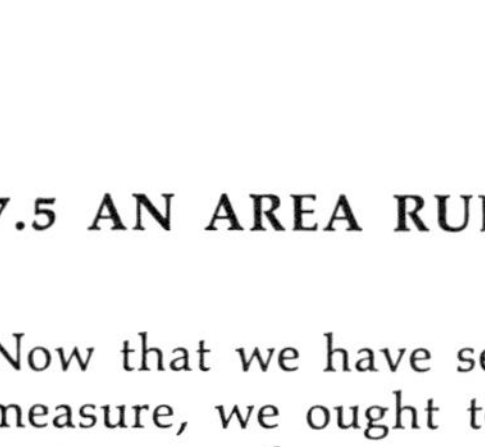

Figure 7.4.1

7.5 AN AREA RULER

Now that we have settled on using the square as unit of area measure, we ought to consider the problem of constructing an instrument for measuring area which is the counterpart for the ruler in linear measure. Such an instrument would have to have a series of square units pieced together with a suitable scale for counting those squares. Such an instrument would look very much like a sheet of graph paper, but how will we arrange the scale? Where will the "end of the ruler" be? How will we order the squares? Figures 7.5.1 and 7.5.2 are just two suggestions as to how this might be done.

25	26	27	28	29	30	31	32
17	18	19	20	21	22	23	24
9	10	11	12	13	14	15	16
1	2	3	4	5	6	7	8

Figure 7.5.1

31	30	29	28	27	26
32	13	12	11	10	25
33	14	3	2	9	24
34	15	4	1	8	23
35	16	5	6	7	22
36	17	18	19	20	21

Figure 7.5.2

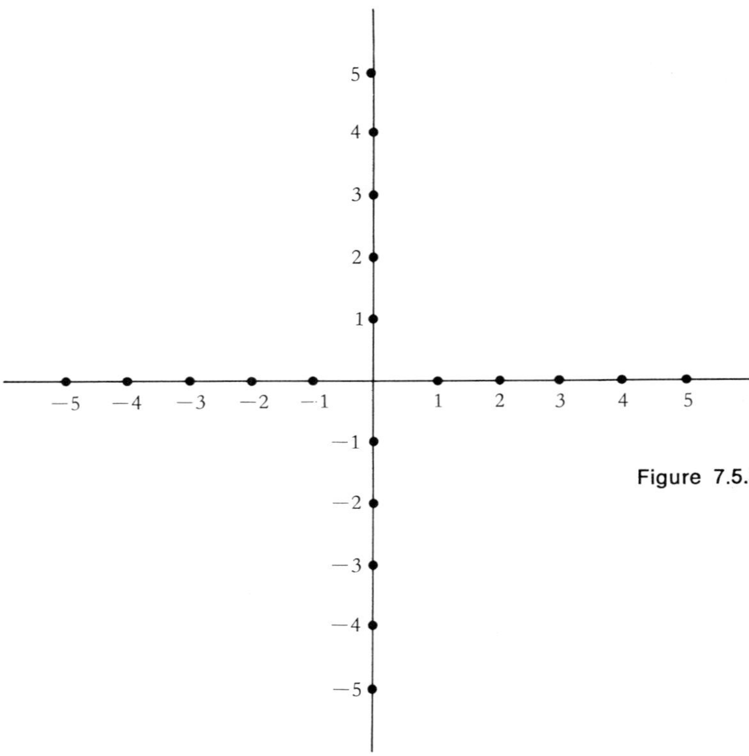

Figure 7.5.3

Experiment 6

1. If you were to use the rulers above to measure the area of a square, where would you have to place the 1 of ruler *A*? of ruler *B*?
2. Which ruler do you think would be easier to use to measure a rectangle? triangle? circle? Explain why.
3. Which ruler do you think more nearly parallels its one-dimensional counterpart? Why?
4. Why are neither of these "rulers" totally acceptable?
5. What is a simple way to count large arrays of squares?

An alternative to the two area rulers above is to designate the intersection of one vertical and one horizontal line as the "beginning" of the ruler and number the units on each line beginning at the point of intersection (Fig. 7.5.3).

Just as the ruler was idealized into a number line, so the area ruler can be idealized by considering the two intersecting lines.

Of course, you probably have recognized by now that what we have constructed is nothing more than the Cartesian plane with a coordinate system. We are able to determine the area of geometric shapes by placing them on a Cartesian plane and then "counting" the number of squares which are covered by the shape. Now consider the following exercise set.

Exercise 7.5.1

1. The regions in Fig. 7.5.4 have been placed on a Cartesian plane. Estimate their areas. For which placement was it easiest to determine the area? Can you suggest other placements which are even easier?
2. Estimate the areas of the polygons in Fig. 7.5.5. Then, suggest a placement which makes the counting of the squares easier.

As the preceding exercise set suggested, our counting task can be made easier by a judicious placement of the polygon on the Cartesian plane. For example, we see it makes sense to place a rectangle so that one pair of its sides is parallel to the x-axis and then the other pair would be parallel to the y-axis. (Do you know why?) Perhaps an even better way would be to place one vertex at the origin, as in Fig. 7.5.6. Now what would be the easiest way to determine its area? You *could* simply count the number of rows of squares covered and the number of columns covered and multiply the two numbers together. But since each square is 1 unit in length, the number of squares in a row equals the number of units in the length of the rectangle. Likewise, the number of squares in a column equals the width. Thus, multiplying the number of squares in a row by the number of squares in a column is the same as multiplying the length of

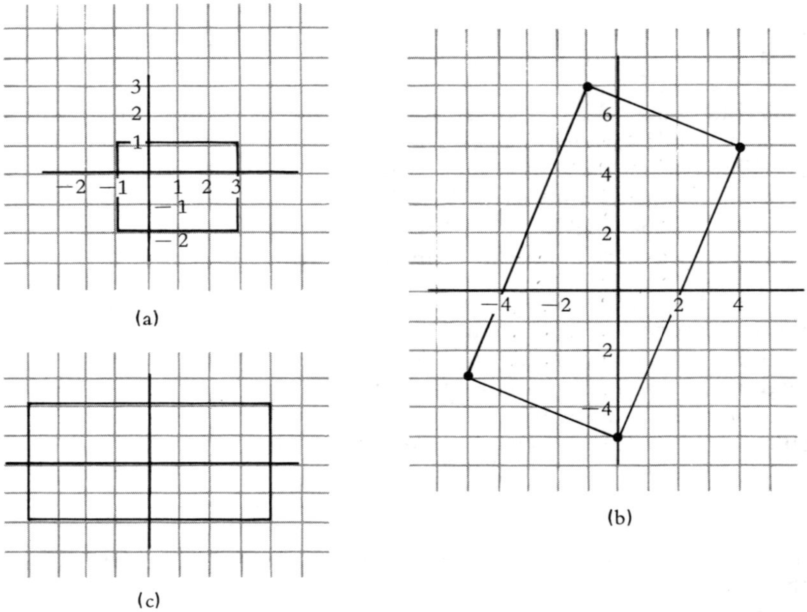

Figure 7.5.4

the rectangle by its width. Thus, we have a basic formula for determining the *area of a rectangle;* namely, $A = l \times w$. This formula is simply an expression which indicates a very efficient way of counting the square units "covered" by the rectangle of length l and width w.

comment . . .

It is essential that students realize two fundamental ideas with regard to area.

1. The problem of determining the area of a region is simply a counting problem.
2. An area formula is simply an efficient way of performing this counting process.

Now let us consider the area of a parallelogram. First, we ought to decide how we will place it on our coordinate system. Since the opposite sides are parallel, it would be wise to place one vertex at the origin and one side on the x-axis, as in Fig. 7.5.7. Now, how can we count the square units which the parallelogram covers? It is not as easy as with the rectangle, since the parallelo-

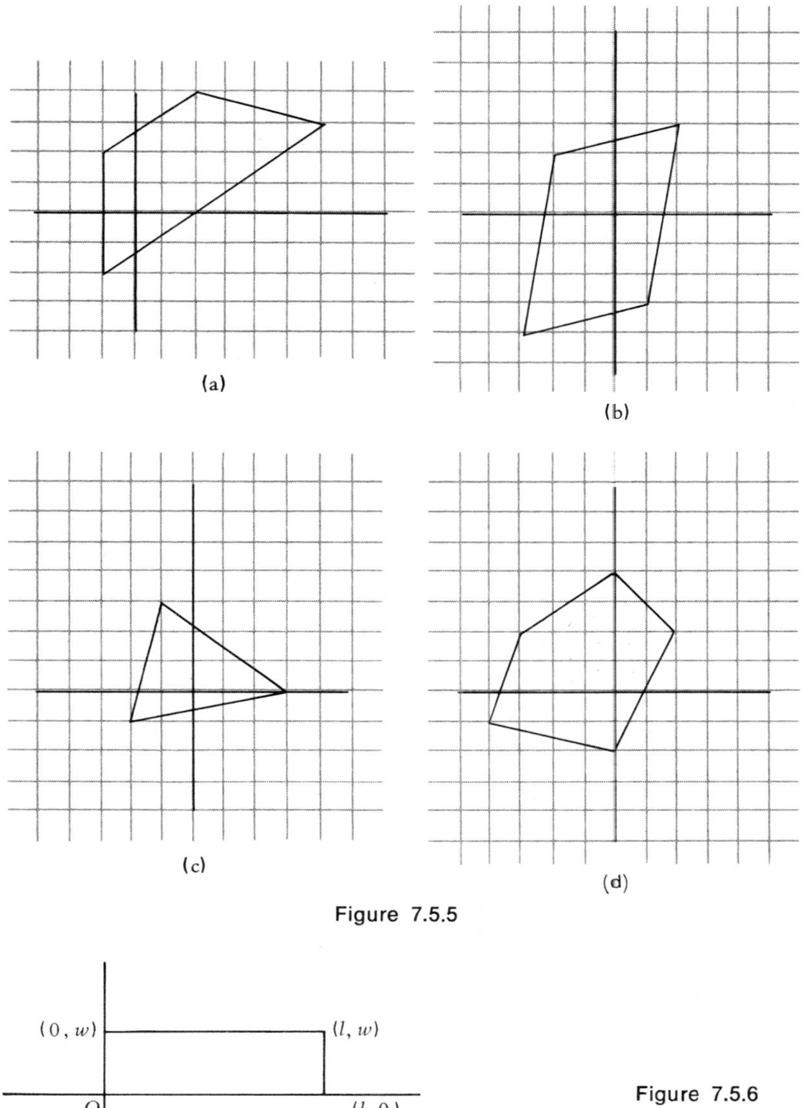

Figure 7.5.5

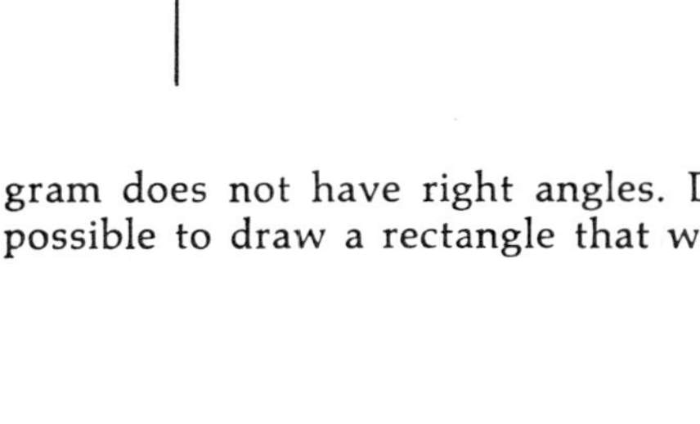

Figure 7.5.6

gram does not have right angles. Do you think it would be possible to draw a rectangle that would have the same area as

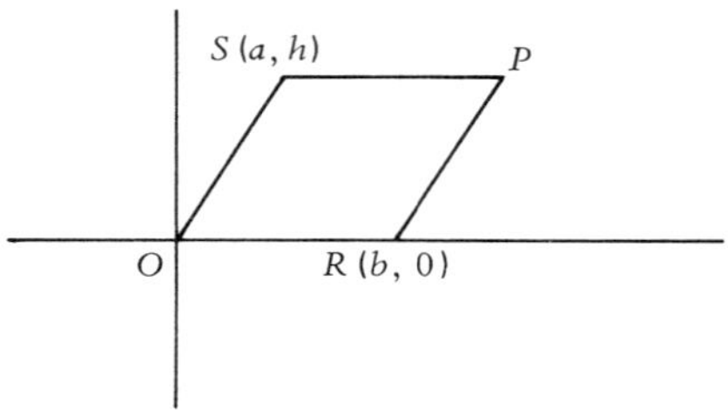

Figure 7.5.7

the parallelogram? Look at Fig. 7.5.8. Do you think that the area of triangle SOT equals the area of triangle PRQ? If these areas are equal, then the area of parallelogram $SQRP$ is the same as the area of rectangle $STQP$. Let us assume this for the moment.

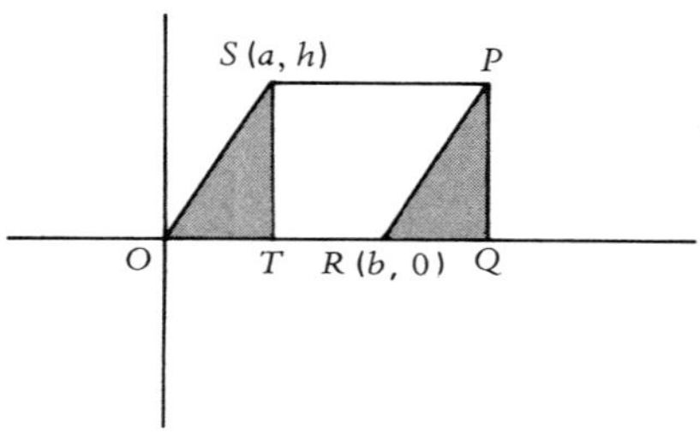

Figure 7.5.8

Then, what are the coordinates of point Q and point P? Do you see that the coordinates of Q are $(a + b, 0)$ and those of P are $(a + b, h)$? The length of $\overline{SP} = |a + b - a| = |b| = b$. The width of the rectangle is the length of $\overline{PQ} = |h - 0| = |h| = h$. Thus, the area of rectangle $STQP$ is bh. If our assumptions that triangles SOT and PQR are equal in area are correct, and if indeed $STQP$ is a rectangle, then the area of a parallelogram is also bh. But how can we be sure that these assumptions are correct? The ideas of congruence, parallelism, and similarity in the next chapters should supply us with the answer. Thus, at this point, we must pause in our discussion of area.

7.6 SUMMARY

1. The measure of the boundary of a two-dimensional region is called the perimeter of the region.
2. Two regions can have equal perimeters, but have different areas.
3. The ratio of the circumference of any circle to its diameter is always the same number which we call pi. Thus, the

circumference can be determined by multiplying the diameter of the circle by pi.

4. Any two-dimensional unit could be used as a unit of area providing it could fit together compactly.
5. The problem of finding the area of a figure is a problem of counting the number of area units which cover the region.
6. An area formula is a statement of an efficient way of counting the area units.
7. The Cartesian coordinate system is a two-dimensional counterpart of the ruler.

Exercise 7.6.1

1. Calculate the area of each of the rectangles in Fig. 7.5.9.

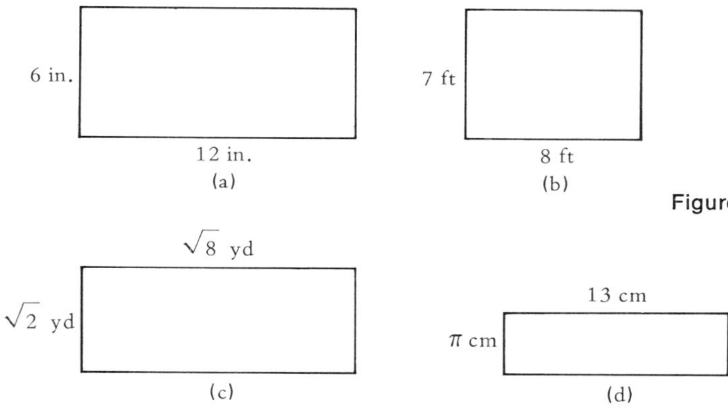

Figure 7.5.9

2. Find the area of a rectangle which is 3 yd long and 2 ft wide.
3. A rectangle has an area of 1325 sq ft and it is 12 ft wide. How long is it?
4. Find the area of a square which is 13.2 ft on a side.
5. How many square feet are in a square yard? How many square inches are in a square yard? Do you have to memorize these numbers or can you calculate them?

CONGRUENT FIGURES

8

8.1 PROLOGUE

We have considered the idea of congruence of line segments and angles. Now we wish to consider the general notion of congruent figures. Geometric figures will be considered to be *congruent* if they have the same size and shape. Each of the simple closed curves in Fig. 8.1.1 have size and shape. It is easy for us to recognize the shapes as we have learned to associate names with each such shape.

Figure 8.1.1

The first curve has the shape of a triangle, the second that of a square, and the third the shape of a pentagon. As to their size, it remains for us to explain what we will mean by that. At any rate, we would agree that these three figures have different shapes and should not be considered congruent.

Now let us consider each of the simple closed curves in Fig. 8.1.2. We would probably observe that each of the curves has the shape of a square but they are not of the same size. We make

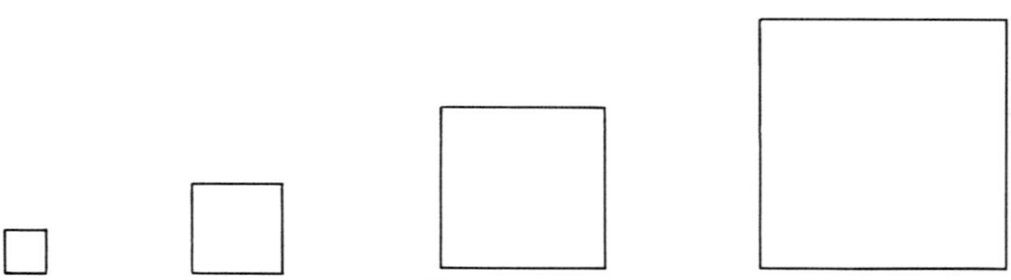

Figure 8.1.2

the observation that they are not the same size even though we still have not identified what we shall mean by "having the same size." This is because we possess an intuitive sense of what size should mean and we cannot imagine an interpretation of size which would classify these figures as having the same size.

Let us continue our investigation by considering the simple closed curves in Fig. 8.1.3.

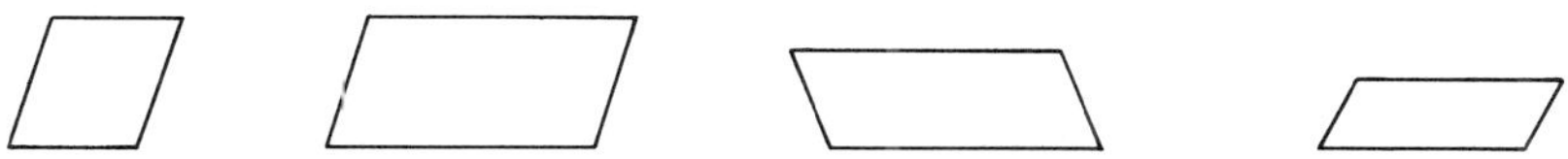

Figure 8.1.3

It appears that each of these curves are parallelograms and we would probably agree that they do not have the same size. Despite the fact that all are parallelograms, we would be reluctant to claim that they have the same shape. Again, they do not quite fit our intuitive notion of "having the same shape."

Similarly, we look at the set of simple closed curves in Fig. 8.1.4; note that they are each triangles but we would deny that they have the same size and shape.

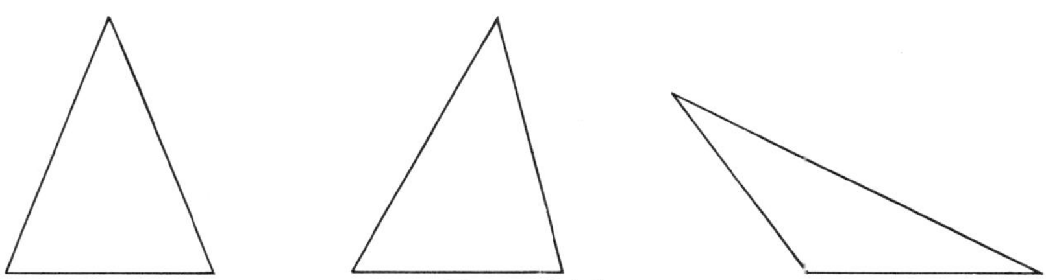

Figure 8.1.4

Describing something as obvious, however, is a very dangerous undertaking. Remember that at one time it was "obvious" to all that the world was flat; it was "obvious" that the earth was the center of the universe; it was "obvious" that man was not intended to fly. It might be interesting if we could revisit this earth a few hundred years from now and discover if there were any beliefs held by today's population as "obvious" which had been shown to be wrong.

It may be suggested that this is all well and good but the observations pertaining to size and shape made about the various sets of figures are not going to be shown to be wrong; rather, they will be shown to be right. As this is the case, our job now is to interpret the obvious so that it may be obvious to all.

Consider the square *ABCD* with diagonal $\overline{AC}$ (Fig. 8.1.5). Suppose we fold the figure over, along the diagonal. We would

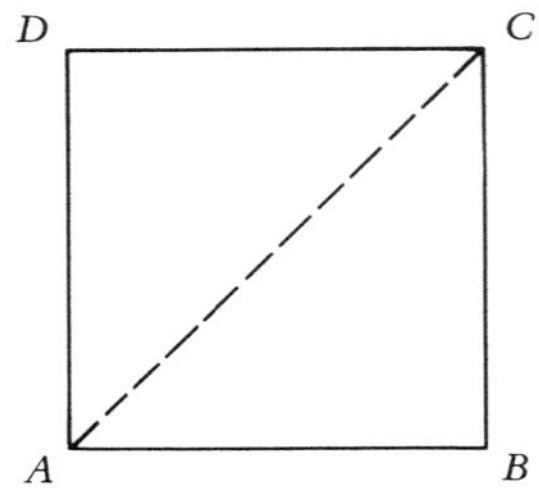

Figure 8.1.5

note that point D matches with point B, side $\overline{DC}$ matches with $\overline{BC}$ in that order, and side $\overline{DA}$ matches with $\overline{BA}$ in that order. We would be most likely to agree that the triangles $\triangle ADC$ and $\triangle ABC$ have the same size and the same shape. Again, this appears "obvious."

Let us look at another square $PQRS$ with line segment $\overline{MN}$ connecting the midpoint of side $\overline{SR}$ to the midpoint of side $\overline{PQ}$ (Fig. 8.1.6). If we fold the figure over along the segment $\overline{MN}$, we find that point R matches with S, point Q matches with P, side $\overline{RQ}$ matches with side $\overline{SP}$ in that order, and that the segments $\overline{RM}$ and $\overline{SM}$ match as do the segments $\overline{QN}$ and $\overline{PN}$. We decide that figures $MRQN$ and $MSPN$ have the same size and shape.

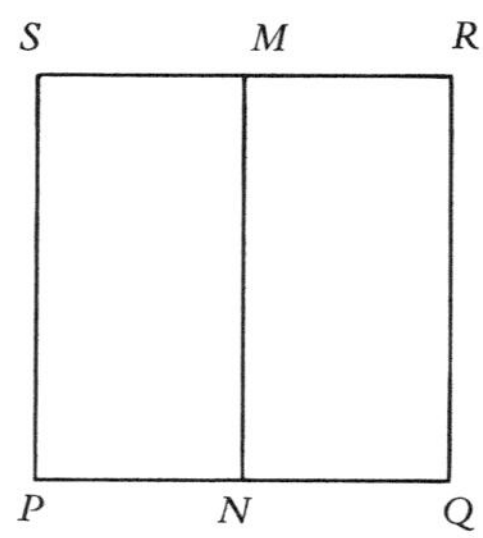

Figure 8.1.6

We begin to formulate an opinion of what it should mean that two real objects have the same size and shape. We would offer the conjecture that two figures would have the same size and shape if they can be made to coincide point by point, by placing one on top of the other. To add supporting evidence to our conjecture, we could point out that a method of producing a figure having the same size and the same shape as a given figure would be to trace the original figure on paper or some other flat material and then use scissors to cut out the new figure.

All right, we now have some notions of what it would mean to say that two physical objects have the same size and shape. Our problem now is to idealize this notion so it can be applied to the idea of polygons and other subsets of space.

Experiment 1

1. Take a plain sheet of paper and fold it in two. Open up the paper and drop a bit of ink in the fold. Now fold the paper again and carefully press it all over, open it and allow the ink to dry. What do you notice about the shape made by the ink?
2. Figure 8.1.7 shows various geometrical shapes. How many ways can you fold these shapes so that one half matches the other half?

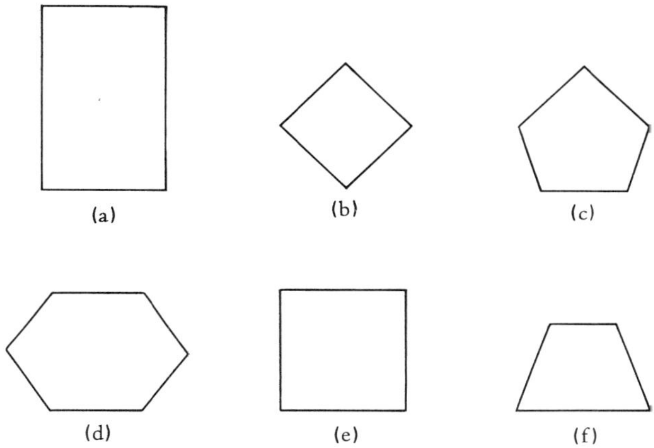

Figure 8.1.7

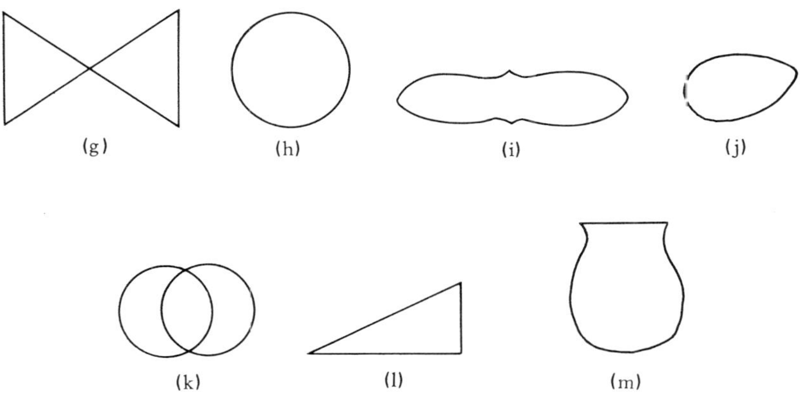

This experiment introduces the idea of symmetry by reflection.

8.1.1 Definition

A figure is symmetric about a line l *called its* axis of symmetry *if for each point P of the figure there exists a point P' of the* *figure (the symmetric partner of P) such that $\overline{PP'}$ is perpendicular to l at some point Q and such that $\overline{PQ} \cong \overline{P'Q}$.*

Example 1

See Figs. 8.1.8 and 8.1.9. Also, a geometric figure may have *symmetry with respect to a point.*

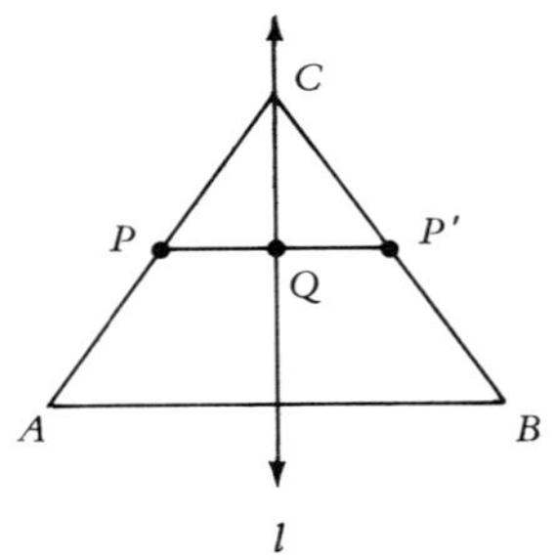

Figure 8.1.8
$\overline{PP'} \perp l$. $\overline{PQ} \cong \overline{P'Q}$.
Triangle *ABC* is symmetric about *l*

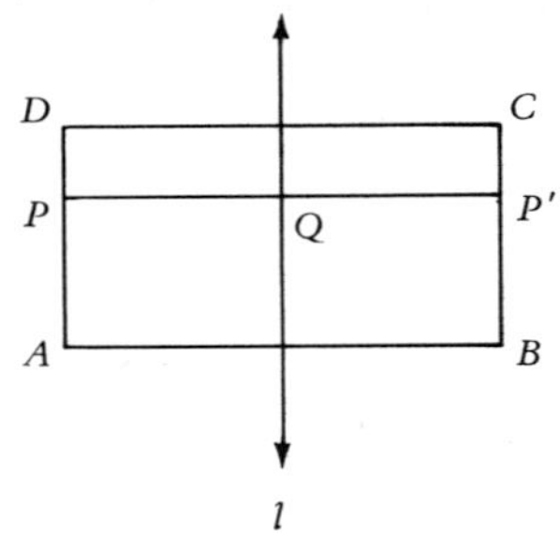

Figure 8.1.9
$PP' \perp l$. $\overline{PQ} \cong \overline{P'Q}$.
Rectangle *ABCD* is symmetric about *l*

8.1.2 Definition

A geometric figure is said to be symmetric with respect to a point *O if and only if for each point A of the figure, there is a point A′ of the figure (the symmetric partner of A) such that O, A, and A′ are collinear and* $\overline{OA} \cong \overline{OA'}$.

Example 2

See Fig. 8.1.10.

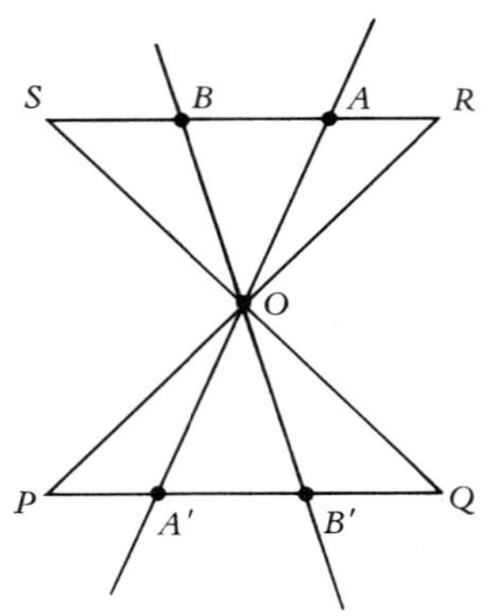

Figure 8.1.10
$\overline{OA} \cong \overline{OA'}$. $\overline{OB} \cong \overline{OB'}$.
Figure *PQORS* is symmetric about point *O*

Exercise 8.1.1

1. Which of the letters of the alphabet are symmetric about a line? Which of the numerals from 0 to 9 are symmetric about a line?

2. a. Which of the figures (a) to (m) of Experiment 1 are symmetric with respect to a point?
   b. Which of the letters of the alphabet and which numerals are symmetric with respect to a point?
   c. Can a figure be symmetric with respect to a line but not symmetric with respect to a point?
   d. Can a figure be symmetric with respect to a point but not symmetric with respect to a line?
   e. Can a figure be symmetric with respect to more than one point?
   f. Can a figure be symmetric with respect to both a point and a line?

Experiment 2

A generalization of the idea of symmetry of a geometric figure is the reflection of a geometric figure in a line of the plane of the particular geometric figure. The distinction is that a symmetrical figure can be reflected in a line that passes through the geometric figure. We now wish to consider the more general situation in which a geometric figure is reflected about a line which may or may not intersect the geometric figure.

Copy the figures of Experiment 1 on one side of a piece of paper. Fold the paper over the vertical line joining the midpoints of the top and bottom of the page. Trace the reflection image of each geometric figure onto the opposite side. With ruler and protractor, check the lengths of the sides and the measures of the angles of the figures and their images.

What conjectures can you make about the effects of this process on (a) collinearity, (b) betweenness, (c) length, (d) angle measurement, (e) area, and (f) perimeter?

8.2 REFLECTIONS

The student undoubtedly has discovered that a figure being symmetrical has a relationship to the concept of congruent figures. If a figure has an axis of symmetry, the two halves of the figure have exactly the same size and shape and thus are congruent. Similarly, it should be apparent that a figure and its reflection image in some line in the plane of the figure have the same size and shape and thus are congruent.

To further develop this idea, suppose we have a piece of paper such as the page of this book and a vertical line which connects the midpoints of the top and bottom of the page. This line will then be parallel to the sides of the page. If we fold the piece of paper on this line, the two halves will be congruent. Of greater interest is the fact that any point P on one half of the page has a reflection partner or image P' on the other half of the page. Thus, there is a one-to-one correspondence of the points of the two half pieces of paper. The points on the vertical line are their own images (Fig. 8.2.1).

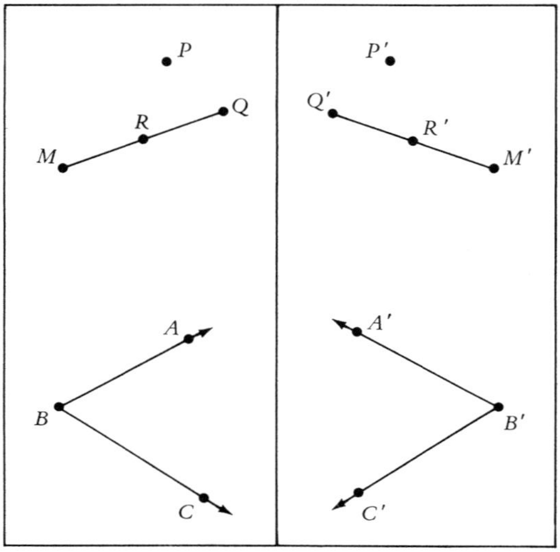

Figure 8.2.1

Among the observations we can make is if points are collinear, then their reflection images appear to be collinear. In addition, the distance between two points seems to be equal to the distance between their images. We can also note that if one point is between two others on a line, its image is between the images of the others. Also, if we have three noncollinear points on one side of the line which determine an angle, it appears that their three images determine an angle of equal measure.

We will now formally define the concept of reflection and state a postulate which indicates the properties which will be assumed about reflections.

8.2.1 Definition

If point P is not on a line l, then the reflection image *of P in (or over) line l is the point P' if and only if l is the perpendicular bisector of $\overline{PP'}$. If P is on line l, the reflection image of P is P itself. Note P, l, and P' all lie in the same plane.*

comment . . .

Line l is the perpendicular bisector of segment PP' if $l \perp \overline{PP'}$ and l intersects $\overline{PP'}$ in a point Q such that $\overline{PQ} \cong \overline{P'Q}$.

8.2.2 Postulate

Every point has exactly one reflection image in a line l. Furthermore, reflection preserves each of the following properties: (a) collinearity, (b) betweenness, (c) distance, and (d) angle measure.

The postulate may be interpreted as follows: (a) If A, B, and C are collinear, then the images in l, A', B', and C' are collinear. (b) If $A - B - C$, then $A' - B' - C'$. (c) If A and B are any two points, then $AB = A'B'$. (d) If A, B, and C determine an angle $\angle ABC$, then A', B', and C' determine an angle $\angle A'B'C'$ and $m° \angle ABC = m° \angle A'B'C'$.

Some immediate consequences of the definition and postulate are the following.

8.2.3 Theorem

a. The reflection image of a line is a line.
b. The reflection image of a ray is a ray.
c. The reflection image of a line segment is a line segment.

More importantly, we may imply that the reflection image of a figure α is a figure β which has the same size and shape as α. If α is a figure with sides and angles such as a polygon, then each side is equal in length to its corresponding image and each angle is equal in measure to its corresponding image.

Any line in a plane can be considered as a line of reflection. Because of the nature of planes and lines we can develop a mathematical meaning for the process of folding. Perhaps the best approach is to think of folding as a kind of function dependent upon a line which assigns points of a plane to points of the plane in a particular way.

8.2.4 Definition

Let $\mathcal{P}$ be a plane and l a line in $\mathcal{P}$. Define the function

$$r_l: \mathcal{P} \to \mathcal{P}$$

as follows: $r_l(P) = P'$ such that P' is the reflection image of P in l. The function r_l is called "a reflection in line l." (See Fig. 8.2.2.)

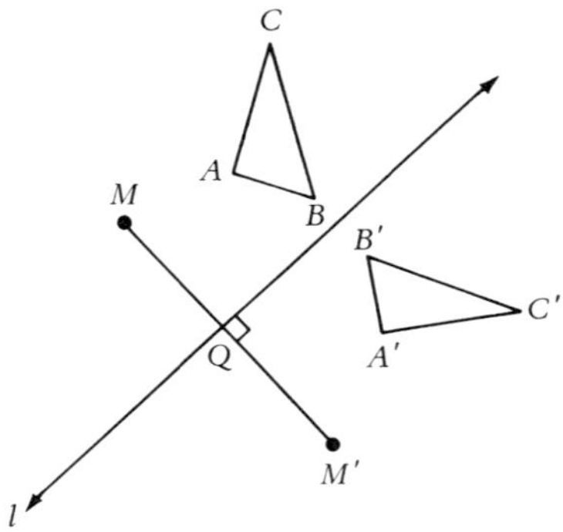

Figure 8.2.2
$l \perp \overline{MM'}$. $\overline{MQ} \cong \overline{M'Q}$.

$r_l(A) = A'$ $\quad r_l(A') = A$
$r_l(B) = B'$ $\quad r_l(B') = B$
$r_l(C) = C'$ $\quad r_l(C') = C$
$r_l(M) = M'$ $\quad r_l(M') = M$
$r_l(Q) = Q$

comment . . .

We can think of line l as a kind of mirror and each point P of the plane is mapped onto its mirror image P'.

comment . . .

Presently, we are limiting the reflections in a line to a plane, although it would be possible to describe a similar situation in space. In such a case, we would be making a reflection in a plane, the plane acting as a mirror assigning each point P of one half-space to its image P' in the other half-space such that the plane is perpendicular to $\overline{PP'}$ and bisects $\overline{PP'}$. Of course, points on the plane would be their own reflection image. This situation will be considered later in this chapter as an experiment.

Returning to the notion of symmetry, we have the following definition.

8.2.5 Definition

A set of points S in a plane is symmetric with respect *to a line l of the plane if and only if the reflection image of each point of S in l is also a point of S. The line l is called a* line of symmetry *for the set.*

Example 1
Let $\triangle ABC$ be an equilateral triangle; then $\triangle ABC$ is symmetric with respect to each of the lines $\overleftrightarrow{AM}$, $\overleftrightarrow{BN}$, and $\overleftrightarrow{CP}$ where they are the angle bisectors of angles A, B, and C, respectively (Fig. 8.2.3).

This example implies that $\overleftrightarrow{AM}$, $\overleftrightarrow{BN}$, and $\overleftrightarrow{CP}$ are also the perpendicular bisectors of sides $\overline{BC}$, $\overline{AC}$, and $\overline{AB}$, respectively.

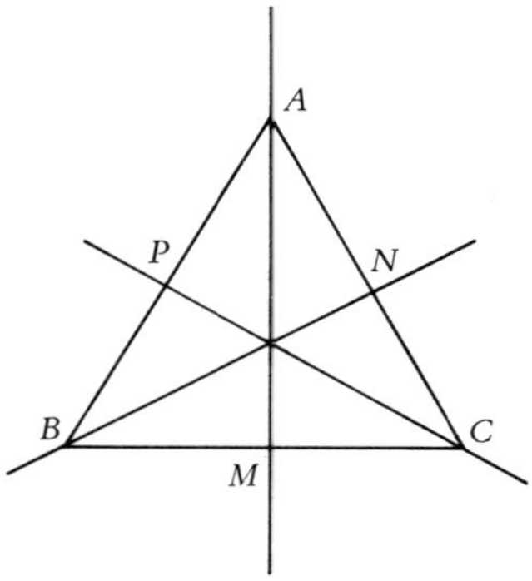

Figure 8.2.3

This suggests the very important property that if a point is equidistant from the endpoints of a line segment, it is on the perpendicular bisector of the line segment. We will be able to establish this fact later.

comment . . .

The picture (Fig. 8.2.3) illustrating this example shows the lines $\overleftrightarrow{AM}$, $\overleftrightarrow{BN}$, and $\overleftrightarrow{CP}$ concurrent in the interior of the triangle. Is this generally true and if so, is it a consequence of the three reflections?

Example 2

Let *ABCD* be a square. Then *ABCD* is symmetric with respect to four lines, the two diagonals and the two lines joining midpoints of pairs of opposite sides (Fig. 8.2.4).

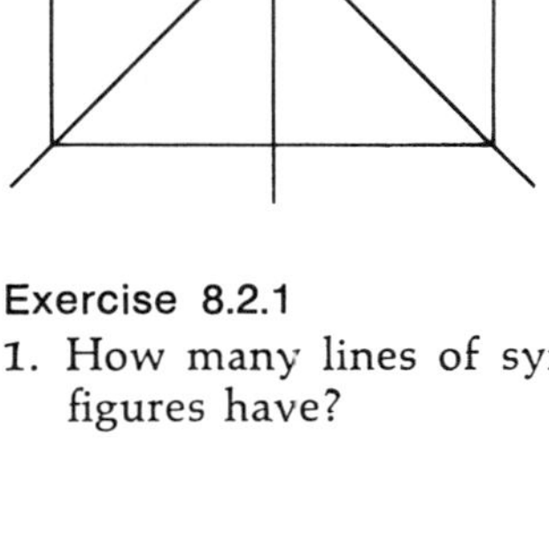

Figure 8.2.4

Exercise 8.2.1

1. How many lines of symmetry does each of the following figures have?

a. line segment
b. ray
c. angle
d. half-plane
e. plane
f. parallelogram
g. rectangle
h. scalene triangle
i. rhombus
j. isosceles triangle
k. pentagon
l. hexagon
m. septagon
n. octagon
o. circle

2. Give arguments supporting each of the following statements.
   a. An isosceles triangle is symmetric with respect to the bisector of its vertex angle.
   b. The bisector of the vertex angle of an isosceles triangle is the perpendicular bisector of the base of the triangle.
   c. Base angles of an isosceles triangle have the same measure.
3. Having established the lines of symmetry of squares, rectangles, parallelograms, rhombi, and isosceles trapezoids, what properties of the sides and angles of these figures can you deduce?

Experiment 3

Once more, copy the figures of Experiment 1 on one side of a piece of paper. Fold the paper over the vertical line joining the midpoints of the top and bottom of the page. Trace the reflection image of each figure. Unfold the piece of paper and then fold it over the horizontal line joining the midpoints of the two sides of the page. Trace the reflection image of each figure. We will note that the original figure and its two reflection images are all of the same size and shape. Is it possible to have a reflection which takes the original figure into the second reflection image? Explain.

Divide a sheet of paper into thirds using two lines. Suppose the figures of Experiment 1 were copied on the first third of the page. If we fold on the first line, the reflection images will be in the middle third of the page. Trace the images and then fold the page on the second line. The second set of figures has reflection images in the last third of the page. Trace the reflection images. As before, the original figure and its reflection images are all of the same size and shape. Is it possible to have a reflection which takes the original figure into the second reflection image? Explain.

8.3 COMPOSITION OF REFLECTIONS

The previous experiment is meant to suggest the possibilities when one reflection of a plane in a line l is followed by a second reflection of the plane in a line m. Any figure which has been twice reflected will have a final image which has the same size and shape as the original figure. There are two possibilities of this successive reflection: (a) when the lines l and m intersect, and (b) when the lines l and m are parallel.

Example 1

See Fig. 8.3.1.

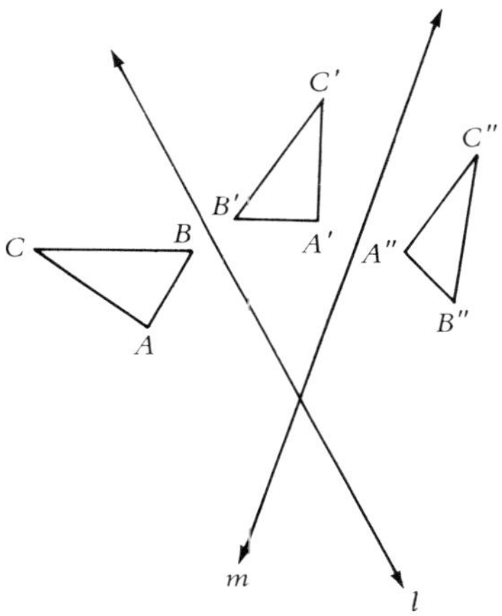

Figure 8.3.1

$r_l(A) = A'$ $\quad$ $r_l(A') = A''$
$r_l(B) = B'$ $\quad$ $r_l(B') = B''$
$r_l(C) = C'$ $\quad$ $r_l(C') = C''$

Example 2

See Fig. 8.3.2.

Example 1 (Fig. 8.3.1) illustrates case (a). In this case, we say that $\triangle ABC$ has been *rotated* into $\triangle A''B''C''$. Example 2 (Fig. 8.3.2) illustrates case (b). Here we say that $\triangle ABC$ has been *translated* into $\triangle A''B''C''$.

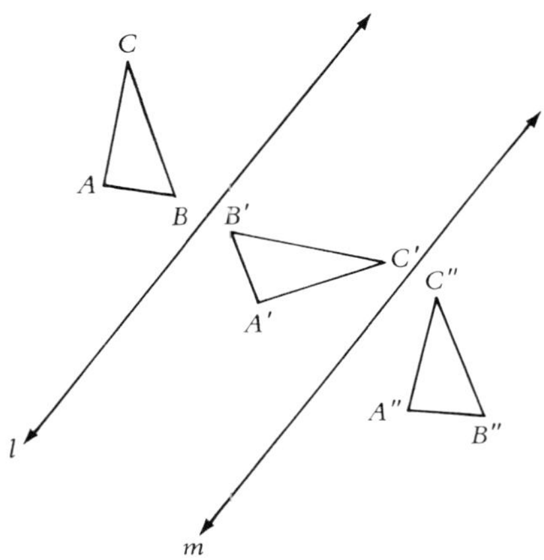

Figure 8.3.2

$r_l(A) = A'$ $\quad$ $r_m(A') = A''$
$r_l(B) = B'$ $\quad$ $r_m(B') = B''$
$r_l(C) = C'$ $\quad$ $r_m(C') = C''$

This process of following a reflection in a line by another reflection in a line is itself a function from the plane onto the plane. It preserves all properties preserved by a reflection, although it itself is not a reflection.

A natural question is: What is the result of following these two reflections in lines by a third reflection or for that matter any series of reflections? The following two examples illustrate two possible outcomes.

Example 3

See Fig. 8.3.3.

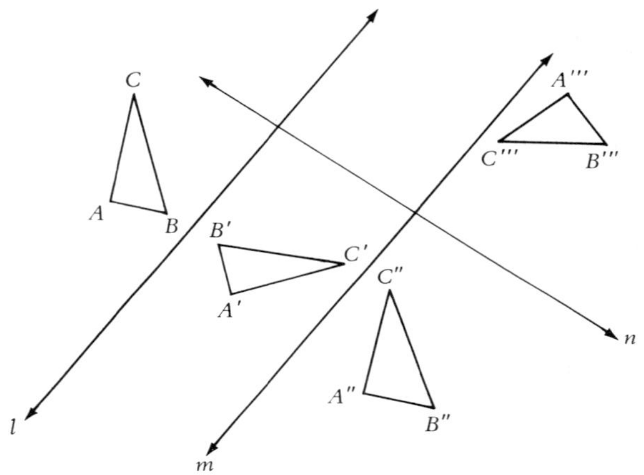

Figure 8.3.3
$r_l(A) = A'$
$r_l(B) = B'$
$r_l(C) = C'$
$r_m(A') = A''$
$r_m(B') = B''$
$r_m(C') = C''$
$r_n(A'') = A'''$
$r_n(B'') = B'''$
$r_n(C'') = C'''$

Example 4

See Fig. 8.3.4.

It is not our intention to laboriously investigate by diagrams all of the possible outcomes of successive reflections. Instead, we will introduce a concept which includes all the possibilities. But first we will wish to make more precise the idea of successive reflections.

8.3.1 Definition

If r_l and r_m are two reflections in line l and m, respectively, we define $r_m \circ r_l$ to be the composition *or* product *of the two reflections. If P is any point of the plane, we have*

$$r_m \circ r_l(P) = r_m(r_l(P))$$

In other words, the composition of two reflections r_l and r_m is accomplished by first reflecting the plane in line l and following this by reflecting the plane in line m. Composition of reflections

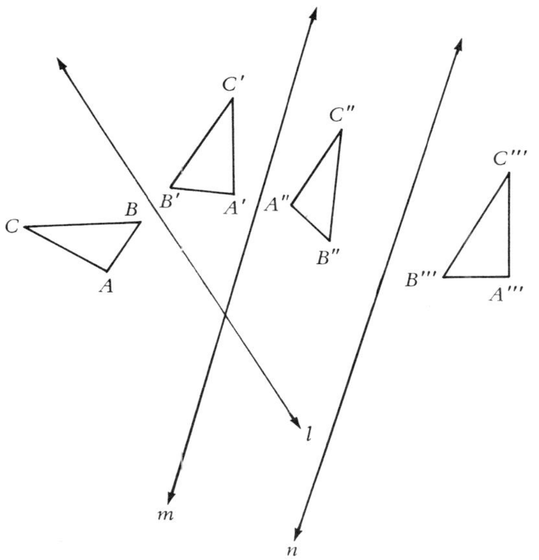

Figure 8.3.4
$r_l(A) = A'$
$r_l(B) = B'$
$r_l(C) = C'$
$r_m(A') = A''$
$r_m(B') = B''$
$r_m(C') = C''$
$r_n(A'') = A'''$
$r_n(B'') = B'''$
$r_n(C'') = C'''$

is not limited to two successive reflections, but can be used to describe a series of reflections. Thus, in the example above, we have illustrated $r_n \circ r_m \circ r_l$.

As we have seen, the composition of reflections need not be a reflection, but the composition does share the properties of a reflection in that it preserves size and shape. We now define the class of all such functions in a plane.

8.3.2 Definition

An isometry *is a function (map) from a plane to itself that preserves distance; that is, σ is an isometry if and only if for each pair of points P and Q in the plane, if $\sigma(P) = P'$ and $\sigma(Q) = Q'$, then $PQ = P'Q'$.*

An immediate consequence of this definition is that all reflections are isometries and all compositions of reflections are isometries. The following are some important properties of isometries which can be deduced from the definition.

8.3.3 Theorem

1. *An isometry preserves collinearity.*
2. *An isometry preserves betweenness.*
3. *An isometry maps line segments into line segments, lines into lines, rays into rays, and circles into circles.*
4. *An isometry preserves angle measure.*

Exercise 8.3.1

Give arguments to support the above four assertions. First try some examples.

comment . . .

An interesting property of isometries is that every isometry is the composition of at most three reflections. Since we will make no particular use of this fact, we will not prove it.

We can now make the following definition.

8.3.4 Definition

Let α and β be any two figures in a plane. We will say that α and β are congruent *(written $\alpha \cong \beta$) if and only if there is an isometry mapping α onto β.*

comment . . .

This definition is consistent with our previous definition of congruent line segments and congruent angles. That is, congruent segments have the same length and congruent angles have the same measure. The converses of these two statements (namely, if two segments have the same length, then they are congruent; and, if two angles have the same measure, then they are congruent) are also true. In general, however, while congruent figures will have many equal measures, having equal measures does not guarantee congruent figures.

We have an instrument that can be used to construct isometric mappings in the plane. It is called a compass and its basic function is to transfer the length of a given line segment. It is illustrated in Fig. 8.3.5.

comment . . .

Interestingly enough, Euclid had the *idea* of a compass, but *not* the actual instrument. Unlike the compass of our experience, Euclid's compass, if it could be built, would collapse each time it was moved. In this way, it could not be used as a distance transferring device. Of course, this was the actual intent; Euclid's compass was meant only to draw circles and arcs of circles, not to act as a measuring device.

Since a compass can be used to transfer the length of a line segment it might quite aptly be called an "isometric machine."

Figure 8.3.5

The process for using the compass to transfer distance is described below.

Construction 1

Given a line segment $\overline{AB}$ and a line $\overleftrightarrow{PQ}$, construct a segment $\overline{PR}$ on $\overleftrightarrow{PQ}$ so that $AB = PR$ (and thus $\overline{AB} \cong \overline{PR}$) (Fig. 8.3.6).

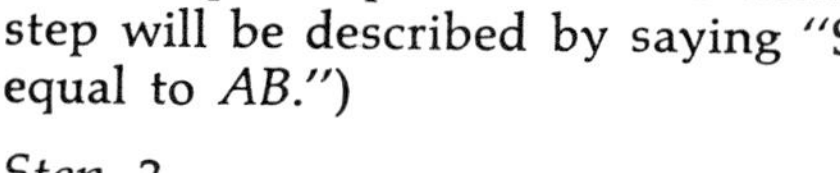

Figure 8.3.6

Step 1

Place the point of the compass on point A. Open the compass so that the pencil point coincides with point B. (From now on, this step will be described by saying "Set the compass to a radius equal to AB.")

Step 2

Keeping the compass set at this radius, place the point of the compass on point P. Draw an arc which cuts $\overleftrightarrow{PQ}$ at some point R (Fig. 8.3.7). Then $\overline{PR}$ is the required segment.

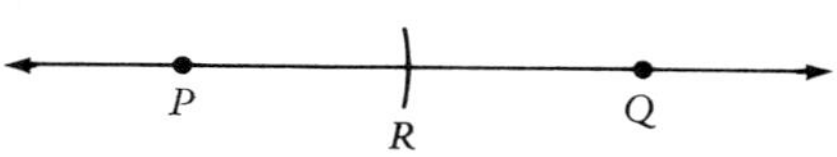

Figure 8.3.7

As can be seen, this is a rather simple process which is a physical illustration of an isometry. Since we have already suggested that isometries not only preserve distances, but preserve angles as well, it must therefore follow that the compass can be used to transfer angle measure. In fact, it should also be possible to use it to transfer the size and shape of other figures such as triangles and other polygons. We shall soon see how this may be done. However, since the compass actually only transfers distance, we shall also need an additional instrument, the straight edge. Basically, we shall use the compass to transfer distance, thus identifying certain points, and we shall use the straight edge to draw certain straight lines which we know to exist through these points.

comment . . .

We frequently use a ruler as our straight edge. However, it should be emphasized that we must use only the straight edge features of the ruler and *not* the uniform markings. Ideally, we should use a stick with no markings on it at all. Likewise, we are not allowed to use any other characteristic of the ruler such as its thickness or the fact that its end is a right angle.

Exercise 8.3.2

1. **8.3.5 Definition**

   Every isometry that is a product of an even number of reflections is called direct. *An* opposite *isometry is the product of an odd number of reflections.*

   8.3.6 Definition

   The sense of a triangle PQR is clockwise *if and only if one moves in the clockwise direction in reading the vertices in cyclic order* $P - Q - R - P$. *It is* counterclockwise *if and only if one moves in the counterclockwise direction in reading the vertices* $P - Q - R - P$. *If triangles PQR and ABC are congruent and both are read in the same sense, they are said to be* directly congruent. *If their senses are not the same, they are* oppositely congruent.

   a. Show that a reflection maps a triangle into a congruent triangle such that the congruent is opposite.
   b. Show that an opposite isometry maps a triangle into an oppositely congruent triangle.
   c. Show that a direct isometry maps a triangle into a directly congruent triangle.
   d. Show that no isometry is both direct and opposite.

2. **8.3.7 Definition**

A plane function α is called a translation through ray PQ *if and only if for each point A, $\alpha(A) = B$ where $\overrightarrow{AB} \| \overrightarrow{PQ}$, $\overline{AB} \cong \overline{PQ}$, and $\overrightarrow{AB}$ and $\overrightarrow{PQ}$ are similarly directed.*

See Figs. 8.3.8 and 8.3.9 as examples.

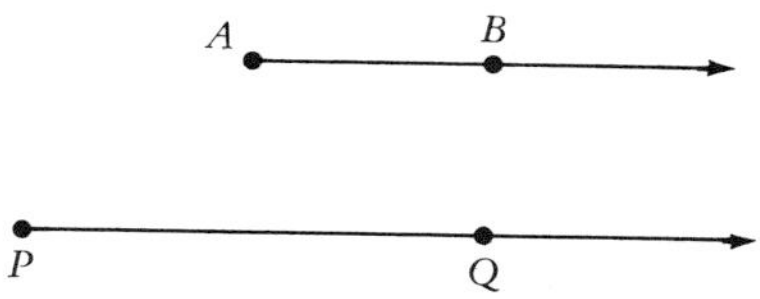

Figure 8.3.8
Similarly directed

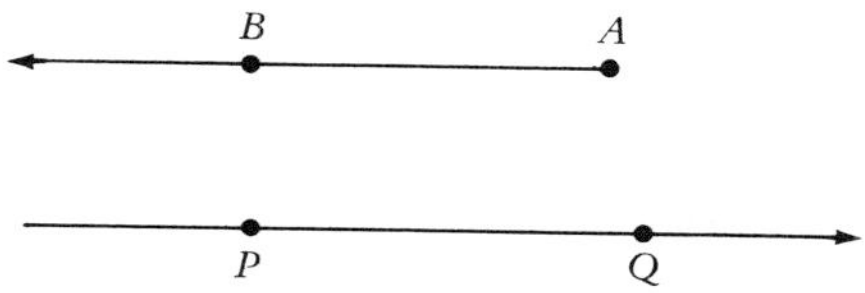

Figure 8.3.9
Oppositely directed

a. Show that every translation is an isometry and can be written as the composition of two reflections in parallel lines l and m.

b. Show that the inverse of a translation through ray $\overrightarrow{PQ}$ is a translation through ray $\overrightarrow{QP}$.

3. **8.3.8 Definition**

Let O be a point and θ the measure of a directed angle. (That is, $\theta = m^\circ \angle BOA$ when *$\angle BOA$ is measured counterclockwise and $\theta = m^\circ \angle BOA$* when *$\angle BOA$ is measured clockwise.) A plane function γ is called a* rotation about O through angle θ *if and only if for each point A, $\gamma(A) = B$ where $A = B = O$ or $\overline{OB} \cong \overline{OA}$ and $m^\circ(\angle BOA) = \theta$. Point O is called the* center of the rotation.

a. Show that if lines l and m intersect at point O with the directed angle from l to m of measure θ, then $\sigma_m \circ \sigma_l$ is a rotation about point O through angle 2θ.

b. Show that every rotation is an isometry and can be written as the composition of two reflections in intersecting lines l and m.

8.4 APPLICATIONS OF ISOMETRIES

Earlier we observed that the following statement seemed to be true:

8.4.1 Theorem

If a point is equidistant from the endpoints of a line segment, then it is on the perpendicular bisector of the line segment.

This may now be explained as follows.

Suppose $\overline{AB}$ is a segment and P is a point such that $AP = BP$. Then consider the bisector of angle APB which intersects $\overline{AB}$ at M (Fig. 8.4.1). Now, since $\overrightarrow{PM}$ is an angle bisector, we know that $\angle APM \cong \angle BPM$. Thus, by our definition of congruence, we know there is an isometry σ which maps $\angle APM$ onto $\angle BPM$. Thus, ray $\overrightarrow{PA}$ is mapped onto $\overrightarrow{PB}$, and since $AP = BP$, A must be mapped onto B. Thus, since σ is an isometry, $AM = BM$. Thus, M is the midpoint of $\overline{AB}$. Also, under σ, $\overrightarrow{MP}$ is mapped onto $\overrightarrow{MP}$ and $\overrightarrow{MA}$ is mapped onto $\overrightarrow{MB}$. Thus, $\angle AMP \cong \angle BMP$. This implies that $\overleftrightarrow{PM}$ is the perpendicular bisector of $\overline{AB}$.

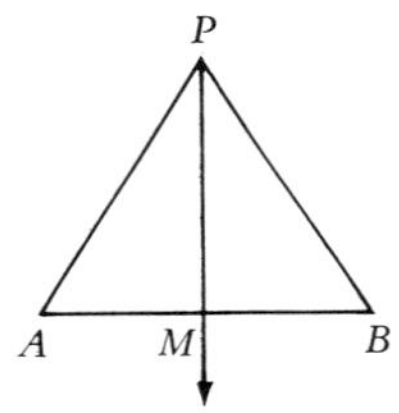

Figure 8.4.1

Now, suppose we consider the problem of transferring the size and shape of a given triangle using our compass and straight edge. How would you propose to do this? The fact that we can only use the compass to transfer distance suggests that perhaps we can transfer the size and shape of a triangle by transferring the lengths of its sides. Thus, consider the following construction.

Construction 2 (SSS Construction)

Given any triangle ABC and line $\overleftrightarrow{PS}$, construct triangle PQR so that $\overline{PQ} \subset \overleftrightarrow{PS}$ and $\overline{AB} \cong \overline{PQ}$, $\overline{AC} \cong \overline{PR}$ and $\overline{BC} \cong \overline{QR}$ (Fig. 8.4.2).

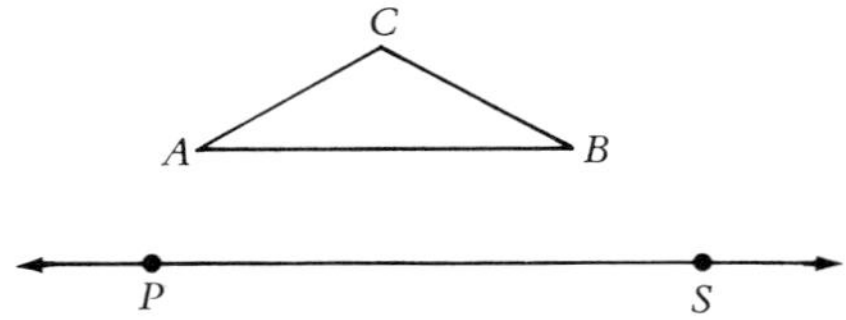

Figure 8.4.2

It is quite obvious what the first step should be since it is a duplication of the last construction:

Step 1

Construct $\overline{PQ}$ on $\overleftrightarrow{PS}$ so that $\overline{AB} \cong \overline{PQ}$ (Fig. 8.4.3). Now the next steps are designed to determine a point R so that $\overline{AC} \cong \overline{PR}$ and $\overline{BC} \cong \overline{QR}$. There are an infinite number of points which satisfy each condition separately, but only a finite number of these points satisfies both conditions. (How many points?)

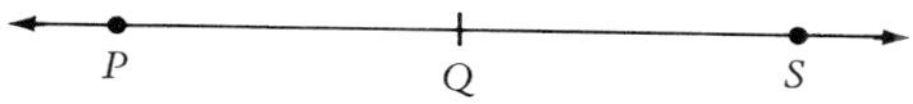

Figure 8.4.3

Step 2

Set the compass to a radius equal to BC, and using Q as center, draw an arc (Fig. 8.4.4).

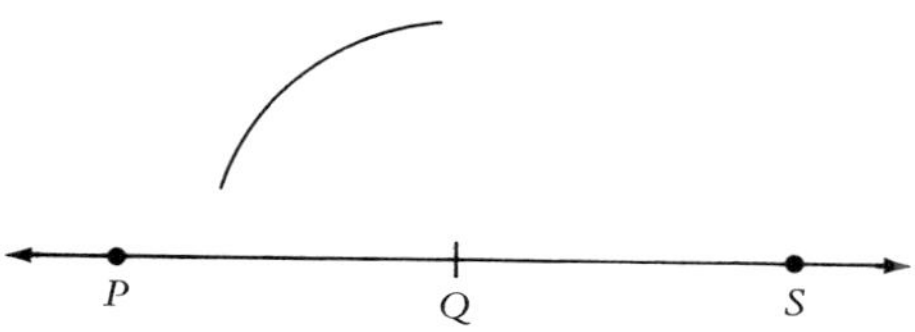

Figure 8.4.4

Now point R must be somewhere on this arc. To find out where, we proceed to Step 3.

Step 3

Set the compass to a radius equal to AC, and using P as center, draw an arc cutting the arc drawn in Step 2 at a point R (Fig. 8.4.5).

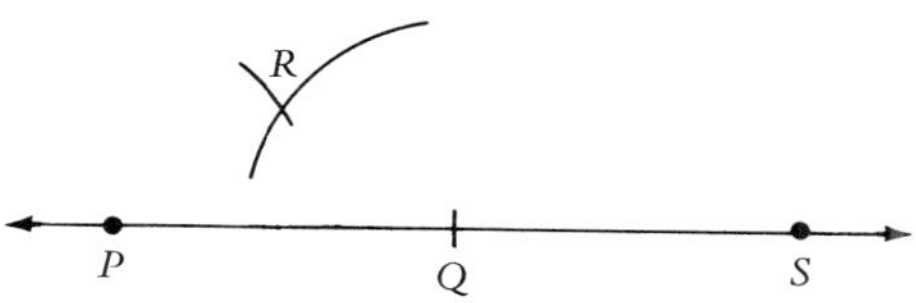

Figure 8.4.5

Since all points on this second arc are at a distance AC from P, and all points on the first arc are at a distance BC from Q, R is a point such that $PR = AC$ and $QR = BC$.

Step 4

Use a straight edge to draw $\overline{PR}$ and $\overline{QR}$. Then $\triangle PQR$ is the required triangle (Fig. 8.4.6).

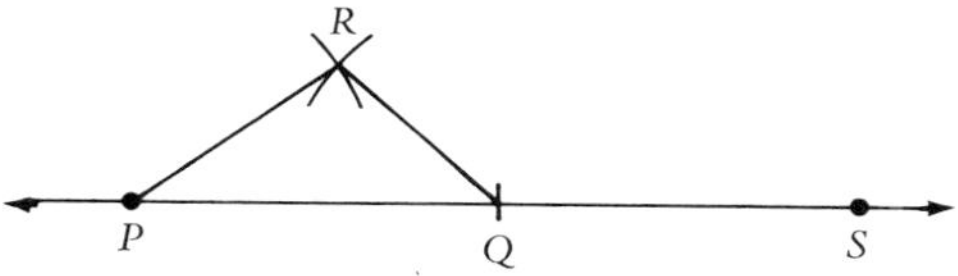

Figure 8.4.6

Thus, we have performed a construction which produces a triangle which has its sides congruent to a given triangle. If we examine the two figures $\triangle ABC$ and $\triangle PQR$, they appear to be the same size and shape. Furthermore, the fact that we were able to perform this using a distance preserving instrument suggests that perhaps there is an isometry which maps $\triangle ABC$ into $\triangle PQR$. If this is true, then we conclude that $\triangle ABC \cong \triangle PQR$. This suggests a very important congruence theorem.

8.4.2 Theorem (SSS Congruence)

If two triangles have three sides of one respectively congruent to three sides of the other, then the triangles are congruent (Fig. 8.4.7).

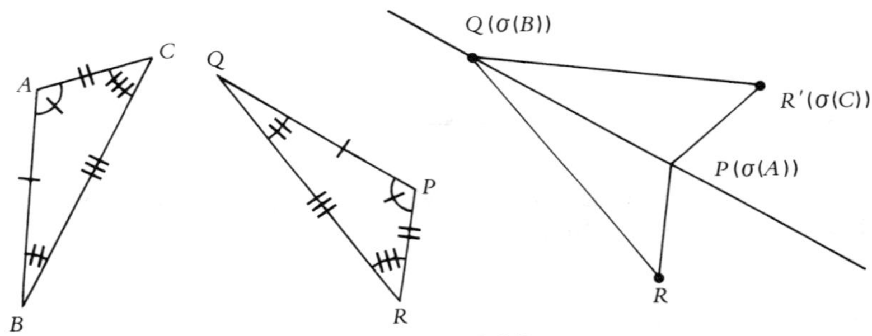

Figure 8.4.7

Suppose triangles ABC and PQR are given with $\overline{AB} \cong \overline{PQ}$, $\overline{AC} \cong \overline{PR}$, and $\overline{BC} \cong \overline{QR}$. To show that $\triangle ABC \cong \triangle PQR$, we must find an isometry σ such that $\sigma(A) = P$, $\sigma(B) = Q$, and $\sigma(C) = R$. Since $\overline{AB} \cong \overline{PQ}$, we know there is an isometry which maps $\overline{AB}$ onto $\overline{PQ}$. Thus, let σ_1 be the isometry that maps A to P, B to Q, and C to R' such that R and R' are on opposite sides of $\overleftrightarrow{PQ}$. Both P and Q are equidistant from R and R'. Thus, $\overleftrightarrow{PQ}$ is the perpendicular bisector of $\overline{RR'}$. Now the reflection in $\overleftrightarrow{PQ}$ maps $\triangle PQR'$ into $\triangle PQR$. The composition of σ_1 and this reflection is an isometry mapping $\triangle ABC$ into $\triangle PQR$ such that $\triangle ABC \cong \triangle PQR$.

We might point out that the isometry to prove that two triangles satisfy the SSS conditions need not be unique. For instance, suppose that both triangles ABC and PQR are equilateral triangles. Then there are six possible one-to-one correspondences between the vertices of the two triangles. Each of these correspondences represent a different isometry. On the other hand, given two triangles ABC and PQR, it does not follow that any one-to-one correspondence between vertices will represent an isometry which will guarantee the congruence of the figures.

If two triangles ABC and PQR are congruent, there will exist a one-to-one correspondence between the vertices, a one-to-one correspondence between the sides, and a one-to-one correspondence between the angles. When we write $\triangle ABC \cong \triangle PQR$, the arrangement of the letters is meant to suggest each of these correspondences. In other words, $ABC \longleftrightarrow PQR$ means that the first letter on the left is matched with the first on the right, the second with the second, and the third with the third, as in Fig. 8.4.8. This correspondence between vertices induces a correspondence between the sides and the angles. The induced correspondence between the sides is

$$\overline{AB} \longleftrightarrow \overline{PQ} \qquad \overline{BC} \longleftrightarrow \overline{QR} \qquad \overline{AC} \longleftrightarrow \overline{PR}$$

and the induced correspondence between the angles is

$$\angle ABC \longleftrightarrow \angle PQR \qquad \angle BAC \longleftrightarrow \angle QPR$$
$$\angle ACB \longleftrightarrow \angle PRQ$$

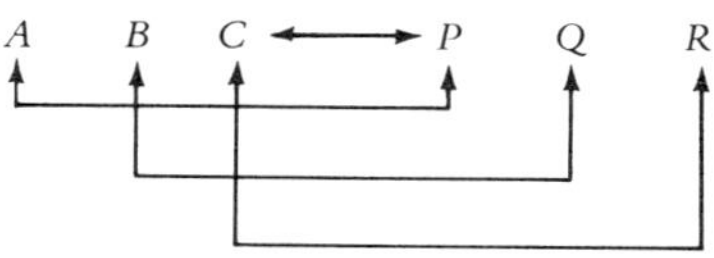

Figure 8.4.8

Thus, when we say that $\triangle ABC \cong \triangle PQR$, we are indicating there exists an isometry of the plane, say σ, such that $\sigma(A) = P$, $\sigma(B) = Q$, $\sigma(C) = R$ with the result that $\overline{AB} \cong \overline{PQ}$, $\overline{BC} \cong \overline{QR}$, $\overline{AC} \cong \overline{PR}$, $\angle ABC \cong \angle PQR$, $\angle BAC \cong \angle QPR$, and $\angle ACB \cong \angle PRQ$.

On the other hand, if we have two triangles ABC and PQR with these six congruent properties, then we know there exists an isometry between the two triangles making them congruent.

Exercise 8.4.1

1. If possible, construct a triangle having each of the sets of three segments in Fig. 8.4.9 as sides. What conditions must exist in order to be able to construct a triangle?

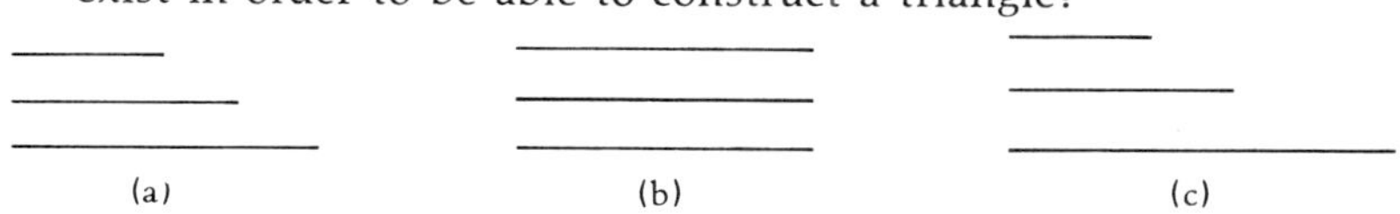

Figure 8.4.9

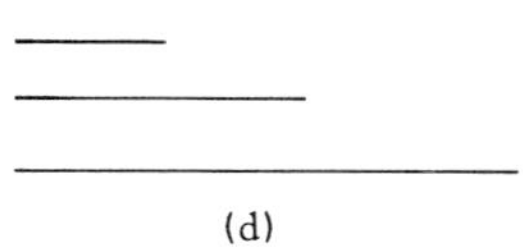

2. Use triangle constructions to duplicate the parallelogram in Fig. 8.4.10. [*Hint:* Draw in a diagonal.]

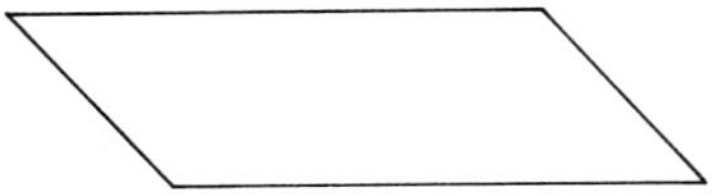

Figure 8.4.10

The SSS congruence theorem shows us that it will not be necessary to check the congruence of all corresponding parts of two triangles to determine if they are congruent. Instead, knowledge that three sides of one triangle are respectively congruent to three sides of the other is sufficient to know that the triangles are congruent. There are other tests for congruence of triangles which do not involve checking the congruence of all corresponding parts of the two triangles. Two of these are the SAS congruence theorem and the ASA congruence theorem.

The SSS theorem and the SSS construction which accompanies it provide us with a method for using the compass to transfer angle measure, for observe we constructed $\triangle ABC \cong \triangle PQR$. In doing so, we also constructed angle P congruent to angle A. That is, we transferred the measure of angle A. This suggests how we can perform the following construction.

Construction 3

Given an angle A and a ray $\overrightarrow{PS}$, construct angle P so that $\angle A \cong \angle P$ (Fig. 8.4.11).

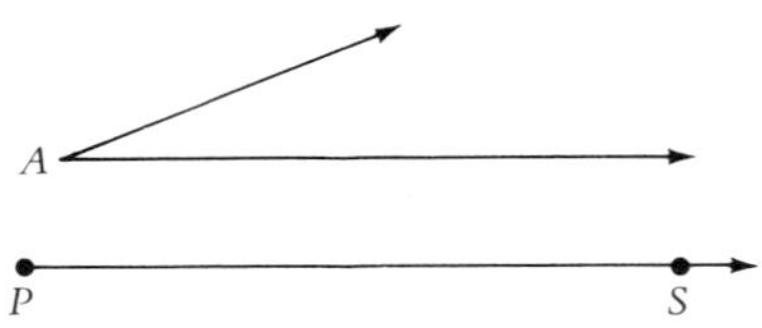

Figure 8.4.11

Step 1

Select any two points B and C distinct from A such that B is on one side of $\angle A$ and C is on the other.

Thus, we have a triangle ABC (Fig. 8.4.12).

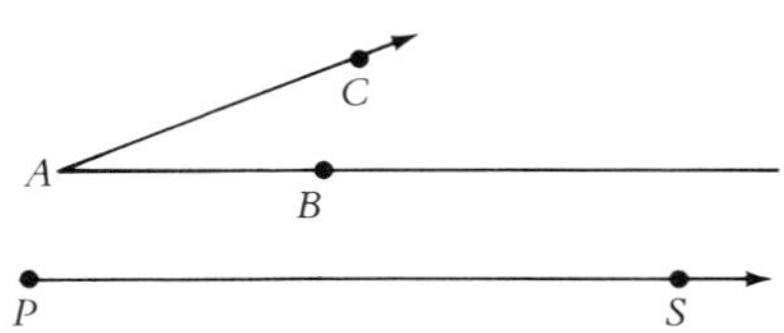

Figure 8.4.12

Step 2

Use the SSS construction to construct $\triangle PQR \cong \triangle ABC$. Then $\angle P$ is the required angle (Fig. 8.4.13).

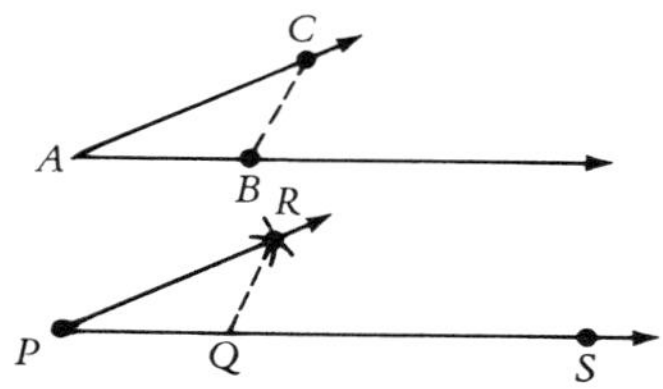

Figure 8.4.13

This is the most general method for transforming angle measure. Actually, the most common method, which was designed for efficiency, is a special case of this. Rather than making any triangle ABC, we make an isosceles triangle with $\overline{BA} \cong \overline{AC}$. This method is given below.

Construction 4

Given $\angle A$ and $\overrightarrow{PS}$, construct $\angle P$ so that $\angle P \cong \angle A$ (Fig. 8.4.14).

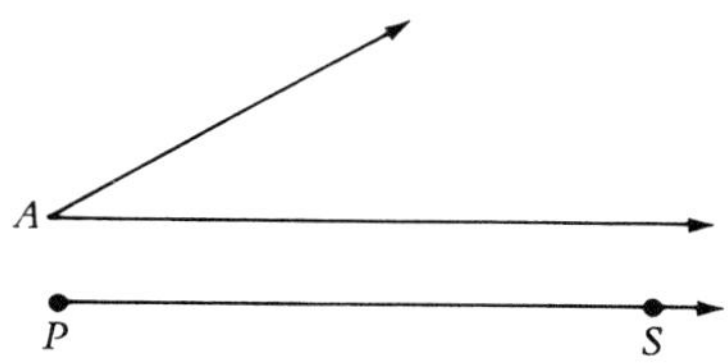

Figure 8.4.14

Step 1

Use A as the center and with a convenient radius, draw an arc which intersects one side of angle A at B and the other side at C.

Thus, we have constructed an isosceles triangle ABC (Fig. 8.4.15).

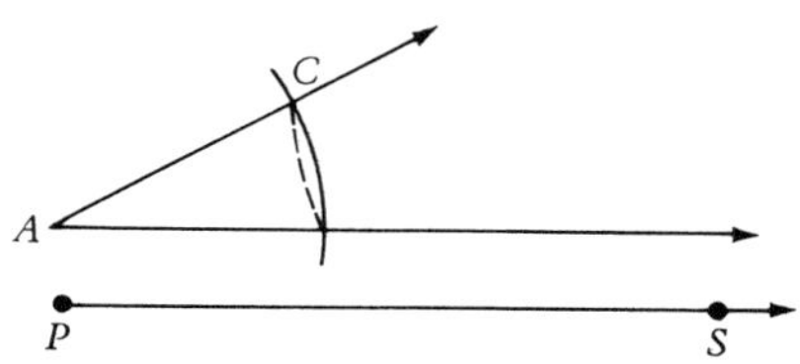

Figure 8.4.15

Step 2

Keeping the same radius, using P as the center, draw an arc which cuts $\overrightarrow{PS}$ at Q.

Here we are transferring the length of side $\overline{AB}$ and $\overline{AC}$, making $\overline{AB} \cong \overline{PQ}$ (Fig. 8.4.16).

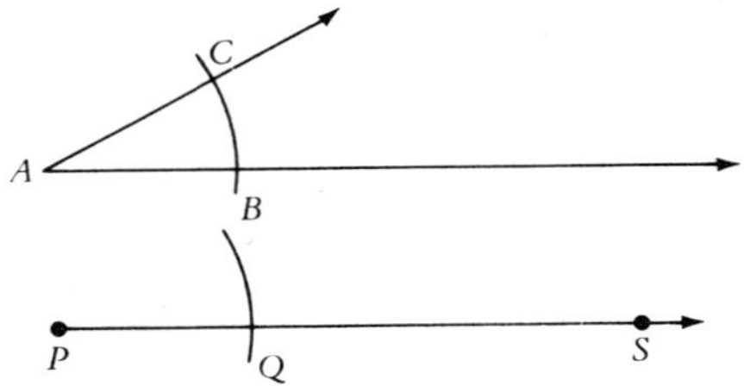

Figure 8.4.16

Step 3

Use Q as the center, and with a radius equal to BC, draw an arc intersecting the first arc at R.

Now we have transferred the length of $\overline{BC}$, making $\overline{BC} \cong \overline{QR}$ and $\overline{AC} \cong \overline{PR}$ (Fig. 8.4.17).

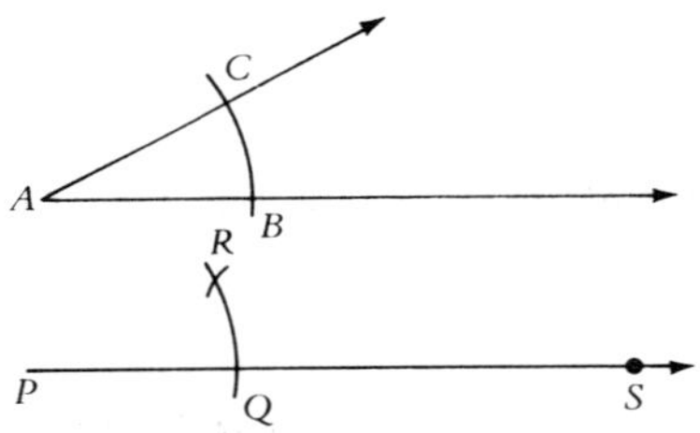

Figure 8.4.17

Step 4

Draw $\overrightarrow{PR}$.

Then $\angle RPQ$ is the desired angle, making $\angle A \cong \angle P$ (Fig. 8.4.18).

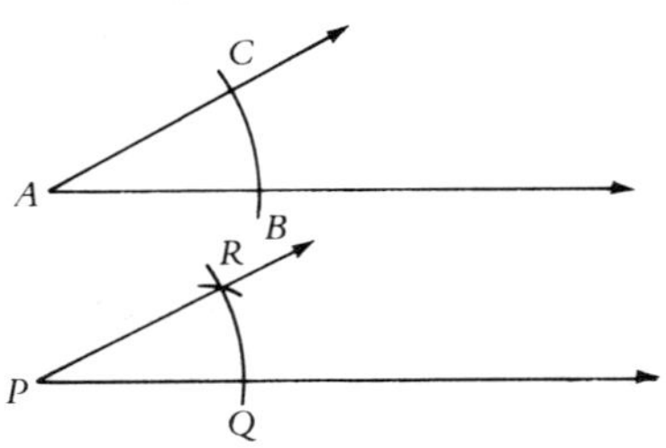

Figure 8.4.18

Now that we have seen how to transfer angle measure as well as segment length, we will return to the problem of transferring the size and shape of a triangle. We have already seen that this can be done by transferring the lengths of the three sides of the triangle. Is it possible to accomplish this same thing not by transferring all three sides, but by transferring only two sides and an angle? There are two possibilities here, one where the angle is included between the two sides (SAS) and one where the angle is opposite one of the sides (SSA). Let us consider the former case first.

Construction 5 (SAS Construction)

Given $\triangle ABC$ and $\overleftrightarrow{PS}$, construct $\triangle PQR$ so that $\overline{PQ} \subset \overleftrightarrow{PS}$ and $\overline{AB} \cong \overline{PQ}$, $\overline{AC} \cong \overline{PR}$ and $\angle A \cong \angle P$ (Fig. 8.4.19).

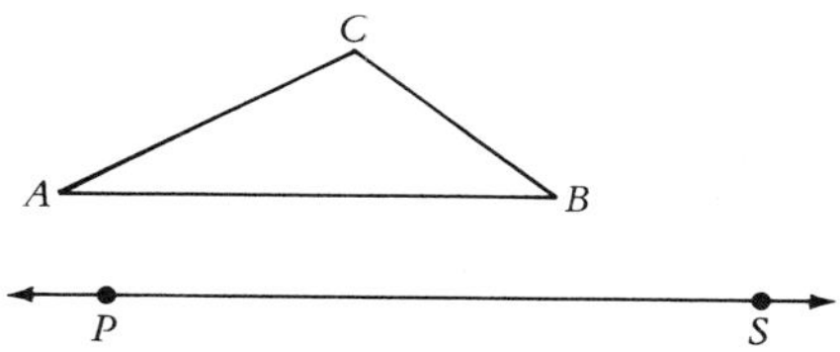

Figure 8.4.19

Step 1

Construct $\angle P$ so that $\angle P \cong \angle A$ and one side of $\angle P$ is on $\overrightarrow{PS}$ (Fig. 8.4.20).

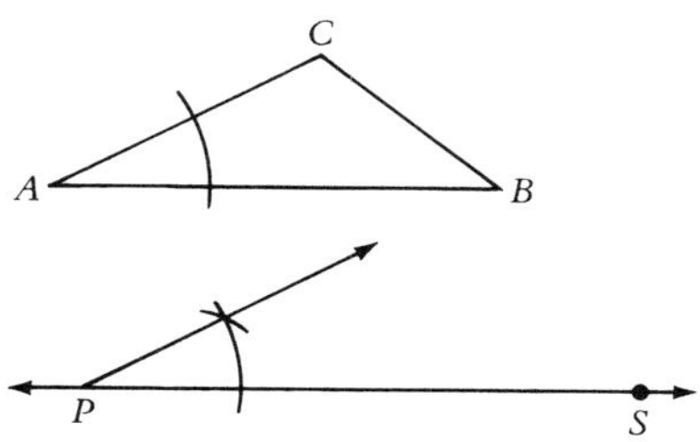

Figure 8.4.20

Step 2

Construct $\overline{PQ}$ on $\overleftrightarrow{PS}$ so that $\overline{AB} \cong PQ$. Also construct $\overline{PR}$ on the other side of $\angle P$ so that $\overline{AC} \cong \overline{PR}$.

Then $\triangle PQR$ is the required triangle (Fig. 8.4.21). Now if you examine the two triangles ABC and PQR it appears as if they have the same size and shape. It seems as though we have transferred the size and shape of $\triangle ABC$ by transferring the length of two of its sides and the measure of the angle included between these sides. Since we performed this by a compass and straight edge construction, it suggests the possibility that perhaps there exists an isometry between the two triangles. This suggests the following congruence theorem.

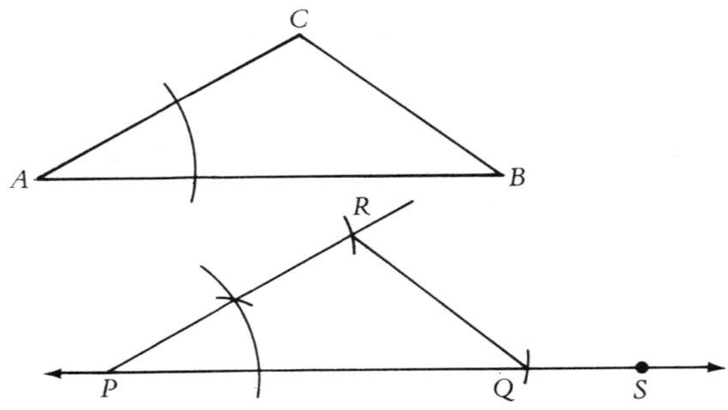

Figure 8.4.21

8.4.3 Theorem (SAS Congruence)

If two triangles have two sides and the included angle of one congruent respectively to two sides and the included angle of the other, then the triangles are congruent.

We offer the following argument to prove this theorem. Suppose two triangles ABC and PQR are given with $\overline{AB} \cong \overline{PQ}$, $\overline{AC} \cong \overline{PR}$, and $\angle BAC \cong \angle QPR$ (Fig. 8.4.22). To show that $\triangle ABC \cong \triangle PQR$, it will be necessary to find an isometry σ such that $\sigma(A) = P$, $\sigma(B) = Q$, and $\sigma(C) = R$. Since $\angle BAC \cong \angle QPR$, we know that there is an isometry σ_1 which maps $\angle BAC$ to $\angle QPR$. This map also sends $\overrightarrow{AB}$ onto one of the rays of $\angle QPR$ and $\overrightarrow{AC}$ onto the other ray of $\angle QPR$. This can occur in one of two ways. If σ_1 maps $\overrightarrow{AB}$ onto $\overrightarrow{PQ}$ and $\overrightarrow{AC}$ onto $\overrightarrow{PR}$, then $\sigma_1(B) = Q$ and $\sigma_1(C) = R$ because σ_1 preserves distances. In this case, σ_1 is the required isometry.

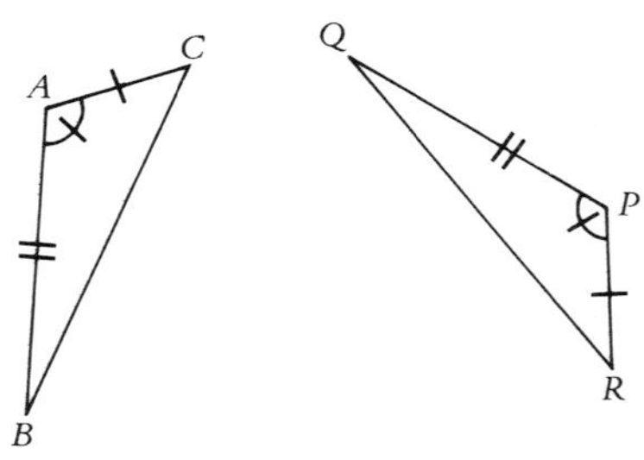

Figure 8.4.22

Now if σ_1 maps $\overrightarrow{AB}$ onto $\overrightarrow{PR}$ and $\overrightarrow{AC}$ onto $\overrightarrow{PQ}$, then a reflection in the bisector of $\angle QPR$ will map $\overrightarrow{AB}$ onto $\overrightarrow{PQ}$ and $\overrightarrow{AC}$ onto $\overrightarrow{PR}$. The composition of σ_1 and this reflection will be the required isometry sending A to P, B to Q, and C to R (Fig. 8.4.23).

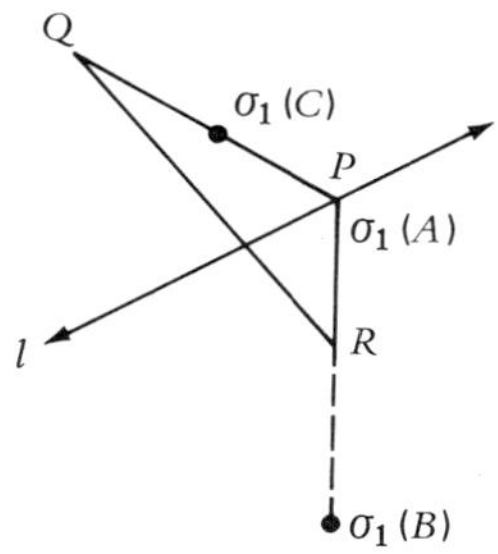

Figure 8.4.23
l is the bisector of $\angle QPR$
$r_l(P) = P$
$r_l(\sigma_1(B)) = Q$
$r_l(\sigma_1(C)) = R$

in our opinion . . .

This so-called transformation approach to congruent figures has many advantages. First of all, young children first become aware of the invariance of size and shape of figures by performing motions of the figures such as reflections, translations, rotations, and combinations of these. Secondly, it is usually necessary to postulate the SAS or SSS congruence statements which have been introduced. Postulates should be simply stated and easy to motivate; these congruence statements are neither. It seems like they should be proven rather than merely accepted.

Exercise 8.4.2

1. Use the angle construction to duplicate each of the angles in Fig. 8.4.24.

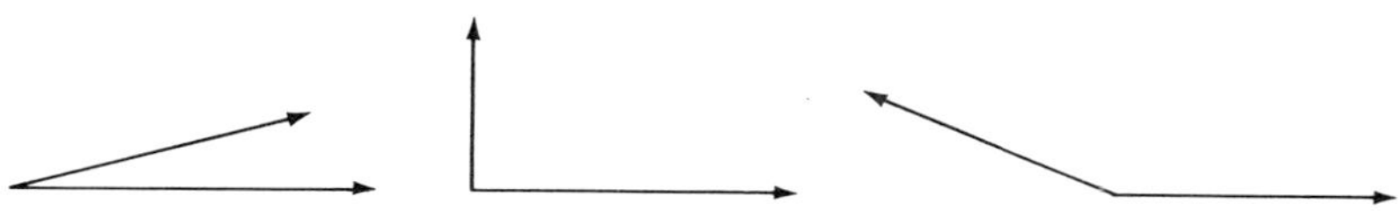

(a) (b) (c)

Figure 8.4.24

2. Use the SAS construction to construct a triangle which has two sides congruent to the given segments and an angle congruent to the given angle and included between the two sides (Fig. 8.4.25).

When we first considered the problem of transferring the size and shape of triangles by transferring the length of the two sides and one angle, we indicated that there were two possibilities. We shall consider the second possibility (SSA) now.

Construction 6

Given $\triangle ABC$ and $\overleftrightarrow{PS}$, construct $\triangle PQR$ so that $\overline{PQ} \subset \overleftrightarrow{PS}$ and $\overline{AC} \cong \overline{PR}$, $\overline{BC} \cong \overline{QR}$ and $\angle A \cong \angle P$ (Fig. 8.4.26).

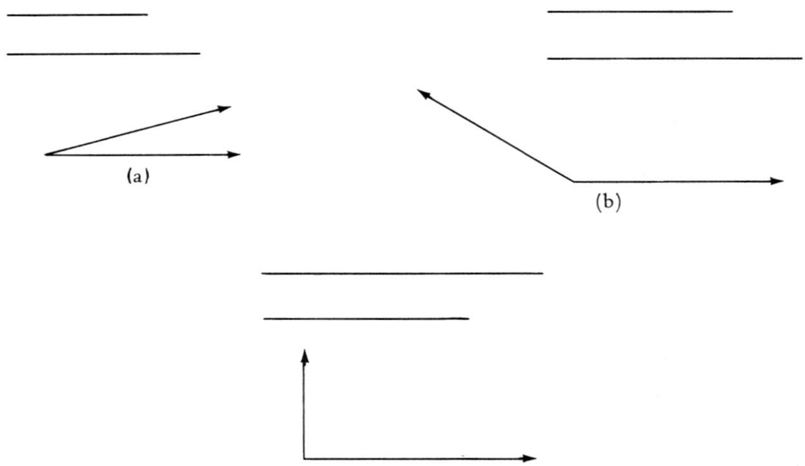

Figure 8.4.25

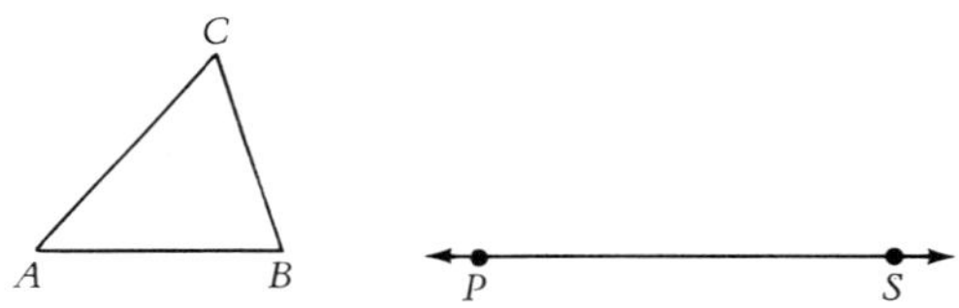

Figure 8.4.26

Step 1

Construct $\angle P$ so that one side of $\angle P$ is on $\overleftrightarrow{PS}$ and so that $\angle A \cong \angle P$ (Fig. 8.4.27).

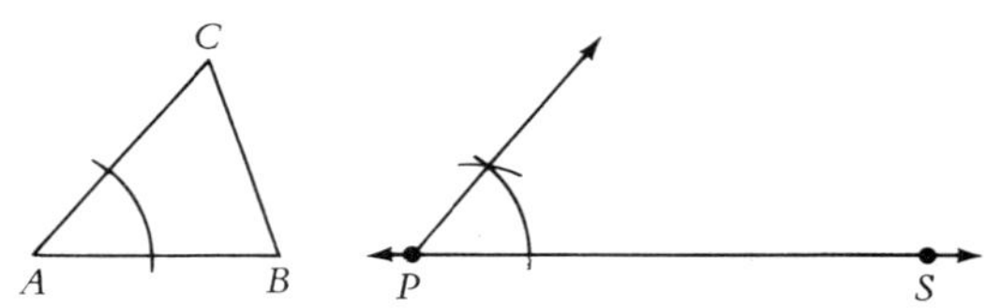

Figure 8.4.27

Step 2

Construct $\overline{PR}$ on the side of P which is not on $\overleftrightarrow{PS}$ so that $\overline{AC} \cong \overline{PR}$ (Fig. 8.4.28).

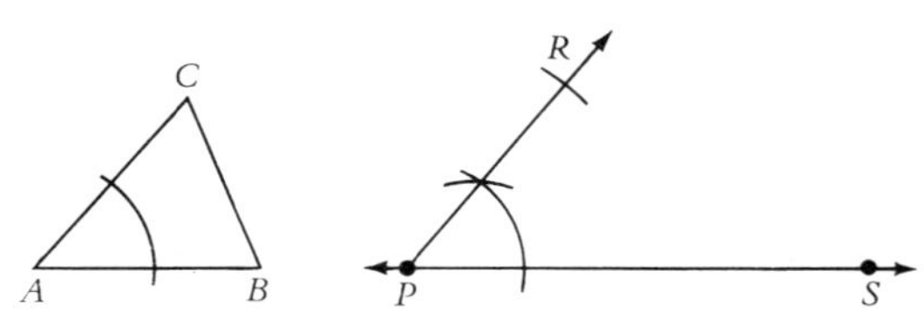

Figure 8.4.28

Step 3

With the compass set at a radius equal to BC, use R as the center, and draw an arc intersecting $\overrightarrow{PS}$ of $\angle P$ (Fig. 8.4.29).

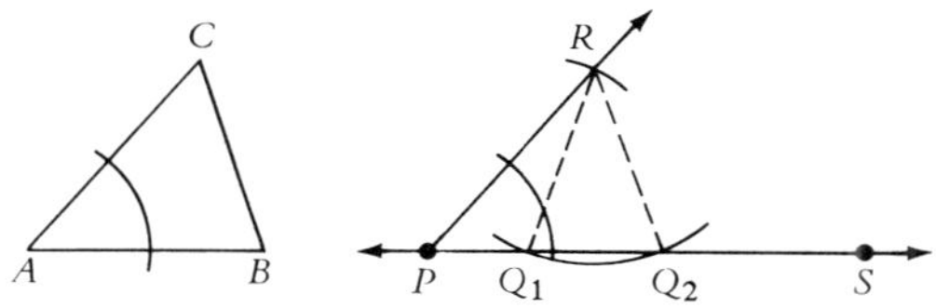

Figure 8.4.29

Now we see that we have run into a problem. This arc intersects $\overleftrightarrow{PS}$ at two points, Q_1 and Q_2. Thus, we have two triangles, $\triangle PQ_1R$ and $\triangle PQ_2R$, with the following properties:

In $\triangle PQ_1R$:	In $\triangle PQ_2R$:
$\overline{AC} \cong \overline{PR}$	$\overline{AC} \cong \overline{PR}$
$\overline{BC} \cong \overline{Q_1R}$	$\overline{BC} \cong \overline{Q_2R}$
$\angle A \cong \angle P$	$\angle A \cong \angle P$

That is, in both cases we have transferred the length of two sides of $\triangle ABC$ and the measure of one of its angles. However, whereas it appears that triangles ABC and PQ_2R are congruent, triangles ABC and PQ_1R are not. That is, it appears possible to transfer the lengths of two sides of a triangle and the measure of the angle opposite one of them and yet not transfer the size and shape of the triangle. Therefore, we must *reject* as a theorem the SSA case.

comment . . .

Up to this point in our development we have used the fact that we could construct a triangle from a given triangle by constructing certain congruent parts of the triangle to motivate the congruence theorems. We should be careful to realize that the fact that a compass (which we referred to as an isometric machine) could be used to perform a construction is not sufficient reason to prove that an isometry exists to map the one figure onto the other. The rejection of the SSA case is an illustration of this for although the compass allowed us to construct ΔPQ_1R, we know there is no isometry which would map ΔABC onto ΔPQ_1R; otherwise they would be congruent.

There is yet another way of transferring the size and shape of a triangle, by transferring only one side and two angles. Once again, we will have two possibilities to consider, one in which the side is included between the two angles (ASA) and one in which the side is opposite one of the angles (AAS). If we consider the (ASA) case first, we will find that we can indeed construct one triangle congruent to another by transferring the measures of these parts. The actual construction will be left as an exercise. The fact that this can be done suggests the following congruence theorem.

8.4.4 Theorem (ASA Congruence)

If two triangles have two angles and the included side of one respectively congruent to two angles and the included side of the other, then the triangles are congruent.

Suppose triangles ABC and PQR are given with $\angle BAC \cong \angle QPR$, $\angle ABC \cong \angle PQR$, and $\overline{AB} \cong \overline{PQ}$. We leave it to the reader to describe the isometry that will make $\triangle ABC \cong \triangle PQR$.

Now that we have established the ASA theorem, suppose we consider the AAS case. Suppose that we have two triangles $\triangle ABC$ and $\triangle PQR$ where $\angle A \cong \angle P$, $\angle B \cong \angle Q$, and $\overline{AC} \cong \overline{PR}$. What can we say about the two triangles? Must they be congruent? (See Fig. 8.4.30.) Since the sum of the measures of the two triangles is 180°, we know that $m°(\angle C) = 180 - [m°(\angle A) + m°(\angle B)]$ and $m°(\angle R) = 180 - [m°(\angle P) + m°(\angle Q)]$, $m°(\angle A) = m°(\angle P)$ and $m°(\angle B) = m°(\angle Q)$. Thus $m°(\angle A) + m°(\angle B) = m°(\angle P) + m°(\angle Q)$, and therefore $m°(\angle C) = m°(\angle R)$. This implies that $\angle C \cong \angle R$. But now observe that we have reduced this case to the ASA case and by our last theorem we know that $\triangle ABC \cong \triangle PQR$. Thus, we have an additional congruence theorem.

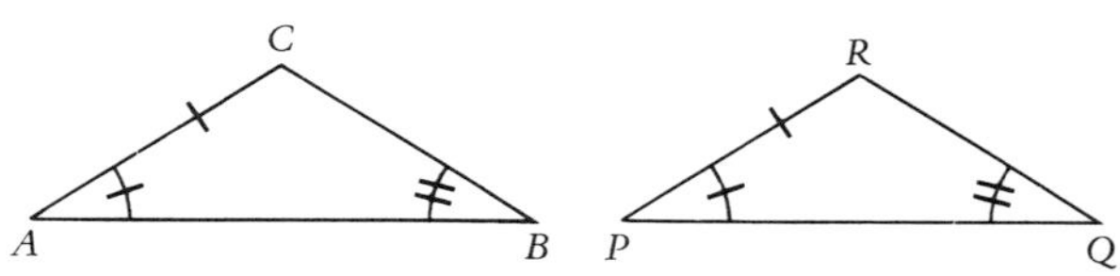

Figure 8.4.30

8.4.5 Theorem (AAS Congruence)

If two angles and a side opposite one of them of one triangle are congruent respectively to two angles and the side opposite one of them of the other triangle, then the triangles are congruent.

As an application of the congruence theorems, we can prove the following result.

8.4.6 Theorem

If two sides of a triangle are congruent, then the angles opposite these sides are congruent (Fig. 8.4.31).

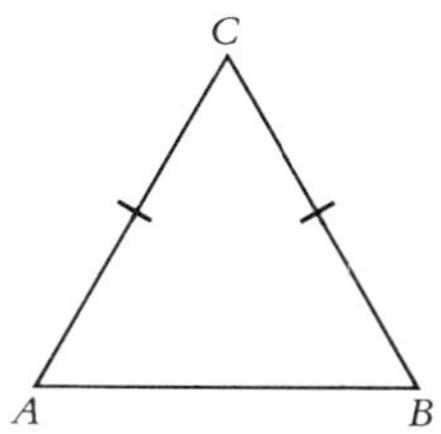

Figure 8.4.31

Suppose $\triangle ABC$ is isosceles with $\overline{AC} \cong \overline{BC}$. Then $AC = BC$ and hence the point C lies on the perpendicular bisector of $\overline{AB}$. If we reflect $\triangle ABC$ in this perpendicular bisector, we have point A mapped to point B, point B mapped to point A, and point C mapped to C. In this way, $\triangle ABC \cong \triangle BAC$ with $\overline{AC} \cong \overline{BC}$, $\overline{BC} \cong \overline{AC}$, and $\overline{AB} \cong \overline{BA}$. As a consequence of this congruence we would have $\angle CAB \cong \angle CBA$.

In a manner similar to that used above we can establish each of the following statements.

1. Every equilateral triangle is equiangular.
2. Every equiangular triangle is equilateral.
3. If two angles of a triangle are congruent, then the sides opposite them are congruent.

Exercise 8.4.3

1. Two triangles PQR and XYZ have $\angle P \cong \angle X$, $\angle Q \cong \angle Y$, and $\overline{PQ} \cong \overline{XY}$. Must the triangles be congruent?
2. Two triangles PQR and XYZ have $\overline{PQ} \cong \overline{XY}$ and $\overline{QR} \cong \overline{YZ}$. Must the triangles be congruent?
3. Give an argument to show that if $\overleftrightarrow{AB}$ and $\overleftrightarrow{CD}$ intersect at X as in Fig. 8.4.32 and if $\overline{CX} \cong \overline{XD}$, $\overline{AX} \cong \overline{AB}$, then $\angle ACX \cong \angle BDX$.

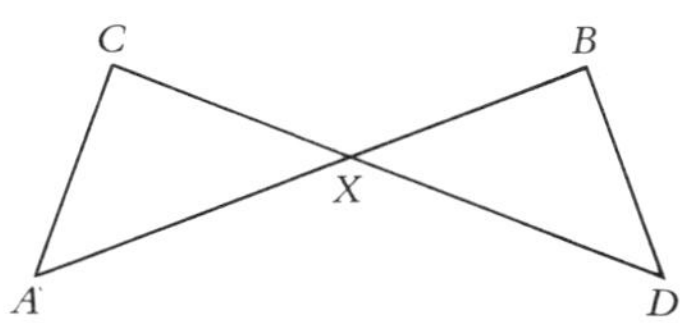

Figure 8.4.32

4. Given Fig. 8.4.33 such that $\overline{AB} \cong \overline{DC}$ and $\overline{AC} \cong \overline{BD}$, give an argument showing that $\triangle ABC \cong \triangle DCB$.

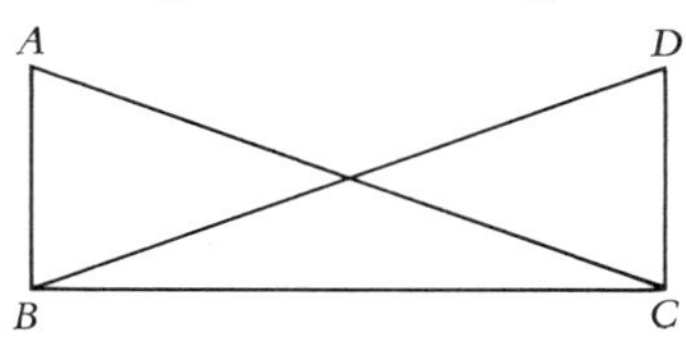

Figure 8.4.33

5. Give an argument showing that the medians to the two congruent sides of an isosceles triangle are congruent.
6. Give an argument showing that the bisector of the base angles of an isosceles triangle are congruent.
7. Give an argument showing that the opposite sides of a parallelogram are congruent.
8. Give an argument showing that if both pairs of opposite sides of a quadrilateral are congruent, then it is a parallelogram.
9. Give an argument showing that the diagonals of a parallelogram bisect each other.
10. Give an argument showing that if the diagonals of a quadrilateral bisect each other, then it is a parallelogram.
11. Give an argument showing that if two opposite sides of a quadrilateral are congruent and parallel, then its a parallelogram.
12. Give an argument showing that if a parallelogram has one right angle, then it is a rectangle.
13. Give an argument showing that if two adjacent sides of a rectangle are congruent, then it is a square.
14. Give an argument showing that if the diagonals of a rectangle are perpendicular, then it is a square.

8.5 MORE EXPERIMENTS

Experiment 4

We have seen that a reflection in a line is the basic isometry of the plane. In space, the situation is quite similar, the basic isometry being a reflection in a plane. Devise a definition for a reflection in a plane and investigate the composition of reflections. Describe a space isometry and parallel the development of all concepts of the chapter, including the notion of symmetry. Investigate each of the following questions.

1. In a plane, an isometry is completely determined by three noncollinear points and their images. How many points of what nature completely determine a space isometry?
2. Various compositions of plane reflections result in transformations which are called translations, rotations, and glide reflections (that is, the composition of a translation and a reflection in a line parallel to the direction of the translation). Describe the space analogs of these transformations. Are there any other possible transformations?
3. What properties are preserved by space isometries?
4. If a plane isometry can be written as the product of at most

three reflections in lines, what similar statement can be made about space isometries?

Experiment 5

The set of all isometries in a plane form an algebraic system known as a *group*. In general, note the following definition.

8.5.1 Definition

Given a set A and a function

$$*: A \times A \rightarrow A$$

with domain the Cartesian product $A \times A$ *and range A, which assigns to each ordered pair (a, b) the element* $a * b$ *of A, we say that the pair (A, *) is a* group *if the following conditions are satisfied.*

1. associativity: *If* $a, b, c \in A$, *then* $a * (b * c) = (a * b) * c$.
2. identity: *There exists* $e \in A$ *such that* $a * e = e * a = a$ *for all* $a \in A$.
3. inverse: *For all* $a \in A$, *there exists* $b \in A$ *such that* $a * b = b * a = e$.

*If, in addition, the following condition is satisfied, we say (A, *) is an* Abelian group.

4. Commutative: *If* $a, b \in A$, *then* $a * b = b * a$.

Find two isometries of the plane which show that the group of all plane isometries does not form an Abelian group.

Another and simpler example of a group can be described as follows. We start with an equilateral triangle and consider all possible symmetries of the triangle. We will think of the triangle as movable, having numbered vertices (Fig. 8.5.1).

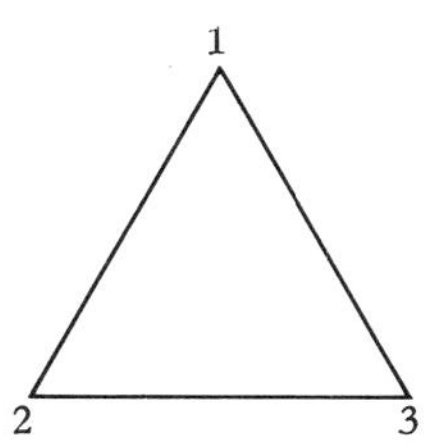

Figure 8.5.1

This triangle can be picked up, turned over, or anything that does not deform it, but then it must be put back in the same position. The possible motions that are performed on the triangle will be elements of a group which can be described by indicating what happens to the vertices.

The first thing we can do to the triangle is to leave it alone. In this case, vertex 1 is sent to vertex 1, vertex 2 to vertex 2, and vertex 3 to vertex 3. We denote this identity motion by I and symbolize it by

$$I = \begin{pmatrix} 1 & 2 & 3 \\ 1 & 2 & 3 \end{pmatrix}$$

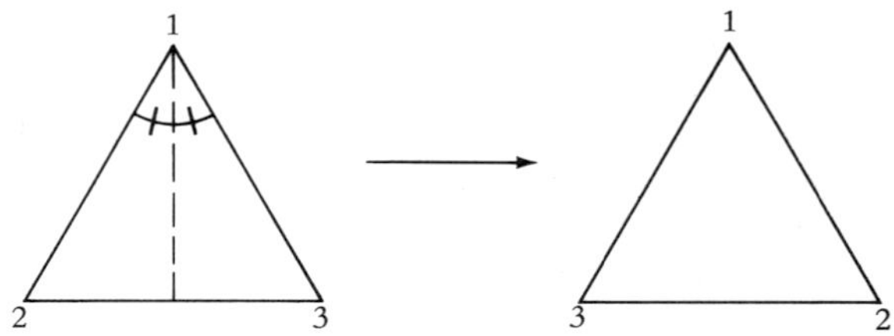

Figure 8.5.2

The triangle can be reflected in each of the bisectors of the angles. For instance, a reflection through the horizontal angle bisector sends 1 to 1, 2 to 3, and 3 to 2 (Fig. 8.5.2). We denote this reflection by

$$D_1 = \begin{pmatrix} 1 & 2 & 3 \\ 1 & 3 & 2 \end{pmatrix}$$

The other two reflections are given by

$$D_2 = \begin{pmatrix} 1 & 2 & 3 \\ 3 & 2 & 1 \end{pmatrix} \quad \text{and} \quad D_3 = \begin{pmatrix} 1 & 2 & 3 \\ 2 & 1 & 3 \end{pmatrix}$$

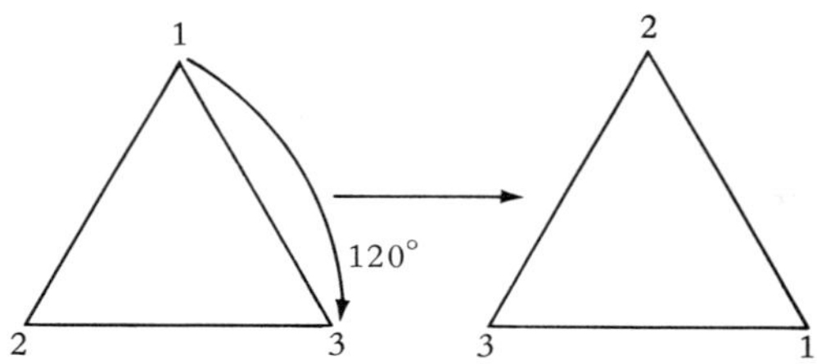

Figure 8.5.3

If we rotate the triangle through 120° in a clockwise direction, this motion sends 1 to 3, 2 to 1, and 3 to 2 (Fig. 8.5.3). We denote this rotation by

$$R_1 = \begin{pmatrix} 1 & 2 & 3 \\ 3 & 1 & 2 \end{pmatrix}$$

If we rotate the triangle through 240°, we get a rotation

$$R_2 = \begin{pmatrix} 1 & 2 & 3 \\ 2 & 3 & 1 \end{pmatrix}$$

It turns out that these six occurrences are the only possible symmetries of the triangle and we can construct a composition table as follows. If X and Y are two symmetries of the triangle, $X \circ Y$ denotes the operation of first performing X and then following this by Y. As an example,

$$D_2 \circ R_1 = \begin{pmatrix} 1 & 2 & 3 \\ 3 & 2 & 1 \end{pmatrix} \circ \begin{pmatrix} 1 & 2 & 3 \\ 3 & 1 & 2 \end{pmatrix} = \begin{pmatrix} 1 & 2 & 3 \\ 2 & 1 & 3 \end{pmatrix} = D_3$$

Note D_2 sends 1 to 3 which is sent to 2 by R_1, the result being that $D_2 \circ R_1$ sends 1 to 2. Similarly, D_2 sends 2 to 2 which is sent to 1 by R_1. Then $D_2 \circ R_1$ sends 2 to 1. Finally, D_2 sends 3 to 1 and R_1 sends 1 to 3 with the result that $D_2 \circ R_1$ sends 3 to 3. Note that $D_2 \circ R_1 = D_3$.

$\circ$	I	R_1	R_2	D_1	D_2	D_3
I						
R_1						
R_2						D_1
D_1						
D_2		D_3				
D_3						

Table 8.5.1

Table 8.5.1 can be filled in showing all the possible results of compositions of symmetries of a triangle. The table is read as follows: Find in the first column the motion D_2, find in the first row the motion R_1. Then find the box which is the intersection of the column containing R_1 and the row containing D_2. In this box we place the motion which is the composition $D_2 \circ R_1$, in this case D_3. Similarly, we have $R_2 \circ D_3 = D_1$. Complete this table. What element of the set of symmetries of an equilateral triangle fulfills the role of an identity? What is the inverse of each motion? Show that this group of symmetries is not Abelian.

Experiment 6

We can describe a similar group by considering the symmetries of a square (Fig. 8.5.4). Describe the various motions which carry the square onto itself in a manner analogous to the motions which carry an equilateral triangle onto itself. Also describe the composition of any two such motions and construct a composition table. Investigate this table to ensure that it is a group. Is it an Abelian group?

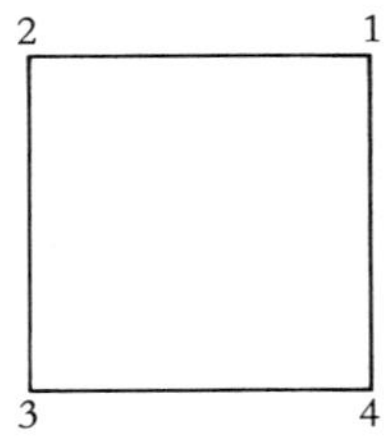

Figure 8.5.4

8.6 SOME MORE CONSTRUCTIONS

As noted earlier, an angle with measure 90 is called a right angle. The two sides of a right angle are said to be perpendicular.

8.6.1 Definition

In general, two lines l and m are said to be perpendicular lines if they intersect to form right angles. To indicate that m is perpendicular to l, we write $m \perp l$ (read "m is perpendicular to l") (Fig. 8.6.1).

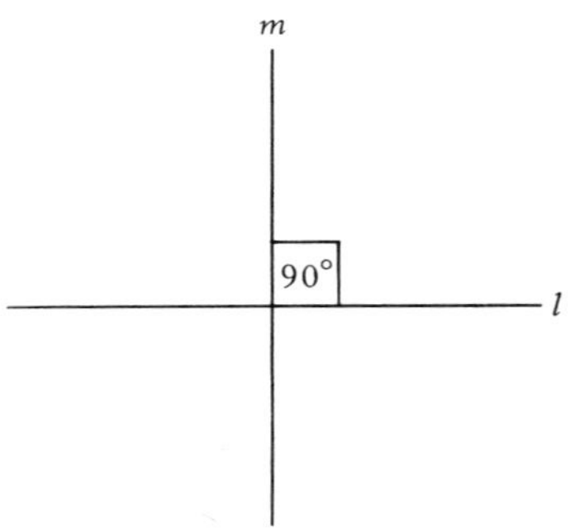

Figure 8.6.1

We can use the compass to construct perpendicular lines.

Construction 7

Given a line l and a point P not on the line l, construct a line m such that $m \perp l$.

Step 1

Pick a point Q so that P and Q are on opposite sides of l.

Step 2

Place the compass point at P and with any fixed radius equal to the distance PQ, draw arcs intersecting l at X and Y (Fig. 8.6.2).

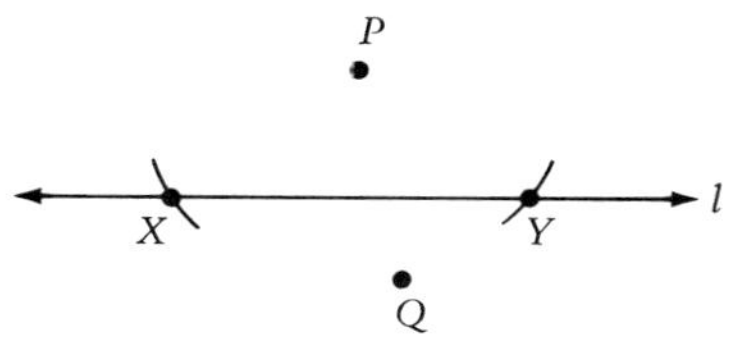

Figure 8.6.2

Step 3

With the compass point at X and with a radius greater than ½ the length $\overline{XY}$, draw an arc. Using the same radius and Y as the center, draw another arc intersecting the first arc at R. Notice that we have located a point R which is equidistant from X and Y (Fig. 8.6.3).

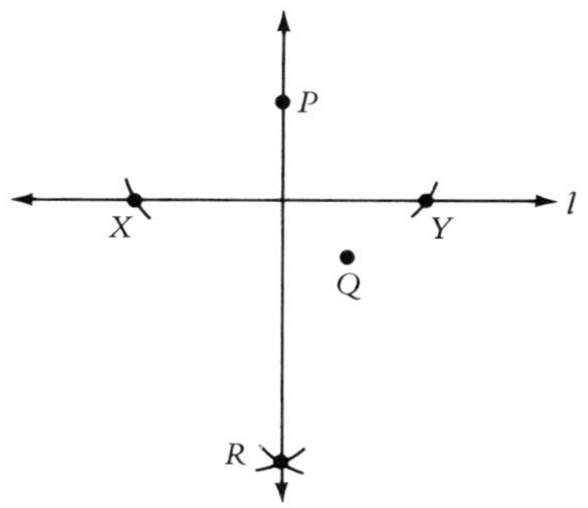

Figure 8.6.3

Step 4

Draw the line $\overleftrightarrow{PR}$. Using the protractor, measure $\angle YMP$ and $\angle XMP$ (M is the point of intersection of lines l and $\overleftrightarrow{PR}$). We find that they are both right angles. Thus, the line $\overleftrightarrow{PR}$ is the required line m which is perpendicular to l.

As an exercise, you should be able to establish that $\overleftrightarrow{PR}$ is the perpendicular bisector of $\overline{XY}$ from previous theorems. We can also use a compass and a straight edge to construct a line parallel to a given line. Basically, we make use of the statement that if two lines are cut by a transversal so that a pair of corresponding angles are congruent, then the lines are parallel. The construction follows.

Construction 8

Given a line l, construct a line m such that $m \| l$.

Step 1

Choose some point P on line l. Draw any line t which intersects l at P (Fig. 8.6.4).

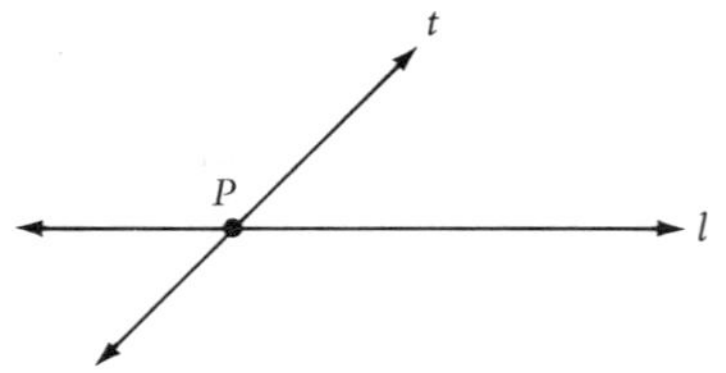

Figure 8.6.4

Step 2

Choose some point other than P on line t and call it Q. Also, choose some point on l other than P and call it X (Fig. 8.6.5).

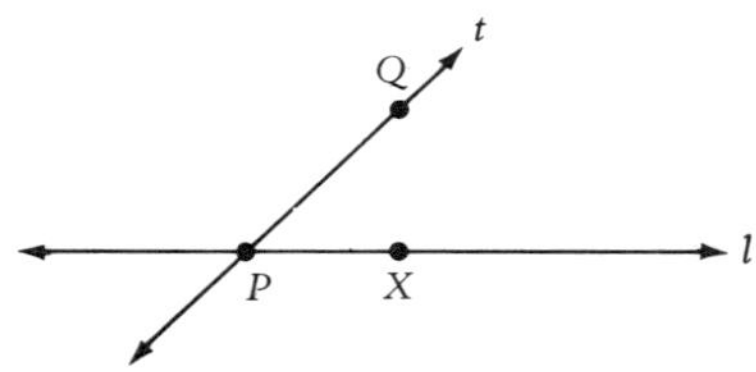

Figure 8.6.5

Step 3

Construct an angle MQN with vertex Q with one side on line t such that this angle and $\angle XPQ$ are congruent corresponding angles (Fig. 8.6.6).

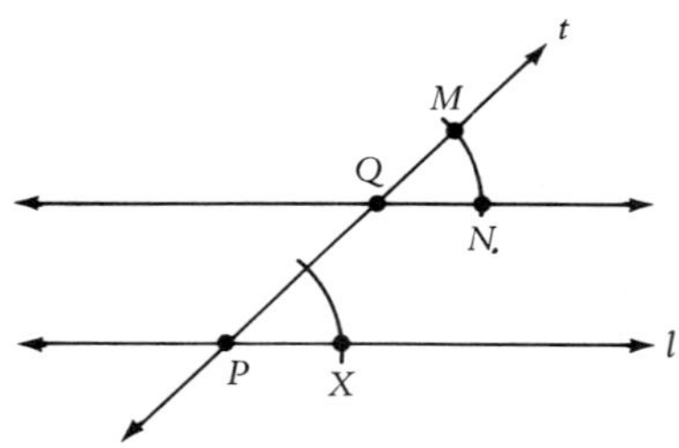

Figure 8.6.6

Then $\overleftrightarrow{NQ} \| l$.

Exercise 8.6.1

1. Using a compass and a straight edge, construct a triangle given
   a. two angles and an included side
   b. two angles and a side opposite
2. Devise a method of bisecting an angle using a compass and a straight edge.
3. Draw a triangle and construct the three angle bisectors using a compass and a straight edge. Where do the angle bisectors meet? Repeat the process for several other triangles. Any conjectures?

4. Consider constructing a triangle by transferring the angle measure of its three angles. What would you conclude about a corresponding (AAA) theorem of congruence?

8.7 SUMMARY

1. A figure is symmetric about a line l called its axis of symmetry if for all points P on the figure, there is a point P' on the figure such that $\overline{PP'}$ is perpendicular to l at a point Q and such that $\overline{PQ} \cong \overline{P'Q}$.
2. A figure is symmetric with respect to a point O if for each point A on the figure, there is a point A' on the figure such that O, A, and A' are collinear and $\overline{OA} \cong \overline{OA'}$.
3. If point P is not on a line l, then the reflection image of P over line l is the point P' if and only if l is the perpendicular bisector of $\overline{PP'}$. If P is on line l, the reflection image of P is P itself.
4. Reflection in a line preserves collinearity, betweenness, distance, and angle measure.
5. Let $\mathcal{P}$ be a plane and l a line in the plane. The function $r_l: \mathcal{P} \to \mathcal{P}$ defined by $r_l(P) = P'$ such that l is the perpendicular bisector of $\overline{PP'}$ if $P \not\in l$, and $r_l(P) = P$ if $P \in l$, is called a reflection in line l.
6. A figure is symmetric with respect to a line l if and only if it coincides with its reflection image.
7. If r_l and r_m are two reflections in lines l and m, respectively, define $r_m \circ r_l$ to be the composition or product of the two reflections. For each point P in the plane, $r_m \circ r_l(P) = r_m(r_l(P))$.
8. An isometry is a function of the points of the plane to itself that preserves distances.
9. If a point is equidistant from the endpoints of a line segment, it is on the perpendicular bisector of the line segment.
10. Let α and β be any two figures, α and β are congruent ($\alpha \cong \beta$) if and only if there is an isometry mapping α onto β.
11. (SAS Congruence) If two triangles have two sides and the included angle of one congruent respectively to two sides and the included angle of the other, then the triangles are congruent.
12. (SSS Congruence) If two triangles have three sides of one congruent respectively to three sides of the other, then the triangles are congruent.
13. (ASA Congruence) If two triangles have two angles and the included side of one congruent respectively to two angles and the included side of the other, then the triangles are congruent.

14. (AAS Congruence) If two angles and a side opposite one of them of one triangle are congruent respectively to two angles and the side opposite one of them of the other triangle, then the triangles are congruent.
15. A set G together with a function $*: G \times G \rightarrow G$ is a group with $(a, b) \rightarrow a * b$ if (a) for every $a, b, c \in G$, $a * (b * c) = (a * b) * c$; (b) there exists $e \in G$ such that for all $a \in G$, $a * e = e * a = e$; and (c) for all $a \in G$, there exists $b \in G$ such that $a * b = b * a = e$.

Exercise 8.7.1

1. **8.7.1 Definition**

   The perpendicular distance from a vertex of a triangle to the line of the opposite side of the triangle is an altitude *of the triangle corresponding to this side.*

   See Figs. 8.7.1 and 8.7.2 as an example.

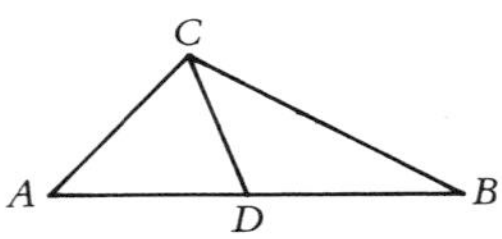

Figure 8.7.1
$\overline{CD}$ is the altitude to $\overline{AB}$ in $\triangle ABC$

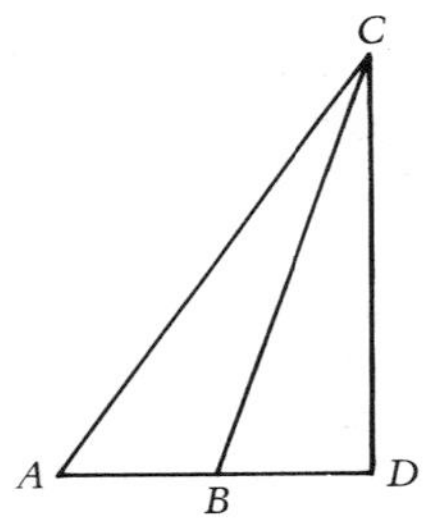

Figure 8.7.2
$\overline{CD}$ is the altitude to $\overline{AB}$ in $\triangle ABC$

   Draw a triangle and construct the three altitudes using a compass and a straight edge. Where do the altitudes meet? Repeat the process for several other triangles. Any conjectures?

2. **8.7.2 Definition**

   In a triangle, a median *is a segment joining a vertex to the midpoint of the opposite side.*

   See Figs. 8.7.3 and 8.7.4 as an example.

   Draw an obtuse triangle and construct its medians. Draw an acute triangle and construct its medians. Draw a right triangle and draw its medians. Repeat the experiment with several other triangles. Any conjectures?
3. Make a conjecture about the median from the vertex of an isosceles triangle to the base of the triangle.
4. a. Show that the product of translations is commutative.

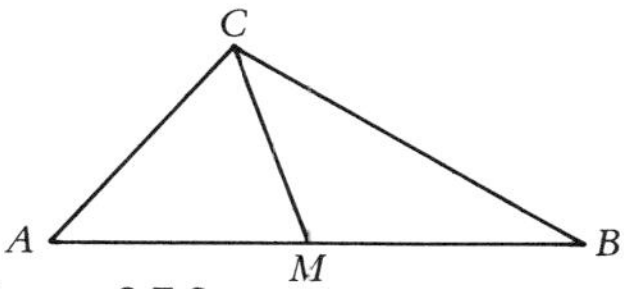

Figure 8.7.3
$\overline{CM}$ is the median to $\overline{AB}$ in $\triangle ABC$

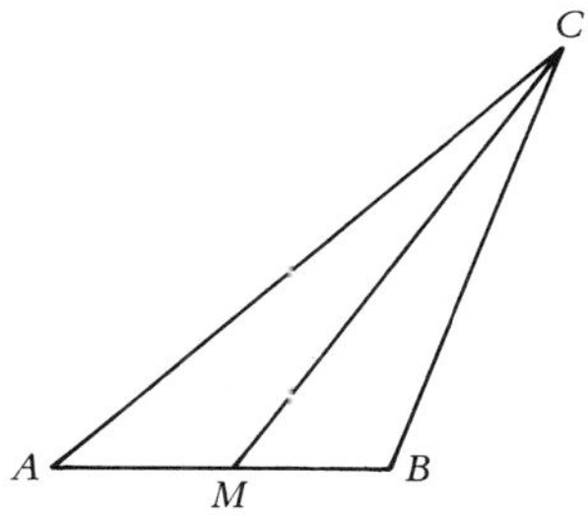

Figure 8.7.4
$\overline{CM}$ is the median to $\overline{AB}$ in $\triangle ABC$

b. Show that every direct isometry is a rotation or a translation.
c. Show that the product of two translations is a translation.
d. Show that the inverse of a rotation about 0 through angle θ is the rotation about O through angle $-\theta$.
e. Show that the product of two rotations about the same center O is a rotation about O.
f. Show that the product of two rotations about the same center O is commutative.

8.7.3 Definition

A glide reflection *is the product of a translation and a reflection in a line parallel to the direction of the translation.*

g. Show that every opposite isometry is either a glide reflection or a reflection.

GEOMETRIC INEQUALITIES

9

9.1 PROLOGUE

Thus far in our discussion, we have limited ourselves to considering conditions which allow us to say that two segments have the same length or that two angles have the same measure. We now consider conditions which allow us to say that one segment is shorter than another or that one angle has a greater measure than another.

In order to do this, we shall want to formalize and summarize some of our intuitive notions about segments and angles of unequal measure. We shall also want to refer to some properties of inequality which you have probably encountered in some earlier mathematics course. For example, we shall want to make extensive use of the *trichotomy property*.

9.1.1 Postulate

(Trichotomy Property) If a and b are two real numbers, then exactly one of the following must be true: $a = b$, $a < b$, or *$a > b$.*

Two other important inequality properties which we shall refer to are the following.

9.1.2 Postulate

(Transitive Property of Less Than) For all real numbers a, b, and c, if $a < b$ and $b < c$, then $a < c$.

9.1.3 Postulate

(Transitive Property of Greater Than) For all real numbers a, b, and c, if $a > b$ and $b > c$, then $a > c$.

With the aid of these properties, we shall investigate and develop important relationships concerning segments and angles of unequal measure. During our investigation we shall find it very convenient and natural to employ a new type of logical reasoning to justify some of our ideas. When that occurs, we will pause briefly to discuss this type of reasoning, for you will soon discover that it is a very useful and important technique to master.

9.2 ANGLES AND SEGMENTS OF UNEQUAL MEASURE

If you look at Fig. 9.2.1, you can see a line segment $\overline{AB}$ with C a point between A and B. It seems quite obvious that under such conditions, $AC < AB$. That is, as long as point C is between points A and B, the measure of $\overline{AC}$ will be less than the measure of $\overline{AB}$. We can summarize this idea by the following statement.

Figure 9.2.1

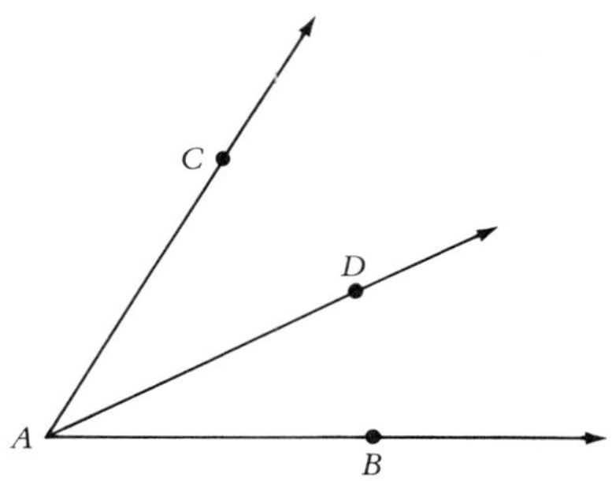

Figure 9.2.2

9.2.1 Postulate

If C is between A and B, then $AC < AB$.

We can consider an analogous situation with angles by considering $\angle BAC$ with $\overrightarrow{AD}$ in the interior of $\angle BAC$ as in Fig. 9.2.2. Here it appears as if $m^\circ \angle BAD$ is less than $m^\circ \angle BAC$. This also can be expressed in the following statement.

9.2.2 Postulate

If $\overrightarrow{AD}$ is in the interior of $\angle BAC$, then $m^\circ \angle BAD < m^\circ \angle BAC$.

[*Note:* From here on, we shall simply say $\overrightarrow{AD}$ is between $\overrightarrow{AB}$ and $\overrightarrow{AC}$.] These two principles allow us to identify segments and angles of unequal measures as long as the segments and angles are located in very restricted ways.

What if the segments under consideration are in different lines of the same plane, or even in different planes? How can we determine if they are unequal? We can perhaps seek an answer by recalling how we determined if two segments were congruent in the last chapter. At that time, we defined two segments to be congruent if and only if there was an isometry mapping one segment onto the other.

Now suppose we have two segments, $\overline{AB}$ and $\overline{CD}$, with $AB < CD$. We use C as the center. With our compass we mark off a segment on $\overline{CD}$ with a radius of AB (Fig. 9.2.3). Where will the compass mark intersect $\overleftrightarrow{CD}$? Of course, the compass mark will be in the interior of $\overline{CD}$. Now, recalling from our last chapter that our compass was suggesting the construction of an isometry, we formalize our discovery with the following statement.

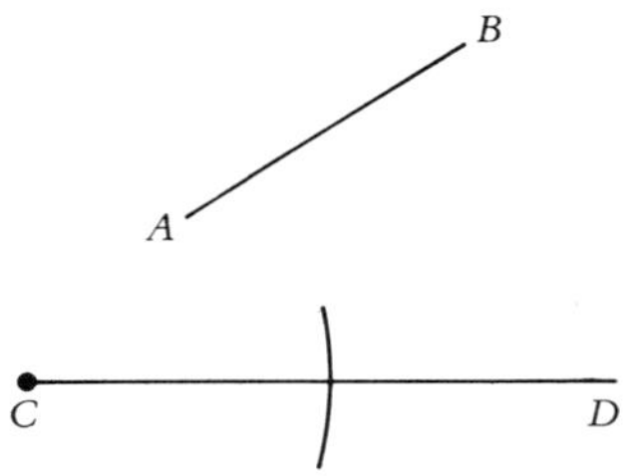

Figure 9.2.3

9.2.3 Definition

For any two segments $\overline{AB}$ and $\overline{CD}$, $AB < CD$ *if and only if there exists an isometry* σ *such that* $\sigma(A) = C$ *and* $\sigma(B)$ *is between* C *and* D *[that is,* $C - \sigma(B) - D$*].*

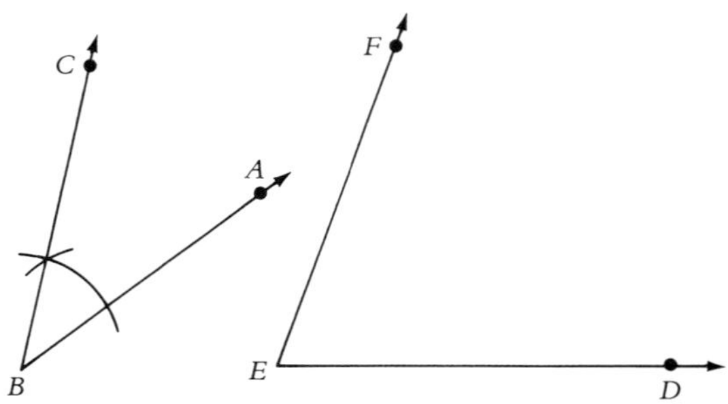

Figure 9.2.4

Now let us consider the analogous situation for angles. Suppose we have two angles $\angle ABC$ and $\angle DEF$ with $m^\circ \angle ABC < m^\circ \angle DEF$ (Fig. 9.2.4). Then suppose we duplicate $\angle ABC$ so that vertex B is at E, $\overrightarrow{BA}$ is on $\overrightarrow{ED}$, and $\overrightarrow{BC}$ is on the F side of $\overleftrightarrow{DE}$ (Fig. 9.2.5). Where do you observe the image of $\overrightarrow{BC}$ to be? It seems to be in the interior of $\angle DEF$. Once again, this compass construction suggests the following statement.

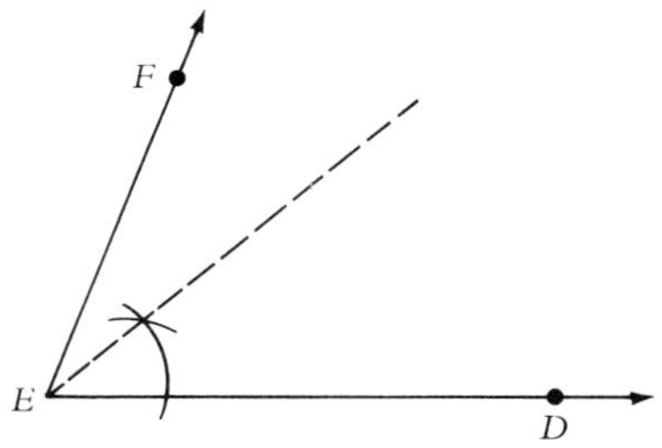

Figure 9.2.5

9.2.4 Definition

For any two angles ABC and DEF, $m^\circ \angle ABC < m^\circ \angle DEF$ if and only if there exisits an isometery σ such that $\sigma(\overrightarrow{BA}) = \overrightarrow{ED}$ and $\sigma(\overrightarrow{BC})$ is between $\overrightarrow{ED}$ and $\overrightarrow{EF}$.

Experiment 1

Draw any triangle ABC. Extend segment $\overline{AB}$ through point B. Select a point D such that $A - B - D$, forming $\overrightarrow{ED}$. Now $\angle CBD$ is called an *exterior angle* of $\triangle ABC$. Now, using your compass, duplicate $\angle C$ so that the image of $\overrightarrow{CB}$ is $\overrightarrow{BC}$ and so that the image of $\overrightarrow{CA}$ is on the D side of $\overleftrightarrow{BC}$. What can you conclude about $m^\circ \angle C$ and $m^\circ \angle CBD$? Now extend $\overline{CB}$ through B and select a point E so that $C - B - E$. $\angle EBA$ is also an exterior angle of $\triangle ABC$. What is true of $\angle EBA$ and $\angle DBC$? Now duplicate $\angle A$ so that the image of $\overrightarrow{AB}$ is $\overrightarrow{BA}$ and the image of $\overrightarrow{AC}$ is on the E side of $\overleftrightarrow{AB}$. What do you conclude about $m^\circ \angle A$ and $m^\circ \angle EBA$ and hence about $m^\circ \angle A$ and $m^\circ \angle DBC$? Repeat your experiment using various triangles of different sizes and shapes. What do you observe? Can you make a conjecture about an exterior angle of a triangle? Do you think $\angle DBF$ would be called an exterior angle of $\triangle ABC$? Explain your answer. What do you think would be a good definition of an exterior angle of a triangle?

9.3 UNEQUAL SIDES AND ANGLES IN TRIANGLES

The preceding experiment has involved a relationship between an exterior angle of a triangle and the two interior angles which are not adjacent to the exterior angle. From now on we shall refer to these interior angles as the *remote interior angles*. The results of the experiment seem to suggest the property that an exterior angle of a triangle is greater in measure than either of the remote interior angles. Although the experiment may lead us to believe that this is always true, we perhaps are a bit skeptical, wondering if perhaps there are triangles where it is not true. How do we know for sure that the ray of the angle we duplicated will always be in the interior of the exterior angle? Look at the diagram in Fig. 9.3.1. Put somewhat differently, the question reduces to whether or not we can always find a point F in the interior of $\angle DBC$ such that $\angle C \cong \angle CBF$. Thus, all we need to do is prescribe a way of finding such a point F. (Of course, if there is one point, there is an infinite number of them, but all we need is one.) Euclid suggested locating point F in the following way:

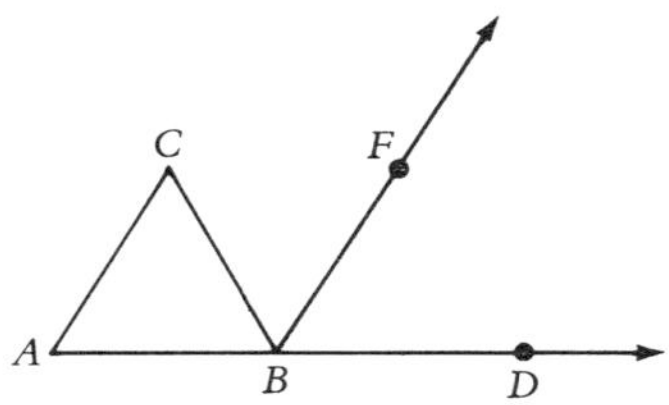

Figure 9.3.1

Select M, the midpoint of $\overline{BC}$. Then extend $\overline{AM}$ through M and let F be the point on $\overrightarrow{AM}$ such that $A - M - F$ and $\overline{MF} \cong \overline{AM}$ (Fig. 9.3.2). Now it is obvious that F is in the interior of $\angle DBC$. What remains to be seen is that $\angle C = \angle CBF$.

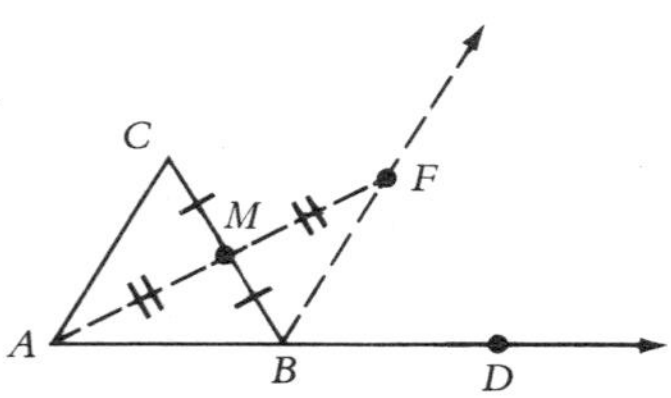

Figure 9.3.2

comment . . .

Actually, a great mathematician, David Hilbert, pointed out that it is not really so obvious that F so constructed is in the interior of $\angle DBC$. However, he did show that with careful analysis, it was true.

We can show that this is true by proving $\triangle AMC \cong \triangle FMB$. The reader should be sure he can do this.

Thus, we have demonstrated that $m^\circ \angle C < m^\circ \angle DBC$. To show that $m^\circ \angle A < m^\circ \angle DBC$, we use a similar argument which uses the exterior angle at B which is vertical to $\angle DBC$. The reader should see if he can do this. To summarize, look at the following.

9.3.1 Theorem

(Exterior Angle Theorem) An exterior angle of a triangle is greater in measure than either of the remote interior angles.

Exercise 9.3.1

1. Name the exterior angles of the triangle in Fig. 9.3.3 and also the remote interior angles relative to each exterior angle.

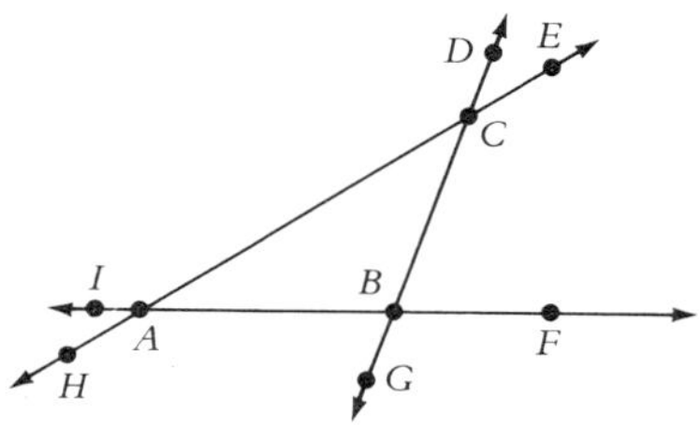

Figure 9.3.3

2. Define what an exterior angle of a quadrilateral should be. Is there an inequality relationship existing between an exterior angle of a quadrilateral and remote interior angles of the quadrilateral?

There is an immediate consequence of this exterior angle theorem. This helps us to establish the idea that through a point not on a line one and only one line may be drawn perpendicular to the line. Let us consider this idea carefully. Suppose we are given a line l and a point P not on l (Fig. 9.3.4). Our statement is suggesting two ideas. First, it says that it is possible to construct at least one line through P perpendicular to l, and, second, it says that it is not possible to construct more than one line through P perpendicular to l. We could have established the first idea earlier using the construction of Fig. 9.3.5. The reader should see if he can prove $\overleftrightarrow{PR} \perp l$, with the congruences, $\overline{PX} \cong \overline{PY}$ and $\overline{XR} \cong \overline{RY}$, given by the construction.

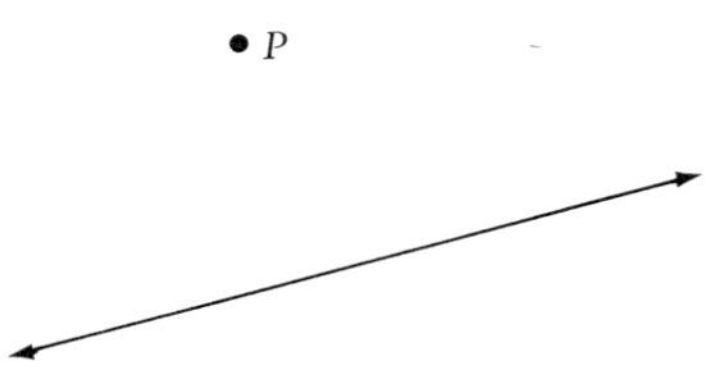

Figure 9.3.4

It is the second idea with which we are concerned at the present time. We wish to argue that it is not possible to draw more than one line through P perpendicular to l. Intuitively, this seems to make sense, but how can we explain it more formally?

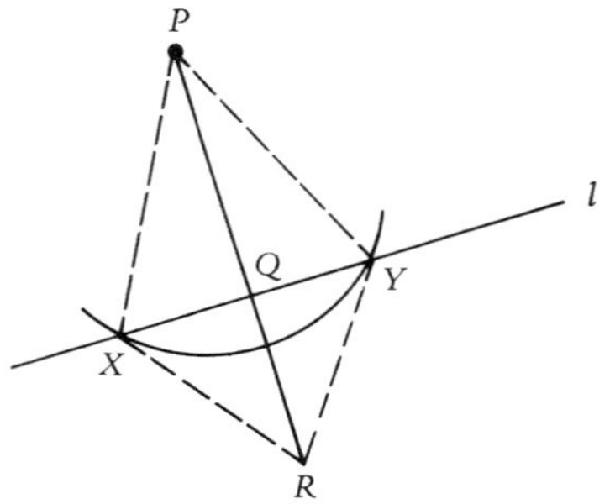

Figure 9.3.5

What would it mean if we could draw more than one perpendicular through P? What would our diagram look like? We have to use our imagination a bit but perhaps the diagram would look like Fig. 9.3.6. Suppose that $\overleftrightarrow{PQ} \perp l$ and $\overleftrightarrow{PR} \perp l$ and that $\overleftrightarrow{PQ}$ and $\overleftrightarrow{PR}$ are distinct lines. Then we would have a triangle PQR. Now how can we use our exterior angle theorem? Can you identify an exterior angle in the diagram? $\angle PRS$ is such an angle. What does our theorem say about $m^\circ \angle PRS$ and $m^\circ \angle RQP$? Since $\angle RQP$ is an interior angle of $\triangle PQR$ and since it is remote from exterior angle PRS, we can conclude that $m^\circ \angle RQP < m^\circ \angle PRS$. But this is not possible because $\angle RQP$ and $\angle PRS$ are both right angles. Thus, it is not possible for $\overleftrightarrow{PQ}$ and $\overleftrightarrow{PR}$ to both be perpendicular to l. That is, it is not possible to have more than two lines perpendicular to the same line from the same point not on that line. We summarize with the following.

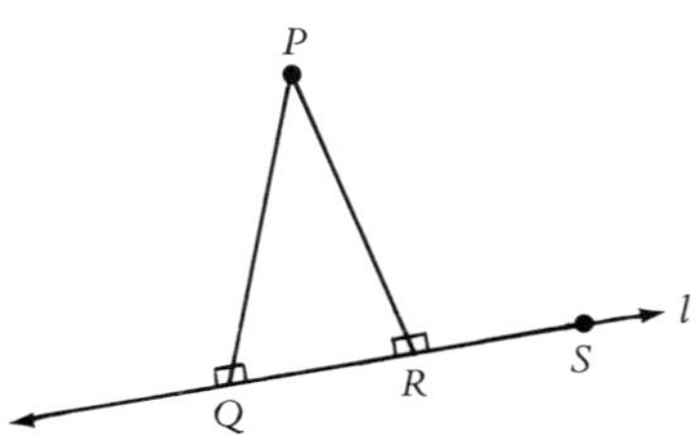

Figure 9.3.6

9.3.2 Theorem

Through a point not on a line one and only one line may be drawn perpendicular to the line.

Perhaps you are aware that the last argument seemed quite different from many of our previous arguments. The technique we used is often referred to as *indirect reasoning*. Undoubtedly, we have all employed this type of reasoning many times. To prove that this is the case, see if you can correctly answer the following question: How old was one of the authors of this book at the time of its writing? Circle the correct age.

(a) 2 years (b) 34 years (c) 508 years

You certainly did not have much difficulty answering that question. But isn't it strange that even though you probably don't know the authors of this book, you were able to conclude that one of them was 34 years old when he was writing the book? Not strange at all; you simply used indirect reasoning in reaching your conclusion. Perhaps you are used to calling it the "process of elimination" technique. But whatever you call it, it is the same kind of reasoning which was used in the last geometrical argument. What were the essential ingredients of your thinking in answering the question? First, the question implied that one of the selections was the correct answer. Second, you were able to eliminate all of the possibilities except one. You knew the author could not have been 2 years old for he would have been too young and he could not have been 508 years old for he would have been too old. Therefore, he must have been 34 years old.

Now reconsider our geometrical argument and notice the similarities. We first recognize that either there is more than one line through P perpendicular to l or there is not. These are the only two possibilities. So, we proceed to eliminate one of these possibilities. There cannot be more than one because if there were we would have a triangle with one right angle as an exterior angle and another right angle as a remote interior angle and thus we would have one right angle less than another right angle which is impossible. Thus, we have eliminated that one possibility. Therefore, the remaining possibility is the only alternative. Thus, it must be true that there is not more than one line through P perpendicular to l.

In general, all forms of indirect reasoning must contain two major steps: first, a listing of all the possible alternatives and, second, an elimination of all of these possibilities except one. The usual way of eliminating the alternatives is by demonstrating that they would lead to a contradiction of some acknowledged fact.

Suppose we try this type of reasoning to solve the following problem: In Fig. 9.3.7 we have triangle ABC with $\overline{AB} \not\cong \overline{BC}$. What can you conclude about the angles opposite these sides, $\angle A$ and $\angle C$? What are the possible alternatives? Either angles A and C are congruent or they are not congruent. Now, can we

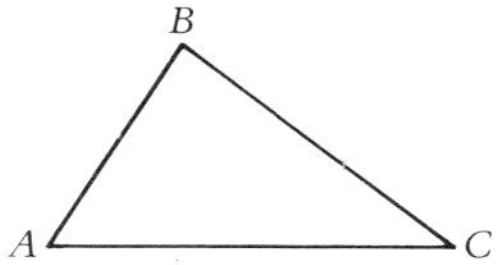

Figure 9.3.7

eliminate one of these two alternatives? Let us suppose that $\angle A \cong \angle C$. What does that imply? We know that if two angles of a triangle are congruent, the sides opposite these angles are congruent. Thus, the supposition that $\angle A \cong \angle C$ implies that $\overline{AB} \cong \overline{BC}$ and this contradicts our hypothesis that $\overline{AB} \ncong \overline{BC}$. Thus, $\angle A$ and $\angle C$ must not be congruent, which is our only other possibility. Thus, we can make the following conclusion.

9.3.3 Theorem

If two sides of a triangle are not congruent, then the angles opposite these sides are not congruent.

Although we shall frequently be using this type of reasoning, we shall seldom use it exclusively in the absence of direct reasoning. For example, in the last statement, indirect reasoning led us to a conclusion but direct reasoning can enable us to further investigate the idea.

In Fig. 9.3.7, suppose we not only know that $\overline{AB}$ and $\overline{BC}$ are not congruent but we also know that $AB < BC$. What can we conclude about the measures of $\angle A$ and $\angle C$? The Law of Trichotomy reminds us that there are three distinct possibilities: either $m^\circ \angle A = m^\circ \angle C$, $m^\circ \angle A < m^\circ \angle C$, or $m^\circ \angle A > m^\circ \angle C$. Indirect reasoning already eliminated the first alternative. Suppose we attempt to eliminate one of the remaining two. Which one shall we eliminate? The diagram itself seems to suggest that $m^\circ \angle A > m^\circ \angle C$. Suppose we try to eliminate the other alternative, $m^\circ \angle A < m^\circ \angle C$. What would this tell us about the diagram? The point is, we are not sure what this implies about the diagram because we do not really have any theorems about triangles with angles of unequal measure. So, even though assuming $m^\circ \angle A < m^\circ \angle C$ may lead to a contradiction, we are not able to deduce it. So, at this point it does not seem as though indirect reasoning can help us. So let us try a different approach. Let us consider the hypothesis that $AB < BC$. What does this mean? This means that there is an isometry σ such that $\sigma(B) = B$ and $B - \sigma(A) - C$. That is, with a compass, we can locate a point D between B and C such that $\overline{AB} \cong \overline{BD}$. See Fig. 9.3.8. Now, connecting A and D with a segment, we have two triangles, $\triangle ABD$ and $\triangle ADC$. Since $\overline{AB} \cong \overline{BD}$ we know $m^\circ \angle BAD = m^\circ \angle BDA$. But, $\angle BDA$ is an exterior angle of $\triangle ADC$ and $\angle C$ is interior and remote. Thus, $m^\circ \angle BDA > m^\circ \angle C$. Thus, we can

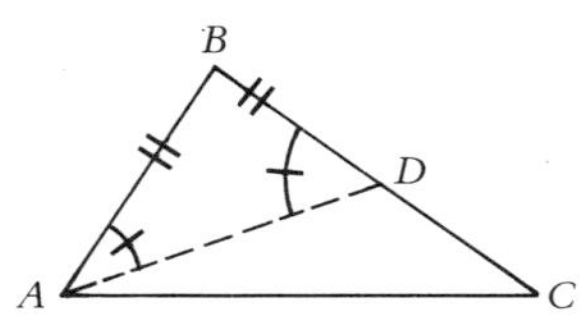

Figure 9.3.8

also say $m^\circ \angle BAD > m^\circ \angle C$. Now we also observe that $\overrightarrow{AD}$ is between $\overrightarrow{AB}$ and $\overrightarrow{AC}$. Therefore, $m^\circ \angle BAC > m^\circ \angle BAD$. Therefore, by the transitive property of greater than, we have $m^\circ \angle BAC > m^\circ \angle C$ or simply $m^\circ \angle A > m^\circ \angle C$.

We can therefore strengthen Theorem 9.3.3 by saying the following.

9.3.4 Theorem

If two sides of a triangle are not congruent, then the angles opposite these sides are not congruent and the angle which is opposite the side with the greater measure has a measure greater than that of the other angle.

This theorem now allows us to establish another important idea. Suppose we have a line l and a point P not on l. What path would you take if you wanted to go from P to l in the shortest possible distance? See Fig. 9.3.9. It appears as if the shortest path would be along the line which goes through P and is perpendicular to l. Let us see how indirect reasoning can give us this result. Suppose we have $\overleftrightarrow{PQ} \perp l$. Suppose $\overleftrightarrow{PS}$ is any other line through P intersecting l (Fig. 9.3.10). We would like to demonstrate that $PQ < PS$. What are the possible alternatives? The law of trichotomy tells us that either $PQ = PS$, $PQ < PS$, or $PQ > PS$. To establish our conclusion we must eliminate alternatives 1 and 3. Suppose $PQ = PS$. Then we know that $m^\circ \angle PSQ = m^\circ \angle SQP$. Since $\overleftrightarrow{PQ} \perp l$, it follows that $m^\circ \angle SQP = 90$ and hence $m^\circ \angle PSQ = 90$. Thus, $\overleftrightarrow{PS} \perp l$ and we have more than one line through P perpendicular to l, which is impossible. Thus, we have eliminated

Figure 9.3.9

Figure 9.3.10

one alternative. To eliminate the possibility that $PQ > PS$, we observe that if $PQ > PS$, our last theorem tells us that $m^\circ \angle PQS < m^\circ \angle PSQ$. Since $\overleftrightarrow{PQ} \perp l$, we know $m^\circ \angle PQS = m^\circ \angle PQT$. Thus, $m^\circ \angle PQT < m^\circ \angle PSQ$. But $\angle PQT$ is an exterior angle and $\angle PSQ$ is remote and interior; so $m^\circ \angle PQT < m^\circ \angle S$ and $m^\circ \angle PQT > m^\circ \angle S$ at the same time. But this contradicts the law of trichotomy. Thus, PQ cannot be greater than PS. Since we have eliminated two of the three alternatives, we must conclude that $PQ < PS$.

Thus, we have the following theorem.

9.3.5 Theorem

The shortest possible distance from a point to a line is the length of the perpendicular from the point to the line.

Exercise 9.3.2

1. Give an indirect argument to support the statement that if a triangle has a right angle, then the other two angles are acute.
2. Consider a triangle with two angles of unequal measure. Can you make a conjecture about the opposite sides? What three alternatives does the trichotomy law give us concerning these sides? Can you eliminate two of these possibilities? Write your conjecture as a theorem and support it by indirect reasoning.

9.4 MORE ON INEQUALITIES

Experiment 2

Consider a triangle ABC. Can you make a conjecture about the relationship between the sum of the lengths of two sides of the triangle and the length of the third side? Can you make any generalization to other polygons? Draw several triangles in which the measure of one angle is successively greater but the adjacent sides remain unchanged in length. What happens to the third side?

Let us now examine the following problem alluded to in our experiment. Suppose A, B, and C are three points in a plane and are noncollinear. Our experimenting would lead us to believe that the distance from A to B plus the distance from B to C is greater than the distance from A to C. In symbols, we represent this statement by $AB + BC > AC$. (Recall that AB refers to the distance from A to B.)

As we have learned, three coplanar, noncollinear points A, B, and C form a triangle. Thus, another way of making our assertion is to claim that the sum of the measures of two sides of a triangle is greater than the measure of the third side (Fig. 9.4.1).

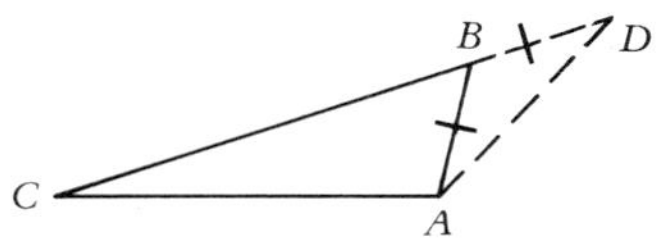

Figure 9.4.1

Let D be some point in the plane such that $D - B - C$ and $\overline{BD} \cong \overline{AB}$. Since B is in the interior of $\angle DAC$, it follows that

$$m^\circ \angle DAC > m^\circ \angle DAB$$

We constructed $\triangle ABD$ in such a way that it is an isosceles triangle with $\overline{BD} \cong \overline{AB}$. This means that

$$\angle DAB \cong \angle ADB$$

from which it follows that

$$m^\circ \angle DAB = m^\circ \angle ADB$$

Thus, we have

$$m^\circ \angle DAC > m^\circ \angle ADB$$

by substitution.

From the second problem of Exercise 9.3.2, it follows that

$$CD > AC$$

On the other hand,

$$CD = DB + BC = AB + BC \quad \text{(Why?)}$$

Thus,

$$AB + BC > AC$$

Thus, we have concluded the following.

9.4.1 Theorem

The sum of the measures of any two sides of a triangle is greater than the measure of the third side.

Exercise 9.4.1

1. Decide for which of the following sets of measures it would be possible to construct a triangle so that its sides have measures equal to those indicated. Justify your answers.
   a. 6 in., 7 in., 5 in.
   b. 13 ft, 1 ft, 2 ft
   c. 10 cm, 20 cm, 10 cm
   d. 2 yd, 3 ft, 5 ft

2. Try to discover a method for constructing a triangle using a compass and a straight edge, given two angles and a side in the triangle that is opposite one of the angles. Conclude that given a correspondence between two angles, if two angles and a side opposite one of them in one triangle are congruent respectively to the corresponding parts of the second triangle, then the correspondence is a congruence (AAS Congruence Theorem). Complete the following argument showing that the previous statement is true. Let $ABC \longleftrightarrow DEF$ be a correspondence between two triangles, such that $\angle BAC \cong \angle EDF$, $\angle ABC \cong \angle DEF$, and $\overline{AC} \cong \overline{DF}$ (Fig. 9.4.2).

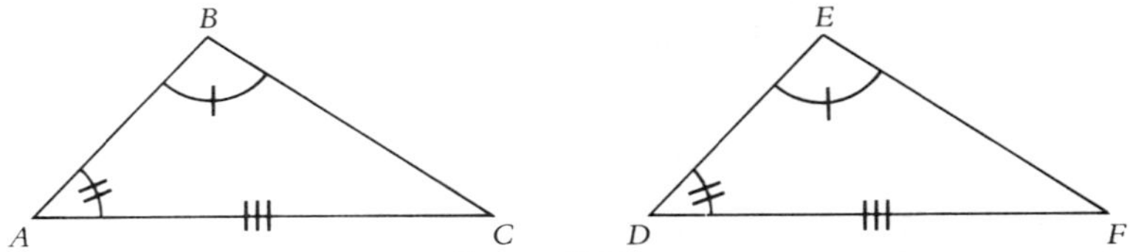

Figure 9.4.2

On $\overrightarrow{AB}$, let X be the point so that $\overline{AX} \cong \overline{DE}$ (Fig. 9.4.3). It will then follow that $\triangle AXC \cong \triangle DEF$ by the SAS principle of congruence. Now suppose that X is not the same point as B. Consider the cases $A - X - B$ and $A - B - X$ and compare $\angle AXC$ with $\angle ABC$ (Fig. 9.4.3).

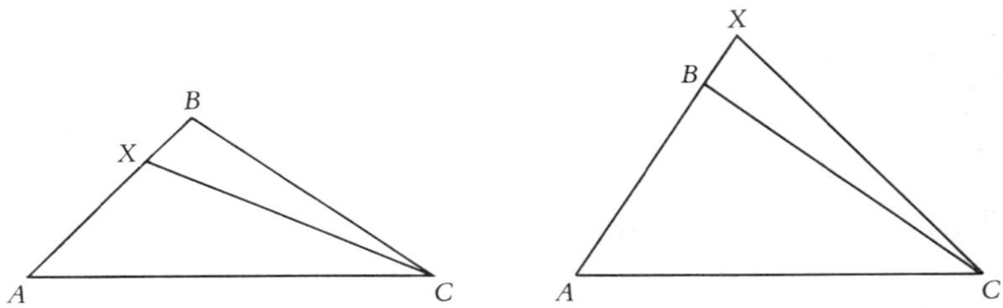

Figure 9.4.3

3. Try to discover a method for constructing a right triangle using a compass and a straight edge, given the hypotenuse and one other (leg) side of the triangle. Conclude that given a correspondence between two right triangles, if the hypotenuse and one other side of one triangle are congruent respectively to the corresponding parts of the second triangle, then the correspondence is a congruence (Hypotenuse–Leg Congruence Theorem). Complete the following argument showing that the previous statement is true. In $\triangle ABC$ and $\triangle DEF$, let $m^\circ \angle BCA = m^\circ \angle EDF = 90$. Let $CBA \longleftrightarrow DEF$ be a correspondence

such that $\overline{BA} \cong \overline{EF}$ and $\overline{CB} \cong \overline{DE}$ (Fig. 9.4.4). Choose a point Q on line $\overleftrightarrow{DF}$ such that $Q - D - F$ and $\overline{QD} \cong \overline{AC}$. Show that $\triangle DEQ \cong \triangle CBA$. From this it will follow that $\overline{DQ} \cong \overline{AC}$, and $\triangle \overline{EQF}$ is an isosceles triangle. Now establish that $\triangle DEF \cong \triangle DEQ$ by AAS principle of congruence. It will then follow that $\triangle DEF \cong \triangle CBA$. (Why?)

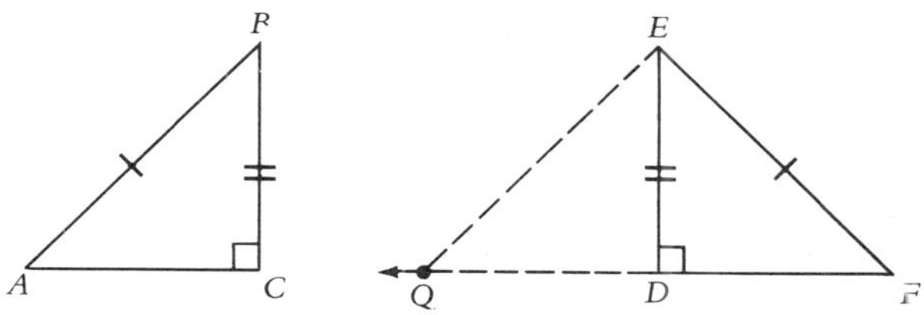

Figure 9.4.4

4. Make conjectures about each of the following questions.
   a. Of all plane figures with a given perimeter, which has the greatest area?
   b. Of all plane figures with a given area, which has the least perimeter?
   c. Of all triangles with a common base and perimeter, which has the greatest area?
   d. Of all triangles with a common base and area, which has the smallest perimeter?
   e. Of all triangles with a given perimeter, which has the greatest area?
   f. Of all triangles with a given area, which has the least perimeter?

9.5 SUMMARY

1. $AB < CD$ if and only if there exists an isometry σ such that $\sigma(A) = C$ and $C - \sigma(B) - D$.
2. $m^\circ \angle ABC < m^\circ \angle DEF$ if and only if there exists an isometry such that $\sigma(\overrightarrow{BA}) = \overrightarrow{ED}$ and $\sigma(\overrightarrow{BC})$ is between $\overrightarrow{ED}$ and $\overrightarrow{EF}$.
3. Given $\triangle ABC$ with a point D on $\overrightarrow{AC}$ such that $A - C - D$, $\angle DCB$ is called an exterior angle of $\triangle ABC$. $\angle ABC$ and $\angle BAC$ are called remote interior angles relative to $\angle BCD$.
4. The measure of an exterior angle of a triangle is larger than either remote interior angles.
5. The perpendicular to a line from a point not on this line is unique.
6. The sum of the measures of two sides of a triangle is greater than the measure of the third side.

7. If there exists a correspondence between two right triangles such that the hypotenuse and a leg of one triangle are respectively congruent to the corresponding parts of the second triangle, the correspondence is a congruence (hypotenuse–leg).

Exercise 9.5.1

1. Given $\triangle PQR$ with $PQ = 5$ in., $QR = 14$ in., and $PR = 11$ in., which angle is the largest? Which angle is the smallest?
2. Given $\triangle PQR$ with $m^\circ \angle QPR = 34^\circ$, $m^\circ \angle PQR = 76^\circ$, and $m^\circ \angle QRP = 70^\circ$, which side is the largest? Which side is the shortest?
3. Draw various quadrilaterals with both diagonals. What is the relationship between the sum of the lengths of the diagonals and the perimeter of the quadrilateral? Can you give an argument supporting your conjecture?
4. Let l be a line in a plane and P and Q two points on the same side of the line. Find the point R on l such that $PR + RQ$ will be as small as possible.
5. Draw various triangles along with the three altitudes. What is the relationship between the perimeter of the triangle and the sum of the three altitudes? Can you give an argument supporting your conjecture?

PARALLELISM

10

10.1 PROLOGUE

Although the intention of the authors of this book is to treat the study of geometry informally, it would not do to ignore one of the most interesting stories of mathematics. In the day of so-called "modern mathematics," it is well to recall the man who is perhaps most entitled to the name "Father of Modern Mathematics." This, of course, would be the ancient Greek mathematician Euclid.

Euclid was the first mathematician to our knowledge to offer a postulational development of mathematics. In his famous *Elements* (written between 330 and 320 B.C.), he organized into a deductive system all the geometry known in his day. He chose a small set of simple geometric facts which he called axioms or postulates and attempted to demonstrate all remaining geometric facts as a consequence of them.

The following statements are paraphrased versions of Euclid's five geometrical statements which were to be the basis of his geometry. We have chosen to state them in the language of this book for the convenience of the reader.

Let the following be postulated:

I. Two points determine a line segment.
II. Any line segment can be extended continuously to form a line.
III. Given any point and any line segment, there exists a circle with the point as the center and the line segment as the radius.
IV. All right angles are congruent.
V. If two straight lines in a plane are cut by a transversal making the sum of the measures of two interior angles on the same side of the transversal less than 180°, then the two straight lines will meet on that side of the transversal.

There were many flaws in Euclid's attempt to establish geometry as a deductive system. For instance, he seemed not to realize the need for undefined terms. But the most serious fault was that his postulates were insufficient to establish the geometry he intended. Many of his proofs were based on arguments not justified by his postulates.

We will not dwell on these various flaws, but instead direct our attention to the history of his fifth postulate, called the Parallel Postulate. This postulate was considered a flaw in his geometry for two thousand years when, in fact, it was evidence of his greatness as a modern mathematician. We notice that the first four postulates are rather simple statements while the Fifth Postulate is quite complicated. Since postulates were originally considered self-evident truths, the Fifth Postulate almost from the beginning was considered suspect. It was felt that such a statement should be proven, not assumed to be true.

Thus, it happened that for approximately the next two thousand years very competent mathematicians tried to prove that the Fifth Postulate was a consequence of the first four postulates.

Each attempt to prove the Parallel Postulate introduced assumptions that turned out to be equivalent to the postulate, thus assuming what was to be proven. Each of the following is an example of such equivalent assumptions.

1. Through a given point only one line can be drawn parallel to a given line.
2. There exists a pair of coplanar straight lines everywhere equally distant from one another.
3. If in a quadrilateral, three angles are right angles, the fourth angle is also a right angle.
4. There exists a triangle having the sum of the measure of its three angles equal to 180°.

In 1733, an Italian priest, Girolano Saccheri (1667–1733) published his great work, *Euclides Vindicates* (*Euclid Vindicated*). It was Saccheri's intention to erase all errors in Euclid and in particular to remove the blemish of the Parallel Postulate. His approach was as follows: Given a line segment $\overline{AB}$, construct segments $\overline{AC}$ and $\overline{BD}$ such that $\overline{AC} \cong \overline{BD}$ and $\overline{AC} \perp \overline{AB}$, $\overline{BD} \perp \overline{AB}$. Then join C and D by a line segment forming the figure known today as a *Saccheri quadrilateral* (Fig. 10.1.1).

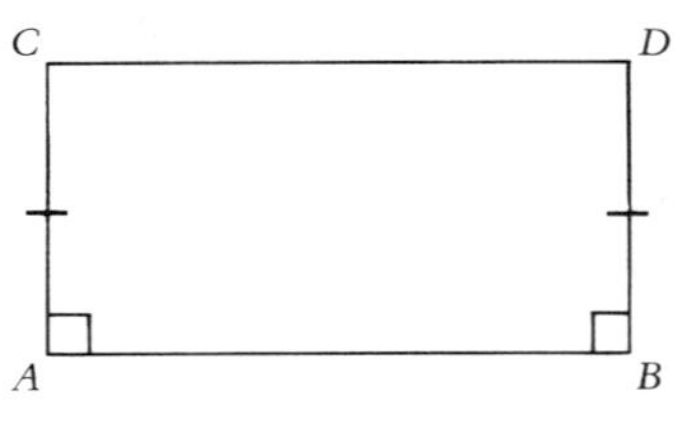

Figure 10.1.1

Without using the Parallel Postulate, Saccheri was able to prove that $\angle ACD \cong \angle BDC$. These two angles were called *summit angles* and Saccheri observed that one and only one of the following three statements was true.

1. *Right Angle Hypothesis:* The summit angles are right angles.
2. *Obtuse Angle Hypothesis:* The summit angles are obtuse angles.
3. *Acute Angle Hypothesis:* The summit angles are acute angles.

Saccheri was able to show that the Parallel Postulate was a consequence of the Right Angle Hypothesis. Thus, he believed that he could prove the Parallel Postulate by using the technique of indirect reasoning. His plan was to eliminate the Obtuse and Acute Angle Hypotheses by showing that they each led to a contradiction. He was able to show that the Obtuse Angle Hypothesis contradicted the infinite length of a line. It was Saccheri's intention to also reach a contradiction by assuming the Acute Angle Hypothesis which would thus establish the Parallel Postulate. Despite elaborate efforts and long work, Saccheri was unable to reach such a contradiction. In the process, Saccheri made many important discoveries and his contribution to mathematics would be remembered with greater respect if it were not that he somewhat lacked intellectual integrity. Being unable to find a contradiction, he reported that the Acute Angle Hypothesis was absolutely false essentially because he wanted it to be false. What Saccheri failed to realize was that there would be no logical contradiction and that, in fact, he had created a new geometry.

Probably the first man to realize that a geometry existed which denied Euclid's Parallel Postulate was Karl Friedrick Gauss (1777–1855) of Germany, certainly one of the greatest mathematicians of all time. In a letter to a friend in 1824, Gauss stated:

> The assumption that the sum of the three angles is less than 180° leads to a curious geometry, quite different from ours (the Euclidean), but thoroughly consistent, which I have developed to my entire satisfaction. . . . The theorems of this geometry appear to be paradoxical and to the uninitiated, absurd; but calm, steady reflection reveals that they contain nothing at all impossible.

Gauss insisted that his friend consider this a private communication and in no way to be made public. Gauss had plans to have his works published after his death. This might seem strange to us; here was surely a major discovery in mathematics. But we must consider the temper of the times. Most prominent mathematicians of the day, dominated by the philosophy of Immanuel Kant, had come to the conclusion that the mystery of the Fifth Postulate could never be solved. Kantian philosophy treated space not as empirical but as intuitive. Space was regarded as something always existing in the mind and not as a concept resulting from external experience. Gauss could have expected much criticism if he put forth the view that geometry becomes an experimental science once it is applied to physical space and that its postulates and their consequences need only be accepted if convenient and if they agree reasonably well with experimental data.

In 1832, the Hungarian mathematician, Wolfgang Bolyai (1775–1856), best remembered for producing his son Johann (1802–1860), published a two-volume treatise on geometry which included an appendix by his son describing Johann's investigations of the Parallel Postulate. The younger Bolyai chose to regard the Parallel Postulate in the form which is known as Playfair's Axiom and states: "Through a given point one and only one line can be drawn parallel to a given line." The denial of the Postulate then implies either that no parallel to the line can be drawn through the point or that more than one such parallel can be drawn. These two statements are respectively equivalent to Saccheri's Obtuse Angle Hypothesis and Acute Angle Hypothesis. Bolyai was able to reject the first statement because it contradicted the infinite extent of a line.

Assuming the second statement, Bolyai's geometry developed and no contradiction appeared and he began to become aware of the significance of what he was doing. He was impressed with the propositions which did not depend on any parallel postulate at all, but which were common to all geometries regardless of what assumptions were made about parallels. He regarded such propositions as the basis of an *absolute geometry*. In a letter to his father outlining his findings, he stated: "Out of nothing I have created a strange new universe."

The elder Bolyai, who had long worked on the Parallel Postulate himself, insisted that his son publish his results at once and so the Appendix appeared in 1832.

Young Bolyai was in for two great disappointments. In a letter from Gauss to Wolfgang Bolyai in March, 1832, Gauss announced:

> . . . the entire content of the work, the path which your son has taken, the results to which he is led, coincide almost exactly with my own meditations which have occupied my mind for from thirty to thirty-five years.

In 1848, Bolyai learned that Nikolai Lobachevsky (1793–1856) had also discovered the new geometry and even published his results in 1829. Lobachevsky, referred to as the "Copernicus of Geometry" became the first mathematician to publish a revolutionary geometry built on an assumption in direct conflict with the Parallel Postulate. His geometry was completely valid, but so contrary to common sense that he called it "imaginary geometry." Thus, even though there is strong evidence that Bolyai actually preceded Lobachevsky in his discoveries, he was forced to share with two men the glory of the invention of what Gauss called *non-Euclidean geometry*.

The next stage in this history of a century of revolution in geometry occurred in 1854 at the University of Göttingen when C. F. B. Riemann (1826–1866) delivered a lecture "on the Hypotheses which lie at the Foundations of Geometry." He urged a global view of geometry as a study of any kind of space of any number of dimensions. Riemann's view of geometry was more non-Euclidean than that of Lobachevsky, feeling that geometry need not even necessarily deal with points or lines or space in the ordinary sense, but rather with sets of ordered n-tuples that are combined according to certain rules.

The next important development in our story came in 1868 when the Italian Eugenio Beltrami (1835–1900) published a paper establishing that this new geometry was as consistent as Euclidean geometry. A postulate system is said to be *consistent* if it is free of contradiction; that is, it can never happen that a statement and its negation are both true. Beltrami was able to develop a model of the new geometry within the framework of Euclidean 3-space. Thus, if there was any contradiction in the new geometry, it would also be a contradiction in Euclidean geometry. Beltrami, Gauss, Bolyai, and Lobachevsky had thus succeeded where Saccheri had failed; they had vindicated Euclid.

The invention of the new geometries indicated that Euclid was correct in suggesting the Parallel Postulate. It could not be proven as a consequence of his other postulates.

Gauss had applied the name non-Euclidean geometry to the geometry based on the negation of the Parallel Postulate and Euclid's other postulates; that is, the geometry of more than one parallel to a given line or Saccheri's Acute Angle Hypothesis. Another new geometry which has come also to be called a non-Euclidean geometry has been invented where the infinite extent of straight lines is not assumed. In this case, there is a geometry in which no parallels exist (the Obtuse Angle Hypothesis).

The two non-Euclidean geometries and Euclid's geometry can be characterized also as follows:

1. *Euclidean geometry* (exactly one parallel): The sum of the measures of the angles of a triangle is 180°.
2. *Lobachevskian geometry* (more than one parallel): The sum of the measures of the angles of a triangle is less than 180°.
3. *Riemannian geometry* (no parallels): The sum of the measures of the angles of a triangle is more than 180°.

David Hilbert (1862–1943), the great German mathematician, described the invention of non-Euclidean geometry as "the most suggestive and notable achievement of the last (19th) century." And, remember this was the century of Darwin's Theory of

Evolution. Geometry, until the works of Gauss, Bolyai, and Lobachevsky, had been considered a physical science, a description of the spatial relations of our world. Euclid's postulates were considered to be obvious truths, the resulting theorems factual statements about the physical world. The invention of non-Euclidean geometries shows rather that the choice of postulates created a world which could be studied, but that this world need not be the physical world.

The invention of non-Euclidean geometries and their consistencies affected the basic notion of truth in mathematics. What was once considered to be a source of absolute truth had finally been recognized as a creation of man much as the work of a novelist. It becomes meaningless to ask, "Which is the true geometry?" in the study of mathematics. Instead, we find it more convenient to consider which geometry is most suitable for a particular application. It can be shown that each of the three geometries fits some small part of the physical world equally well. Each geometry in certain ways approximates the portion of the world with which we are familiar. Thus, we cannot say our world is Euclidean or that it is non-Euclidean.

Since the invention of these two non-Euclidean geometries, other geometries have been invented. Any geometry with postulates not equivalent to Euclid's postulates has come to be called a non-Euclidean geometry. The mathematician Bernhard Riemann invented a whole class of non-Euclidean geometries which were used by Einstein in developing his general theory of relativity.

In 1872, Felix Klein (1849–1925), in his inaugural address when he became professor at Erlangen, described geometry as the study of those properties of figures that remain unchanged under a particular group of transformations. For instance, plane Euclidean geometry is the study of such properties of figures including areas and lengths which are unchanged under the group of transformations made up of reflections, translations, and rotations. These so-called rigid transformations are equivalent to Euclid's unstated axiom that figures remain unchanged when moved about in a plane.

Since that time, the branch of mathematics called *topology* has been developed which is the most general investigation of spaces. It can be truthfully said that the invention of non-Euclidean geometry spurred the greatest progress in mathematics perhaps of all time. It began what must be described as a new Golden Age of Mathematics which continues to this day. To illustrate this, it has been observed that more mathematics has been invented and discovered in the past twenty-five years than in all time preceding. This continuous stimulus was supplied by the genius of men such as Gauss, Bolyai, Lobachevsky, Riemann, and Klein.

in our opinion . . .

It has been argued that it is not appropriate to discuss non-Euclidean geometries with prospective elementary school teachers because, first, it is too difficult a subject and, second, they would never have cause to use this knowledge. We believe that both these arguments are invalid and quite the contrary, it is both essential and appropriate that they be exposed to this topic. The depth to which this topic is pursued in this book is certainly not beyond the grasp of our readers—what it requires, rather than a super intellect, is an open and imaginative mind˙and if a prospective teacher does not possess these qualities, perhaps he does not belong in the classroom teaching youngsters who certainly have these qualities. Second, we believe that it is quite likely, indeed, that a bright, young, inquisitive student might dare to suggest that all of the experiments and reasoning concerning the sum of the angles of a triangle are not convincing to him. Another student may well suggest that parallel lines do not behave as nicely as the textbook suggests. How can the teacher parry these queries? Perhaps the teacher who is ignorant of the existence of non-Euclidean geometries might suggest to the student that he is being ridiculous and illogical. On the other hand, the teacher who is at least aware of the profundity of the student's remarks can respond to the student in a much more positive and encouraging manner.

Finally, we believe that the whole historical controversy surrounding this topic provides a very interesting and exciting insight into the lives and works of prominent mathematicians.

Experiment 1

We can investigate what is known as finite Bolyai-Lobachevskian geometries. As usual, we will not define the notions of point, line, and space nor will we make any attempt to make precise the relations: lies on, passes through, and so on. We will define two lines to be parallel if they do not intersect.

Then, our geometry consists of a space whose elements are points and certain subsets of points are lines subject to the following axioms.

Axioms

1. For any two points of the space, there is exactly one line containing the two points.
2. For any line in the space, there is a point that does not lie on the line.
3. There are at least two points on every line.

4. There exists at least one line.
5. Through a point not on a line, there are exactly two lines that pass through the point and are parallel to the line.

Investigate this geometry, supposing the additional axiom: There exists at least one line of the space that contains exactly three points. Determine the number of points on each line, the number of lines that pass through each point, the number of points in the space, and the number of lines in the space.

What effect would there be if Axiom 5 were replaced by the following?

5′. Through a point not on a line, there are exactly three lines that pass through the point and are parallel to the line.

10.2 ABSOLUTE GEOMETRY

In this book, we will use the following version of the Parallel Postulate:

10.2.1 Postulate

(Parallel Postulate) Through a given external point there is at most one line parallel to a given line.

Before considering the consequence of the Parallel Postulate, we will examine some features of what is often described as absolute or neutral plane geometry, that is, Euclidean geometry without the Parallel Postulate. We intend to indicate how far we can go in our study without involving the Parallel Postulate.

Remember that two lines are parallel if they are coplanar and do not intersect. Each of the following statements can be deduced without making use of the Parallel Postulate.

1. Two lines in a plane are parallel if they are both perpendicular to the same line.
2. Let l be a line and P a point not on l. Then there is at least one line through P parallel to l. (Note this is not the Parallel Postulate.)
3. Given two lines l and m in a plane cut by a transversal t, if a pair of alternate interior angles are congruent, then l and m are parallel.
4. Given two lines l and m in a plane cut by a transversal t, if a pair of corresponding angles are congruent, then the lines are parallel.
5. The diagonals of a Saccheri quadrilateral are always congruent.
6. Let $ABCD$ and $PQRS$ be Saccheri quadrilaterals with lower

bases $\overline{AD}$ and $\overline{PS}$. If $\overline{AD} \cong \overline{PS}$ and $\overline{AB} \cong \overline{PQ}$, then $\overline{BC} \cong \overline{QR}$, $\angle B \cong \angle Q$, and $\angle C \cong \angle R$.

7. In any Saccheri quadrilateral, the upper base angles are congruent.
8. In any Saccheri quadrilateral, $ABCD$ with lower base $\overline{AD}$, $m° \angle BDC \geqq m° \angle ABD$.
9. If $\triangle ABC$ has a right angle at A, then $m° \angle B + m° \angle C \leqq 90$.
10. Every right triangle has only one right angle and its other two angles are acute.
11. The hypotenuse of a right triangle is of greater length than either of the remaining sides.
12. In any triangle ABC, let $\overline{BD}$ be the perpendicular from B to $\overleftrightarrow{AC}$. If $\overline{AC}$ is a longest side of $\triangle ABC$, then D is between A and C.
13. In any triangle ABC, $m° \angle A + m° \angle B + m° \angle C \leqq 180$.

We will not offer arguments for each of these statements, instead we will select a few and offer an outline of the arguments used to deduce them. The reader should try and complete these arguments and consider the remaining statements as exercises requiring proofs.

10.2.2 Theorem

Two lines in a plane are parallel if they are both perpendicular to the same line.

Outline of proof. Let l, m, and t be three lines in a plane with $l \perp t$ and $m \perp t$. We wish to conclude that $l \| m$. We argue indirectly supposing that l and m are not parallel but instead intersect in a point (Fig. 10.2.1). But in this assumption it appears that from a point P not on a line t there are two perpendiculars. This then is a contradiction.

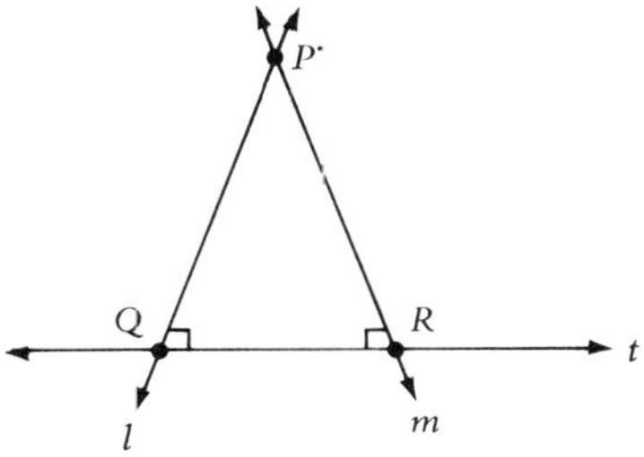

Figure 10.2.1

10.2.3 Theorem

Given two lines l and m in a plane cut by a transversal t, if a pair of alternate interior angles are congruent, then l and m are parallel.

Outline of proof. Again, we argue indirectly supposing the lines l and m are not parallel and in fact intersect in some point P (Fig. 10.2.2). By hypothesis, we have that $\angle PQR \cong \angle SRQ$, but

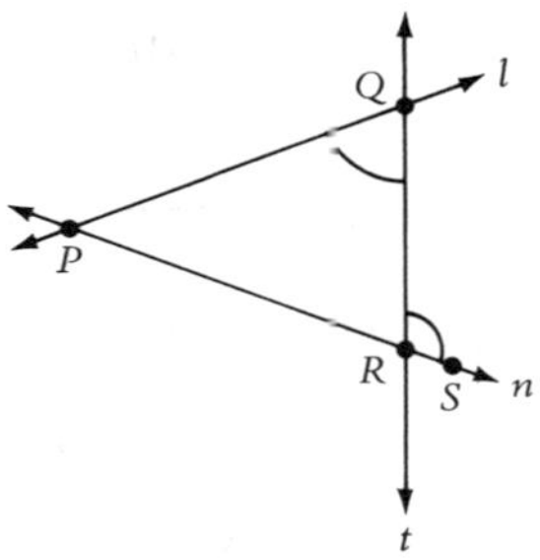

Figure 10.2.2

$\angle SRQ$ is an exterior angle of $\triangle PQR$ and as such must be greater in measure than each remote interior angle. Thus, again, we have a contradiction.

10.2.4 Theorem

The diagonals of a Saccheri quadrilateral are always congruent.

Outline of proof. Let $ABCD$ be a Saccheri quadrilateral with lower base $\overline{AD}$ (Fig. 10.2.3). By recalling the definition of a Saccheri quadrilateral, it can be shown that $\triangle BAD \cong \triangle CDA$ using the SAS Congruence Theorem. Thus, the result follows.

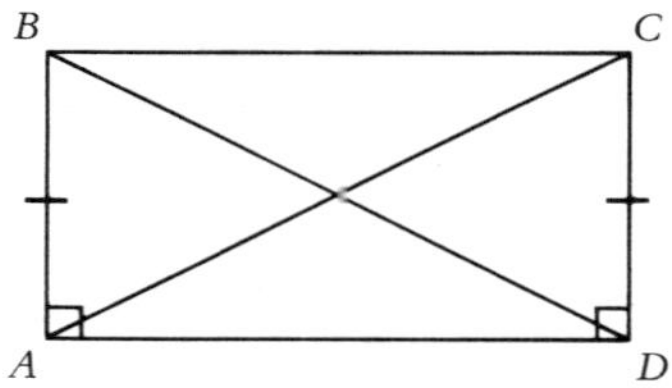

Figure 10.2.3

10.2.5 Theorem

In any triangle ABC, $m^\circ \angle A + m^\circ \angle B + m^\circ \angle C \leqq 180$.

Outline of proof. Some side of the triangle must be a longest side, so we will assume that it is $\overline{AC}$. Let $\overline{BD}$ be the perpendicular to $\overleftrightarrow{AC}$. By statement 12, we know that D lies between A and C (Fig. 10.2.4).

Thus, $\triangle ADB$ and $\triangle BDC$ are right triangles and by statement 9, we may conclude that

$$m^\circ \angle A + m^\circ \angle ABD \leqq 90$$
$$m^\circ \angle C + m^\circ \angle CBD \leqq 90$$

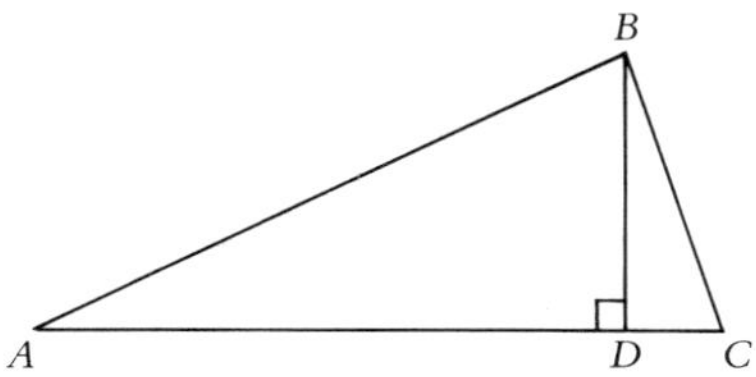

Figure 10.2.4

Adding and noting, since D is in the interior of $\angle ABC$, that $m^\circ \angle B = m^\circ \angle ABD + m^\circ \angle CBD$, we have the conclusion of the statement.

Exercise 10.2.1

Offer arguments to support the assertions of the remaining statements of this section.

10.3 CONSEQUENCES OF THE PARALLEL POSTULATE

We will use the remainder of this chapter to reinforce the understanding and dependency of certain topics introduced earlier. Henceforth, we will be assuming the Parallel Postulate and will use it in deriving new information. In statement 3, we were asked to make an argument showing that if two lines are cut by a transversal and a pair of alternate interior angles are congruent, then the two lines are parallel. As we saw, it was possible to offer an argument in support of this statement without appealing to the Parallel Postulate. Interestingly enough, we need the Parallel Postulate to derive the converse of this statement.

10.3.1 Theorem

If two parallel lines are cut by a transversal, then alternate interior angles are congruent.

Let us consider the following argument which justifies this statement: Let l and m be two parallel lines cut by a transversal t in points P and Q, respectively (Fig. 10.3.1).

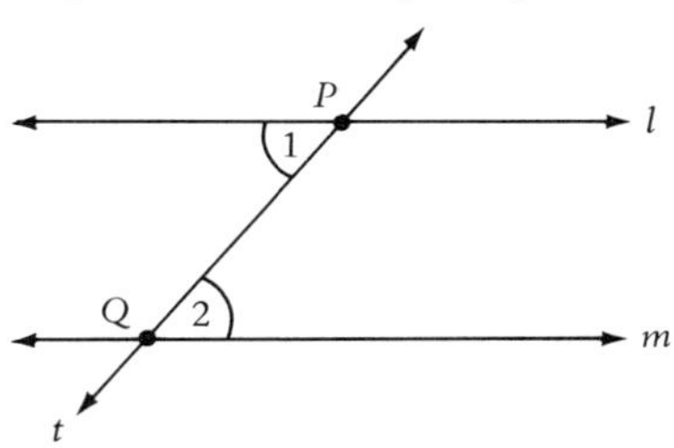

Figure 10.3.1

We will attempt to show that angles denoted in Fig. 10.3.1 by 1 and 2 are congruent. Our argument will be indirect. It is certainly true that these two angles are congruent or they are not congruent. We will suppose that they are not congruent and show that this assumption leads to a contradiction. Thus, the other alternative must be true; namely, the two angles are congruent.

If $\angle 1$ and $\angle 2$ are not congruent, it is possible to construct a line through P, say n, for which alternate interior angles are congruent (Fig. 10.3.2). That is, $\angle 3 \cong \angle 2$. Note since $\angle 1$ and $\angle 3$ are not congruent, lines l and n are distinct. In Theorem 10.2.3, we saw that if two lines are cut by a transversal making a pair of alternate interior angles congruent, then the lines are parallel. Thus, $m \| n$. But $m \| l$ and we have two lines through point P which are parallel to line m. This contradicts the Parallel Postulate. Thus, our assumption that $\angle 1$ and $\angle 2$ are not congruent has led to a contradiction. Hence, $\angle 1$ and $\angle 2$ are congruent.

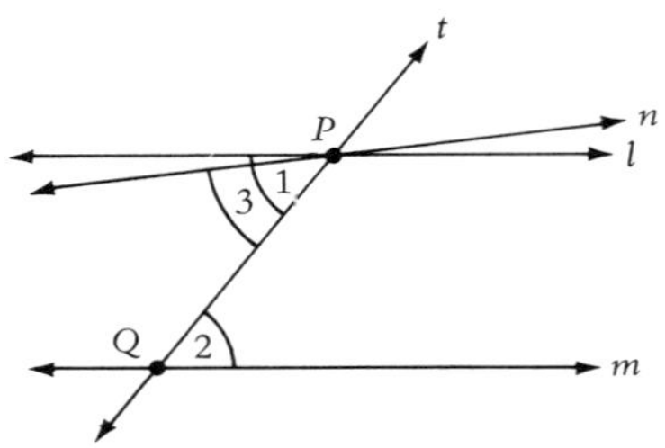

Figure 10.3.2

Exercise 10.3.1

1. Give an argument showing that if two parallel lines are cut by a transversal, then each pair of corresponding angles is congruent.
2. Give an argument showing that if two parallel lines are cut by a transversal, then interior angles on the same side of the transversal are supplementary.
3. Give an argument showing that in a plane, if two distinct lines are parallel to the same line, they are parallel to each other.
4. Give an argument showing that in a plane, if a line is perpendicular to one of two parallel lines, it is perpendicular to the other.
5. In Chapter 6, we gave an argument that the sum of the measures of the angles of a triangle is 180. Examine this argument and determine how we made use of the Parallel Postulate. The use was subtle and not explicitly referred to but its use was implied.
6. Give an argument showing that if a correspondence exists

between two triangles such that a pair of corresponding angles are congruent, then the remaining pair of corresponding angles are also congruent.

7. Give an argument showing that either diagonal of a parallelogram separates the parallelogram into two congruent triangles.
8. Give an argument showing that in a parallelogram, any two opposite sides are congruent.
9. Give an argument showing that in a parallelogram, any two opposite angles are congruent.
10. Give an argument showing that in a parallelogram, any two consecutive angles are supplementary.
11. Give an argument showing that the diagonals of a parallelogram bisect each other.
12. Give an argument showing that if both pairs of opposite sides of a quadrilateral are congruent, the quadrilateral is a parallelogram.
13. Give an argument showing that if two sides of a quadrilateral are parallel and congruent, the quadrilateral is a parallelogram.
14. Give an argument showing that if the diagonals of a quadrilateral bisect each other, then the quadrilateral is a parallelogram.

Consider any triangle ABC such as the one pictured in Fig. 10.3.3. Let D and E be the midpoints of $\overline{AB}$ and $\overline{BC}$, respectively. Form the line segment $\overline{DE}$. Is there any conjecture that can be made about the relationship between $\overline{DE}$ and $\overline{AC}$? Most people would be inclined to predict that $\overline{DE} \| \overline{AC}$ but there is more to it than that. Let F be the point of the ray $\overrightarrow{DE}$ such that $D - E - F$ and $DE = EF$. Construct segment $\overline{FC}$. It is easy to see that $\triangle BDE \cong \triangle CFE$ by SAS Congruence. (Which sides are congruent? Which pair of angles is congruent?) Because of this we see that $\angle ECF \cong \angle EBD$. (Why?) From this it follows that $AB \| CF$. (Why?) Since D is the midpoint $\overline{AB}$ it follows that $AD = DB$ and thus $AD = FC$. (Why?) This means that $ADFC$ is a parallelogram. (Why?) Finally, we can conclude that $DE = \frac{1}{2}AC$. (Why?)

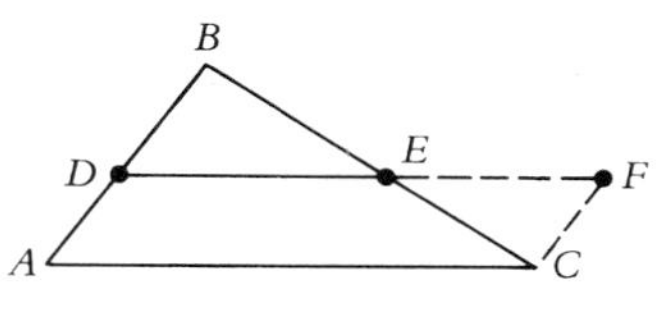

Figure 10.3.3

Summarizing, we have the following.

10.3.2 Theorem

The segment joining the midpoints of two sides of a triangle is parallel to the third side and has length half as long as the third side.

15. Give an argument showing that if a parallelogram has one right angle, then it has four right angles and the parallelogram is a rectangle.
16. Give an argument showing that in a rhombus, the diagonals are perpendicular to one another.
17. Give an argument showing that if the diagonals of a quadrilateral bisect each other and are perpendicular, then the quadrilateral is a rhombus.

Thus far, we have limited our consideration to two parallels cut by a transversal. We gain some useful information by examining the consequence of several lines in a plane being cut by a transversal. For instance, suppose we have three parallel lines l, m, and n in a plane cut by a transversal t (Fig. 10.3.4). The points of intersection of t with the lines l, m, and n will be denoted respectively by P, Q, and R. Suppose that it happens that $\overline{PQ} \cong \overline{QR}$; we say that the three lines intercept congruent segments on the transversal. Suppose now that these same three lines l, m, and n are cut by another transversal in points X, Y, and Z, respectively. What can we say about the segments $\overline{XY}$ and $\overline{YZ}$? They appear to be congruent.

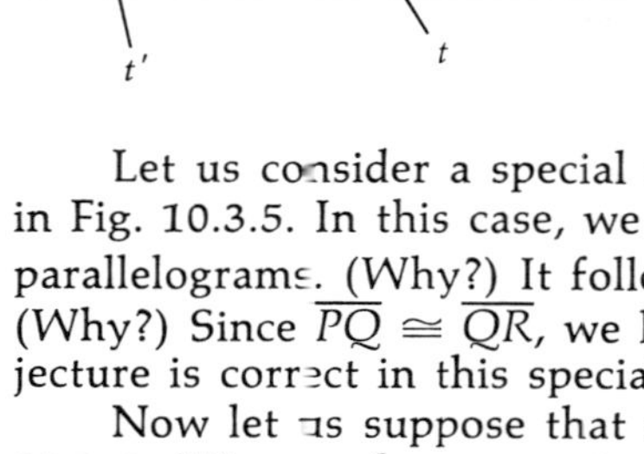

Figure 10.3.4

Let us consider a special case to get started. Suppose $t' \| t$ as in Fig. 10.3.5. In this case, we notice that $XPQY$ and $YQRZ$ are parallelograms. (Why?) It follows that $\overline{XY} \cong \overline{PQ}$ and $\overline{YZ} \cong \overline{QR}$. (Why?) Since $\overline{PQ} \cong \overline{QR}$, we have $\overline{XY} \cong \overline{YZ}$. Thus, our conjecture is correct in this special case.

Now let us suppose that t' is not parallel to t as in Fig. 10.3.6. (We are also supposing that $P \neq X$.) In our special case, we were able to conclude that $\overline{XY} \cong \overline{PQ}$ and $\overline{YZ} \cong \overline{QR}$ because they were pairs of opposite sides of parallelograms. In the case now under consideration, $XPQY$ and $YQRZ$ are not parallelo-

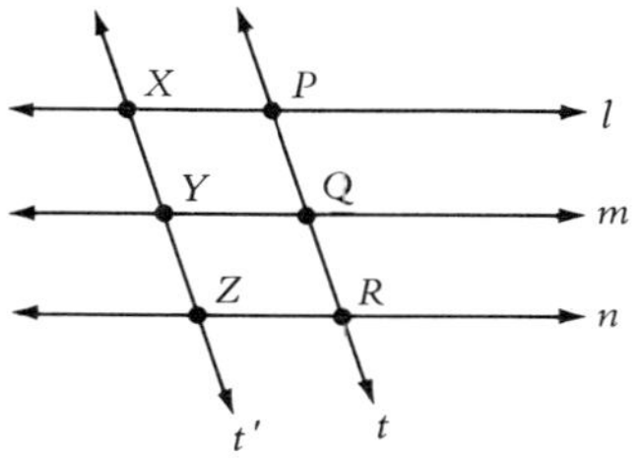

Figure 10.3.5

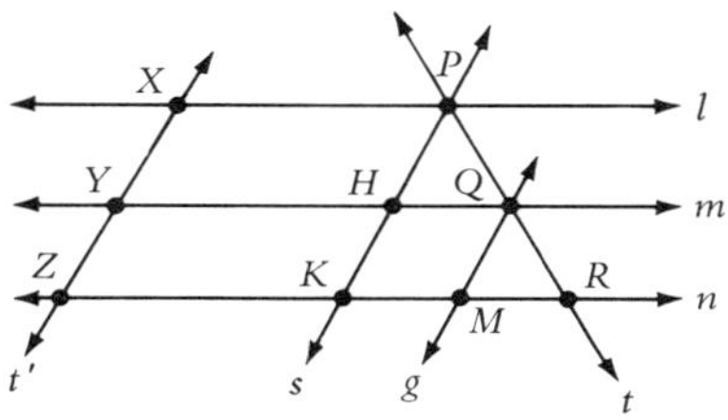

Figure 10.3.6

grams, so we need different evidence. The notion of a parallelogram will help, however, as seen by constructing a new transversal s through P such that $s \| t'$, cutting m and n in points H and K, respectively. Now we can conclude that $\overline{XY} \cong \overline{PH}$ and $\overline{YZ} \cong \overline{HK}$. Suppose next we let g be the line passing through Q such that $g \| t'$ intersecting line n in point M. The triangles PHQ and QMR can be shown to be congruent by ASA. (Explain why.) Since $\overline{PH}$ and $\overline{QM}$ are corresponding parts of congruent triangles, they are congruent. It is also true that $\overline{HK} \cong \overline{QM}$. (Why?) Thus, $\overline{PH} \cong \overline{HK}$ and it follows that $\overline{XY} \cong \overline{YZ}$. Thus, we have justified the following statement.

10.3.3 Theorem

If three parallel lines intercept congruent segments on one transversal, then they intercept congruent segments on any other transversal.

This last statement can be generalized to more than three parallel lines.

10.3.4 Theorem

If three or more parallel lines intercept congruent segments on one transversal, then they intercept congruent segments on any other transversal.

See if you can discover an argument that will justify this statement.

10.4 PARALLEL PLANES

Planes as well as lines can have the property of being parallel. We will say that two planes in space are parallel if they do not intersect. If planes $\mathcal{P}$ and $\mathcal{S}$ are parallel, we will symbolize this by $\mathcal{P} \| \mathcal{S}$. If l is a line which is parallel to a plane $\mathcal{P}$ (that is, it does not intersect $\mathcal{P}$) we denote this fact by $l \| \mathcal{P}$.

Example 1
See Figs. 10.4.1 and 10.4.2.

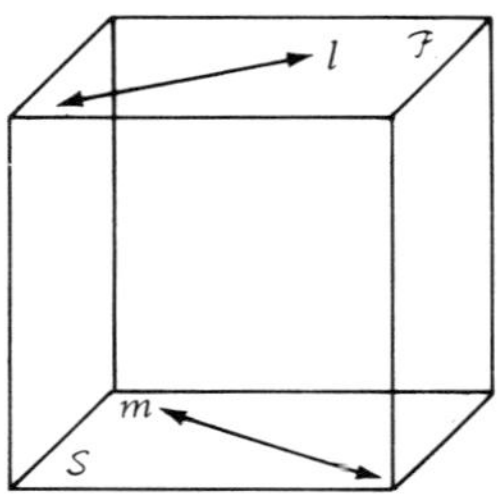

Figure 10.4.1
$\mathcal{P} \| \mathcal{S}$
Note l is not parallel to m (Why?)

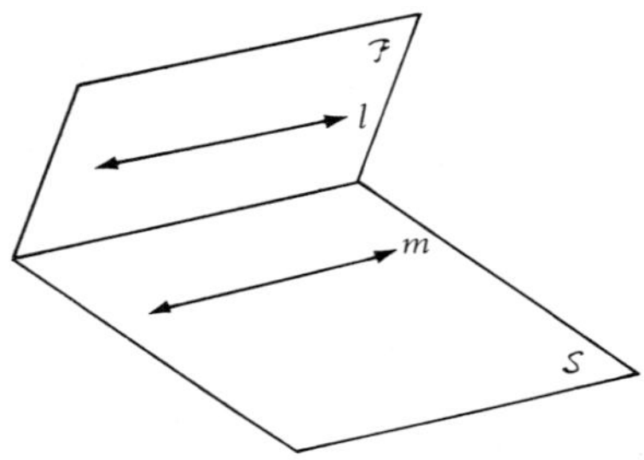

Figure 10.4.2
$\mathcal{P}$ is not parallel to $\mathcal{S}$
$l \| m$ (Why?)

Exercise 10.4.1
In each of the following, draw a picture which illustrates the statement.

1. If a plane intersects two parallel planes, then it intersects them in two parallel lines.
2. If a line is perpendicular to one of two parallel planes, it is perpendicular to the other.
3. Two planes perpendicular to the same line are parallel.
4. If two planes are each parallel to a third plane, they are parallel to each other.
5. Two lines perpendicular to the same plane are parallel.
6. A plane perpendicular to one of two parallel lines is perpendicular to the other.
7. Two parallel planes are everywhere equidistant.

Experiment 2

We will recall that when two lines intersect they form four angles. There is an analogous situation when two planes intersect such as in Fig. 10.4.3. They also seem to form four angles. We call such configurations dihedral angles. The definition of dihedral angles is analogous to our definition of angle which was the union of two noncollinear rays which have a common endpoint.

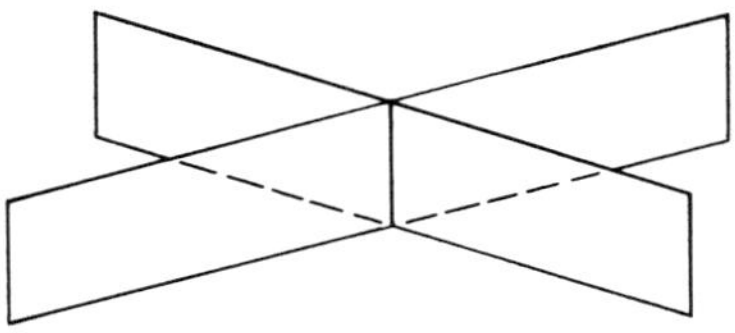

Figure 10.4.3

10.4.1 Definition

A dihedral angle *is the union of a line and two noncoplanar half-planes having the line as their separation. The line is called the* edge *of the dihedral angle. The union of the edge and either half-plane is called a* face *or* side *of the dihedral angle.*

Example 2

See Fig. 10.4.4. $\overleftrightarrow{XY}$ is the edge, $\overleftrightarrow{PXY}$ and $\overleftrightarrow{QXY}$ are the faces or sides. This angle will be symbolized by $\angle P - \overleftrightarrow{XY} - Q$.

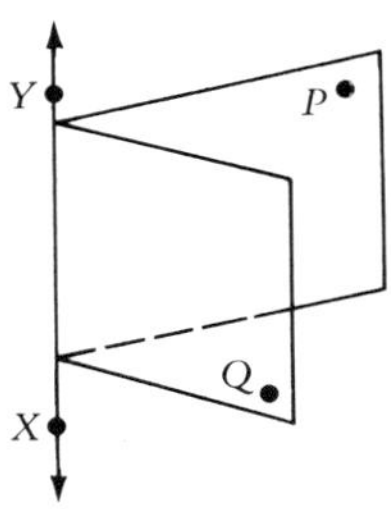

Figure 10.4.4

Exercise 10.4.2

1. Define vertical dihedral angles analogous to vertical angles.
2. Let two planes be cut by a third plane; draw diagrams illustrating interior, exterior, corresponding, alternate interior, and alternate exterior dihedral angles analogously to interior, exterior, corresponding, alternate interior, and alternate exterior angles formed by two lines cut by a transversal, in a plane.
3. Choose any point on the edge of a dihedral angle and pass a plane perpendicular to the edge through this point, intersecting each of the sides in a ray. The resulting angle formed by these rays is called a plane angle of the dihedral angle. (As an example, see Fig. 10.4.5.) Give an argument showing that any two plane angles of a given dihedral angle are congruent.
4. We can now define the *measure* of a dihedral angle to be the

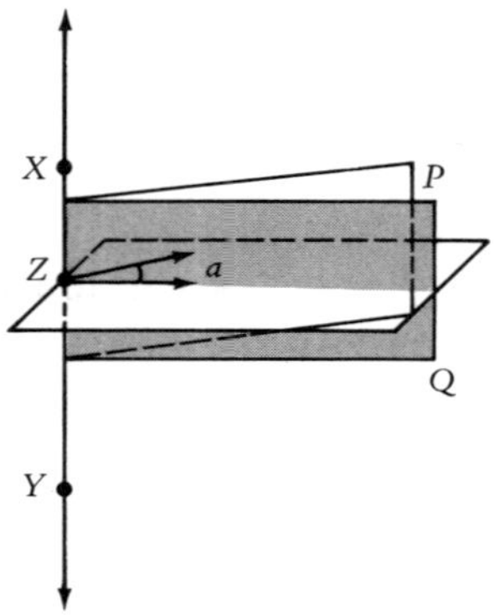

Figure 10.4.5
$\angle a$ is a plane angle of the dihedral angle $\angle P - \overleftrightarrow{XY} - Q$

real number which is the measure of any of its plane angles. A dihedral angle is a right dihedral angle if its plane angles are right angles. Two planes are perpendicular if they form right dihedral angles. Give an argument showing that if a line is perpendicular to a plane, then any plane containing this line is perpendicular to the given plane. Give an argument showing that if two planes are perpendicular, then any line in one of them perpendicular to their line of intersection is perpendicular to the other plane.

10.5 SUMMARY

1. Parallel Postulate: Through a given external point there is at most one line parallel to a given line.
2. If two parallel lines are cut by a transversal, then alternate interior angles are congruent.
3. The segment joining the midpoints of two sides of a triangle is parallel to the third side and has length half as long as the third side.
4. If three or more parallel lines intercept congruent segments on one transversal, then they intercept congruent segments on any other transversal.
5. Two planes in space are parallel if they do not intersect.
6. A dihedral angle is the union of a line and two noncoplanar half-planes having this line as their edge.

Exercise 10.5.1

1. The *median* of a trapezoid is the segment joining the midpoints of its nonparallel sides (Fig. 10.5.1). Give an argument showing that the median of a trapezoid is parallel to the bases and equal in length to half the sum of the lengths of the bases.

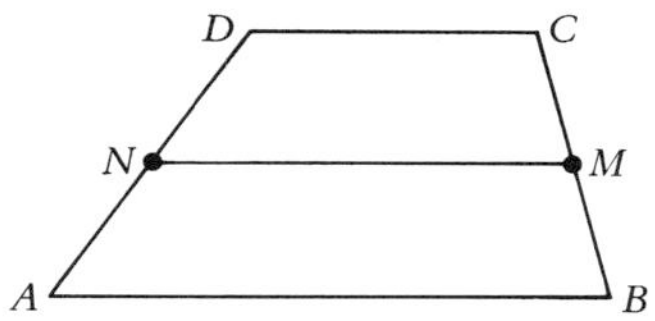

Figure 10.5.1

2. Given $ABCD$ is a square and the points P, Q, R, S dividing the sides as shown in Fig. 10.5.2. Here a and b represent the lengths of the indicated segments. Give an argument showing that $PQRS$ is a square (Fig. 10.5.2).

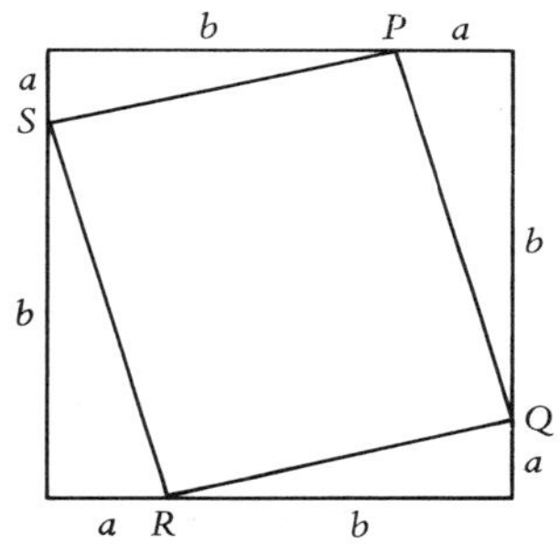

Figure 10.5.2

3. Give an argument showing that the medians of a triangle are concurrent in a point two-thirds the way from any vertex to the midpoint of the opposite side.

SIMILAR FIGURES AND INDIRECT MEASUREMENT

11

11.1 PROLOGUE

In studying plane geometric figures, we have considered the relationship of congruence. We have seen that, roughly speaking, two geometric figures are congruent if they have the same size and shape. One consequence of two geometric figures being congruent is that they share certain measurement properties. For instance, corresponding sides are of equal length and corresponding angles have equal measure. In addition, two congruent figures which have the feature of area will have equal area.

In this book, we consider various measuring processes. In each case, the goal of the measuring process is to assign a number to some geometrical or physical quantity. This assignment can be done directly or indirectly. In making a direct measurement, there is an actual comparison of the object to be measured with some standard unit of measurement.

Such a comparison is not always possible in a direct way. For example, in measuring time, temperature, velocity, weight, and even certain distances, the measure of each of these quantities is actually obtained by measuring some other quantity. We use instruments such as clocks, thermometers, speedometers, and spring scales to determine the time, temperature, speed, and weight indirectly. These measurements are made possible due to mankind's increased sophistication in combining mathematical, scientific, and mechanical knowledge.

Other quantities are measured indirectly using mathematical formulas to translate knowledge obtained by direct measurement into a number for the quantity. As an example, to determine the area of a rectangular plot of land, we determine its length and width and because mathematicians have devised a formula, this will allow us to determine the area.

Sometimes indirect measurement must be used to determine heights such as that of a mountain that are inaccessible to direct measurement. Such measurements are related to a branch of mathematics known as trigonometry. This subject began as an accumulation of mathematical formulas which allowed man to make various indirect measurements. The justifications for the various formulas were theorems of Euclidean geometry, usually those that dealt with the concept of similar figures.

Intuitively, two plane geometric figures are said to be similar if they have exactly the same shape but not necessarily the same size. This being the case, we would properly conclude that congruent figures are also similar figures, but it is not necessarily the case that similar figures are congruent figures.

As examples of similar figures, we would have all equilateral triangles, all squares, all circles, and all line segments (Fig. 11.1.1).

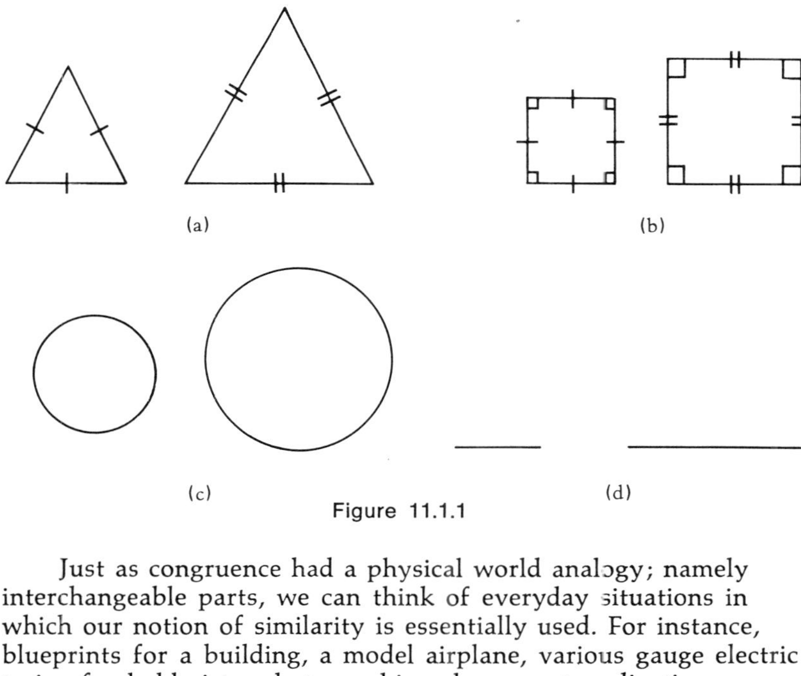

(a) (b) (c) (d)

Figure 11.1.1

Just as congruence had a physical world analogy; namely interchangeable parts, we can think of everyday situations in which our notion of similarity is essentially used. For instance, blueprints for a building, a model airplane, various gauge electric trains for hobbyists, photographic enlargements, adjusting a recipe for a cake meant to serve twelve to one which will serve one hundred, map making, and so on. Try to think of other situations in which the idea of similarity would be involved in the physical world.

Exercise 11.1.1

1. Would all isosceles triangles be similar? Explain.
2. Would all rectangles be similar? Explain.

3. Would all parallelograms be similar? Explain.
4. In space, would all tetrahedra be similar? all spheres? all cubes? Explain.
5. In space, would all pyramids be similar? all cylinders? Explain.

Since congruent geometric figures are also similar, our study of transformations and their invariants is again pertinent. At this point, we discard the demand that our transformations preserve size but insist that shapes be preserved. The set of transformations in which we are interested is sometimes called similarity transformations. We shall see that such transformations increase or decrease all lengths in the same ratio but leave shapes unchanged. For this reason, we will first wish to consider size transformations.

11.2 SIZE TRANSFORMATIONS

Let P be a point in a plane which does not lie on the two distinct lines l and m where $l \| m$. Further, P is in the half-plane determined by l which does not contain m (Fig. 11.2.1).

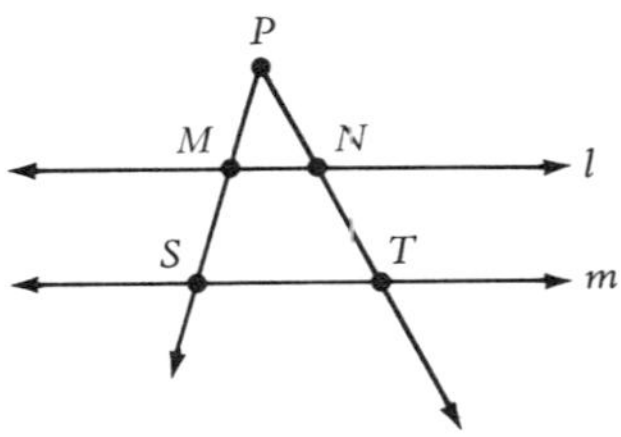

Figure 11.2.1

Draw two rays from P intersecting lines l and m in the points M, N, S, and T as in Fig. 11.2.1. Note that we have two triangles, namely $\triangle PMN$ and $\triangle PST$, and that $\angle PMN$ and $\angle PST$ are corresponding angles as are $\angle PNM$ and $\angle PTS$. Since $l \| m$, these pairs of corresponding angles are respectively congruent. On the other hand, $\angle SPT \cong \angle MPN$ as well. It is clear that these two triangles will not have the same size, but we maintain that they have the same shape and should be considered similar. We can think of this as a method of enlargement. The line segment $\overline{MN}$ has been enlarged, becoming the line segment $\overline{ST}$. We can think of segments $\overline{PM}$ and $\overline{PN}$ being stretched into segments $\overline{PS}$ and $\overline{PT}$, respectively.

Let us consider this enlargement idea a bit more extensively. Let ABC be a triangle and let point P be some point in the exterior of this triangle (Fig 11.2.2). Draw the rays $\overrightarrow{PA}$, $\overrightarrow{PB}$, and

$\overrightarrow{PC}$. Choose any point X on ray $\overrightarrow{PA}$ such that $P - A - X$. Construct the line segments $\overline{XY}$ and $\overline{XZ}$ which are parallel respectively to $\overline{AC}$ and $\overline{AB}$ with Y on ray $\overrightarrow{PC}$ and Z on ray $\overrightarrow{PB}$. Note, as a result of this construction, $\overline{YZ} \| \overline{CB}$. (Why?)

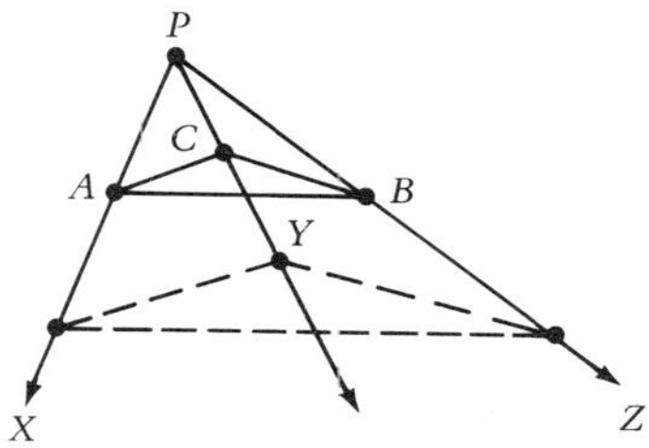

Figure 11.2.2

In the resulting figure, we claim that $\triangle XZY$ is an enlargement of $\triangle ABC$ and that the triangles are similar. We note that the corresponding angles of the two triangles are congruent. (Why?) If we had chosen the point X so that $P - X - A$ and followed the same procedure, it would result that $\triangle XYZ$ would be a miniaturization of $\triangle ABC$.

comment . . .

We leave it to the reader to discover what happens if point P had been chosen either on the triangle ABC or in its interior. The reader should also consider the effect of a similar procedure on a tetrahedron.

We have essentially introduced the notion of a *size transformation*. We make this concept more precise in the following definition.

11.2.1 Definition

Let O be a point and k be a positive real number. For any point A, the size-transformation image of A is the point A' on $\overrightarrow{OA}$ such that

$$OA' = k \cdot OA$$

O is the center *and k the* magnitude *of the transformation. If $k \geqq 1$, we call the size transformation a* dilation. *If $\emptyset < k < 1$, we call the size transformation a* contraction.

Perhaps the most noticeable property of size transformations is that they do not, in general, preserve distances. The exception, of course, is when $k = 1$. Thus, with this exception, a size transformation will *not* send a geometric figure into a congruent figure.

Still, size transformations do possess many of the properties of reflections. In particular, images of collinear points are collinear, betweenness is preserved, and the image of an angle is an angle of the same measure. To summarize, we will assume the following postulates.

11.2.2 Postulate

Size transformations preserve (a) collinearity, (b) betweenness, and (c) angle measure.

11.2.3 Postulate

In a size transformation of magnitude k, the distance between images is k times the distance between their preimages.

We will denote the size transformation with center O and magnitude k by $S(O, k)$. Note that the size transformation $S(O, k)$ is a function with the domain a plane and the range the same plane, such that O is mapped to itself and any other point A is mapped to a point A' such that O, A, A' are collinear and $OA' = k \cdot OA$.

If we consider, then, the effect of a size transformation $S(O, k)$ on a triangle ABC, we find that its image is a triangle $A'B'C'$ with each of the properties listed in Fig. 11.2.3.

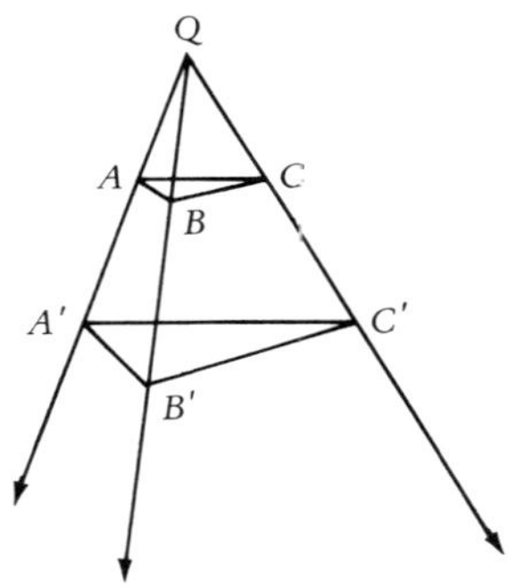

Figure 11.2.3
1. $\angle ABC \cong \angle A'B'C'$
2. $\angle BCA \cong \angle B'C'A'$
3. $\angle CAB \cong \angle C'A'B'$
4. $A'B'/AB = k$
5. $B'C'/BC = k$
6. $A'C'/AC = k$

We will consider $\triangle ABC$ and $\triangle A'B'C'$ to be *similar triangles* if there is a mapping of the plane onto itself which takes A to A', B to B', and C to C' such that the six properties listed with Fig. 11.2.3 are true. Thus, size transformations will become the basis of the concept of similar figures. In this it relates to similarity as reflections relate to congruence. Size transformations will not be the only mappings that will send geometric figures to similar figures. For suppose a size transformation is followed or preceded by a reflection in a line. The result of this composition will be that geometric figures are mapped to similar figures.

This is because reflections preserve all properties that are preserved by size transformations (and, of course, reflections also preserve distances). Thus we are led to making the following definition.

11.2.4 Definition

Let α and β be any two figures in a plane. Then α and β are similar *(written $\alpha \sim \beta$) if and only if there is a composition of reflections or size transformations mapping α to β.*

Since compositions of reflections are isometries, we can conclude that the composition of an isometry and a size transformation will map geometric figures to similar figures. This fact is included in the following definition.

11.2.5 Definition

A similarity *is a mapping of the plane which sends each pair of points A and B into points A′ and B′, such that for some fixed positive real number k, $A'B' = k \cdot AB$. The number k is called the* magnitude *(or* ratio*) of the similarity.*

Each similarity is the composition of an isometry and a size transformation. Any similarity that is not an isometry is either a rotation followed by a size transformation with the same center or it is a reflection followed by a size transformation whose center lies on the line of reflection. The first similarity is a *direct transformation,* while the second is an *opposite transformation.*

Example 1

See Fig. 11.2.4 and Fig. 11.2.5.

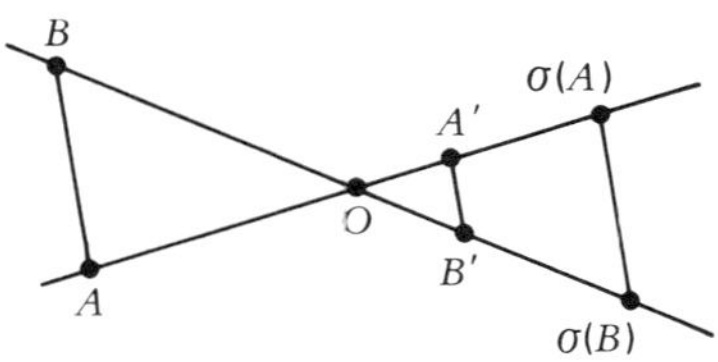

Figure 11.2.4
Direct transformation with $k < 1$, mapping $\overline{AB}$ into $\overline{A'B'}$, O the center, σ a rotation with center O.

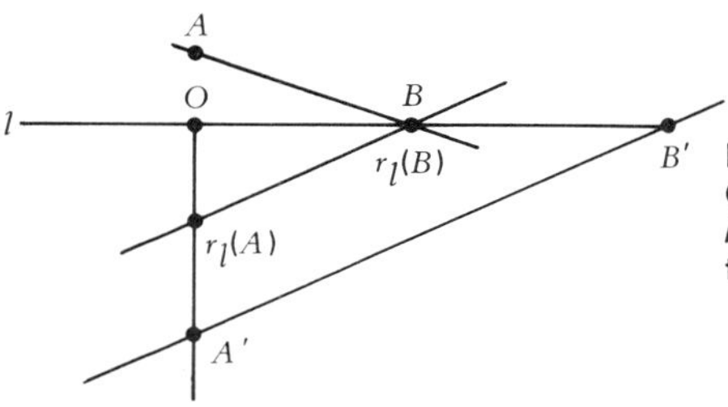

Figure 11.2.5
Opposite transformation with $k > 1$, mapping $\overline{AB}$ into $\overline{A'B'}$ O the center. r_l a reflection in line l

The following is a partial list of properties of similarities which we can deduce.

1. A similarity of magnitude 1 is an isometry.
2. Similarities map line segments into line segments.
3. Similarities preserve angle measures.
4. A similarity maps a triangle into a similar triangle such that corresponding angles are congruent and corresponding sides are in *proportion.* (That is, the ratio of corresponding sides is the magnitude of the similarity.)
5. A similarity maps a polygon into a similar polygon such that corresponding angles are congruent and corresponding sides are proportional.
6. A similarity is completely determined by any three noncollinear points and their images.

We have the following important theorem, used to conclude many results about similar polygons.

11.2.6 Theorem

(SAS Similarity) If two triangles have two sides of one proportional to two sides of the other and the included angles congruent, then the triangles are similar.

The following argument will verify this statement. Suppose we have two triangles ABC and PQR such that $AB/PQ = k$, $BC/QR = k$, and $\angle ABC \cong \angle PQR$. Consider the size transformation $S(O, k)$ with any point O as the center. Apply $S(O, k)$ to $\triangle PQR$. The image of $\triangle PQR$ is $\triangle P'Q'R'$ which is congruent to $\triangle ABC$ by the SAS Congruence Theorem. Thus, there is an isometry σ mapping $\triangle P'Q'R'$ to $\triangle ABC$. Thus, the composition $\sigma \circ S(O, k)$ maps $\triangle PQR$ onto $\triangle ABC$ and the composition is a similarity (Fig. 11.2.6).

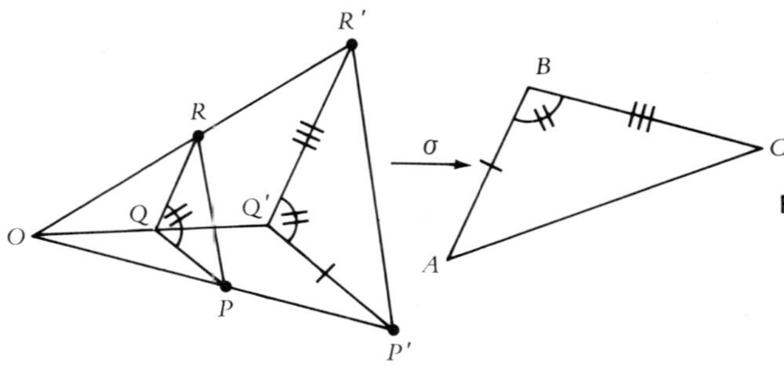

Figure 11.2.6

By using the ASA Congruence Theorem and the SSS Congruence Theorem, we can prove the following theorems.

11.2.7 Theorem

(AA Similarity) If two triangles have two pairs of corresponding angles congruent, then the triangles are similar.

The argument for this theorem is as follows. Suppose we are given $\triangle ABC$ and $\triangle PQR$ with $\angle ABC \cong \angle PQR$ and $\angle ACB \cong \angle PRQ$. Let k be the ratio RQ/CB. Apply the transformation $S(O, k)$ with any center O to $\triangle ABC$. The image of $\triangle ABC$ is a triangle $A'B'C'$ which is congruent to $\triangle PQR$ by the ASA Congruence Theorem. This is so because $\angle ABC \cong \angle A'B'C'$ and $\angle ACB \cong \angle A'C'B'$ and $C'B' = k \cdot CB = (RQ/CB) \cdot CB = RQ$. Thus, $\overline{C'B'} = \overline{RQ}$. Therefore, there is an isometry σ mapping $A'B'C'$ to $\triangle PQR$. The composition $\sigma \circ S(O, k)$ is a similarity between $\triangle ABC$ and $\triangle PQR$.

11.2.8 Theorem

(SSS Similarity) If two triangles have three sides of one proportional to the three sides of the other, then the triangles are similar.

We leave it to the reader to supply the necessary argument to verify this theorem.

Exercise 11.2.1

1. Supply arguments to justify the statements of properties of similarities.
2. Supply arguments to justify each of the following statements.
   a. A size transformation $S(O, k)$ maps $\overline{AB}$ into $\overline{A'B'}$ such that $\overline{AB} \| \overline{A'B'}$.
   b. A size transformation $S(O, k)$ maps a circle of radius r into another circle whose radius $r' = k \cdot r$ and whose center C' is the image of the center C of the given circle.
3. Show that the composition (product) of two size transformations $S(O, k)$ and $S(O, j)$ is commutative; that is, $S(O, j) \circ S(O, k) = S(O, k) \circ S(O, j)$. Show also that $S(O, j) \circ S(O, k) = S(O, jk)$.
4. Show that the inverse of the size transformation $S(O, k)$ is $S(O, 1/k)$.
5. Show that the set of all isometries of a plane form a group under the operation of composition.

11.3 APPLICATION OF SIMILARITIES

With the addition of one preliminary result, we will be able to give an argument justifying the Pythagorean Theorem. We will need the following assertion.

11.3.1 Theorem

The altitude to the hypotenuse of a right triangle divides it into two triangles, each of which is similar to it.

This can be restated as follows: Let $\triangle ABC$ be a right triangle with $m^\circ \angle C = 90$ and $\overline{CD}$ the perpendicular from C to $\overleftrightarrow{AB}$. We claim that $\triangle ACD \sim \triangle ABC$ and $\triangle CBD \sim \triangle ABC$ (Fig. 11.3.1). Clearly, $\angle A \cong \angle A$; also, $\angle ADC \cong \angle ACB$ since these are both right angles. Thus, $\triangle ACD \sim \triangle ABC$ by the AA Similarity Theorem. Also, $\angle B \cong \angle B$ and $\angle CDB \cong \angle ACB$ so again by the AA Similarity Theorem we have $\triangle CBD \sim \triangle ABC$.

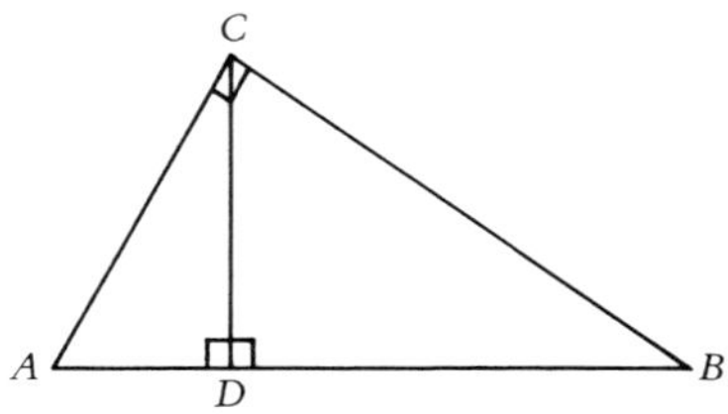

Figure 11.3.1

Now we can show the validity of the Pythagorean Theorem, one of the great theorems of geometry which has innumerable important consequences in all branches of mathematics.

11.3.2 Theorem

(Pythagorean Theorem) In any right triangle, the square of the length of the hypotenuse is equal to the sum of the squares of the lengths of the other two sides.

We will restate this as follows. Let $\triangle ABC$ be a right triangle with $m^\circ \angle C = 90$. If $AB = c$, $AC = b$, and $BC = a$, then $a^2 + b^2 = c^2$ (Fig. 11.3.2).

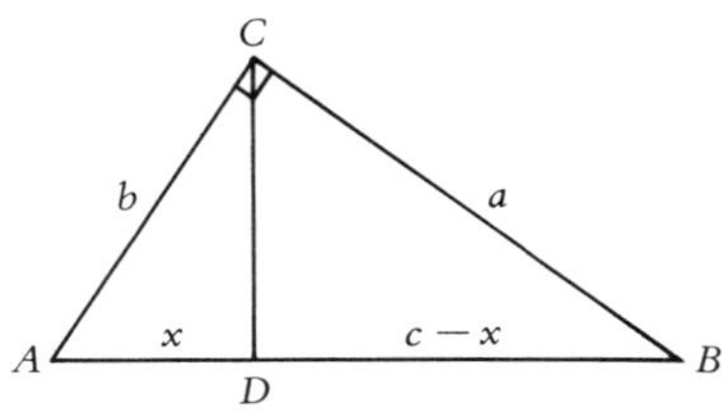

Figure 11.3.2

Let $\overline{CD}$ be the altitude from C to $\overleftrightarrow{AB}$ dividing $\overline{AB}$ into two segments $\overline{AD}$ and $\overline{DB}$ of lengths x and $c - x$, respectively. As we have seen, $\triangle ACD \sim \triangle ABC$ and $\triangle CBD \sim \triangle ABC$.

Since $\triangle ACD \sim \triangle ABC$, $x/b = b/c \Rightarrow x = b^2/c$.

Since $\triangle CBD \sim \triangle ABC$, $(c - x)/a = a/c \Rightarrow c - x = a^2/c$.

Now $c = x + (c - x) = b^2/c + a^2/c = (a^2 + b^2)/c$.

Thus, $c^2 = a^2 + b^2$ as was to be shown.

The converse of the Pythagorean Theorem is also true:

11.3.3 Theorem

Given a triangle whose sides have lengths a, b, and c, if $a^2 + b^2 = c^2$, then the triangle is a right triangle and the angle opposite the side of length c is the right angle.

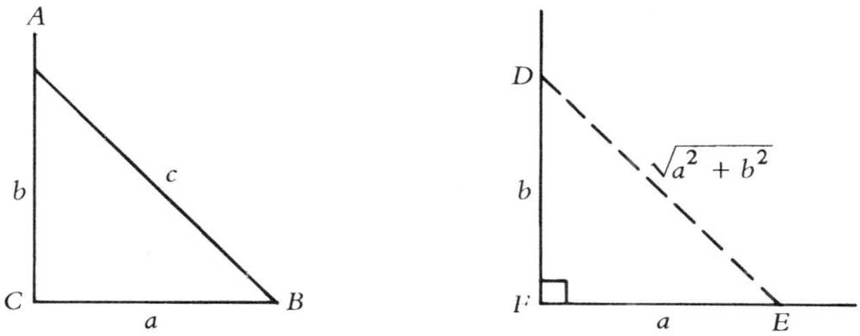

Figure 11.3.3

To see this, let $\triangle ABC$ be such that $a^2 + b^2 = c^2$. Let F be a right angle with D and E points on the sides of $\angle F$ such that $FE = a$ and $FD = b$ (Fig. 11.3.3). Because of the Pythagorean Theorem,

$$DE^2 = a^2 + b^2$$

from which it follows

$$DE = \sqrt{a^2 + b^2} = c$$

Thus, $\triangle ABC \cong \triangle DEF$ by SSS Congruence Theorem.

Since $\triangle DEF$ is a right triangle, it follows that $\triangle ABC$ is a right triangle and that $\angle C$ is a right angle.

comment . . .

The proof given above for the Pythagorean Theorem is not due to Pythagoras (ca. 530 B.C.). But then there are hundreds of different proofs of the theorem, even one originating with President James Garfield. No one knows for sure what argument the Pythagoreans used, but the common belief is that it was a dissection type as suggested in Fig. 11.3.4. Does it suggest a proof to you?

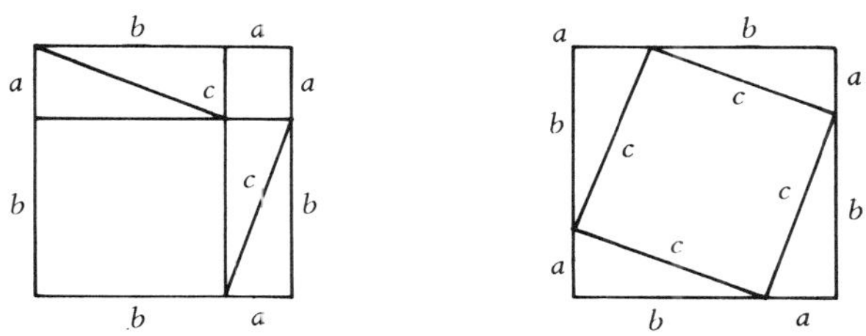

Figure 11.3.4

Experiment 1

Closely related to the Pythagorean Theorem is the problem of determining integers a, b, and c to represent the sides and hypotenuse of a right triangle. Such a triple of numbers is known as a *Pythagorean Triple*. Ancient tradition attributes to Pythagoras himself a solution expressed by the numbers

$$\frac{n^2 - 1}{2}, \qquad n, \qquad \frac{n^2 + 1}{2}$$

where n is an arbitrary odd integer >1.

The problem amounts to finding all systems of positive integers which satisfy the Pythagorean Equation

$$x^2 + y^2 = z^2$$

It is sufficient to seek only so-called *primitive solutions;* that is, solutions consisting of integers without common divisors. Can you explain why this is so? An example of a primitive solution is the set of numbers 3, 4, and 5; another would be the set 5, 12, and 13. Can you find any other primitive solutions?

To facilitate your investigation, we might point out that for any primitive solution x, y, and z, the integers will be relatively prime in pairs (that is, each two of them are without common divisors other than one). From this, we must conclude that x and y cannot be both even, but neither can they be both odd. The truth of the second half of this assertion is based on the fact that the square of an odd integer, when divided by 8, leaves the remainder 1. Suppose both x and y are odd; then

$$x^2 = 1 + 8m \qquad \text{and} \qquad y^2 = 1 + 8n$$

where m, n are positive integers. Thus, $x^2 + y^2 = 2 + 8(m + n)$, implying that z^2 and thus z is even, but the square of an even number is divisible by 4, whereas $2 + 8(m + n)$ is not divisible by 4.

Establish arguments to support the various assertions of the previous paragraph and use the result that one of the pair x and y must be even and the other odd, making z necessarily odd, to seek other primitive solutions of the Pythagorean Equation.

Exercise 11.3.1

1. Give an argument showing that the triangle whose vertices are the midpoints of the sides of a given triangle is similar to the given triangle.
2. Give an argument showing that if a line parallel to one side of a triangle intersects the other two sides in distinct points, then it cuts off a triangle similar to the given triangle.

3. Given parallelogram $ABCD$ with diagonals $\overline{AC}$ and $\overline{BD}$ intersecting at F, show that $BF/FD = CF/FA$.
4. In Fig. 11.3.5, $\overline{BD} \perp \overline{AC}$ and $BE = ED = 2AE = \frac{1}{2}EC$. Show that $\triangle AEB \sim \triangle BEC$, $\triangle DEC \sim \triangle AEB$, and $\overline{AB} \perp \overline{BC}$.

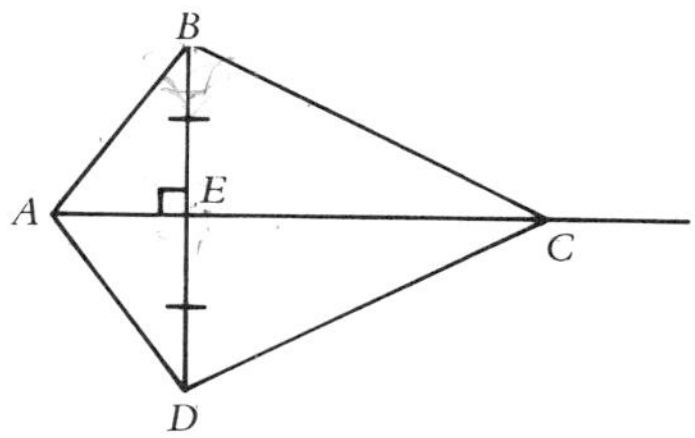

Figure 11.3.5

5. Give an argument showing that in similar triangles corresponding medians have the same ratio as corresponding sides.
6. Given right $\triangle ABC$ with altitude drawn to the hypotenuse and lengths as shown in Fig. 11.3.6, find the unknown lengths.

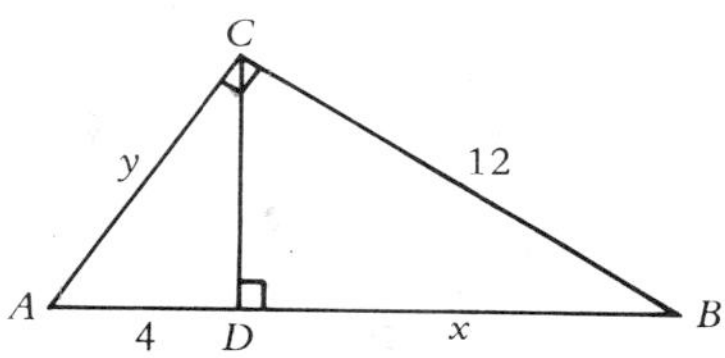

Figure 11.3.6

7. In a right triangle, if the altitude to the hypotenuse is 12 and the hypotenuse is 25, find the length of each leg and of the segments of the hypotenuse.
8. In trapezoid $ABCD$ with parallel sides $\overline{AB}$ and $\overline{CD}$, let the diagonals meet at E. Show that $\triangle ABE \sim \triangle CDE$.
9. Show that the figure formed by joining the midpoints of the sides of a square is a square having half the area of the given square.
10. Show that the median of a trapezoid (the line joining the midpoints of the nonparallel sides) has length equal to half the sum of the bases of the trapezoid.

in our opinion . . .

In keeping with the authors' conviction that mathematics should be presented in an integrated manner, we include the next section on elementary trigonometry. Man's need to compute the measure of angles and segments for which geometry would require scale diagrams led to the development of the science of trigonometry. Historically, trigonometry, which united aspects of the three sciences of arithmetic, algebra, and geometry, was, until the thirteenth century, merely a tool of the science of astronomy. The systematization of plane and spherical trigonometry as a science independent of astronomy was accomplished by the Persian mathematician, Nasir addin at-Tusi, in the second half of the thirteenth century.

Trigonometry is a branch of mathematics very important from a practical point of view. It is important in surveying, the building industry, in navigation, and any occupation which requires the measuring of distances.

11.4 TRIGONOMETRY: AN APPLICATION OF SIMILAR TRIANGLES

In Fig. 11.4.1, we observe that $\triangle ABE$ and $\triangle ACD$ are two right triangles. Since A is common to both triangles, by the AA Similarity Theorem, $\triangle ABE \sim \triangle ACD$. It follows then that the lengths of corresponding sides are proportional. In particular,

$$\frac{BE}{CD} = \frac{AB}{AC}$$

which is equivalent to the useful proportion

$$\frac{BE}{AB} = \frac{CD}{AC}$$

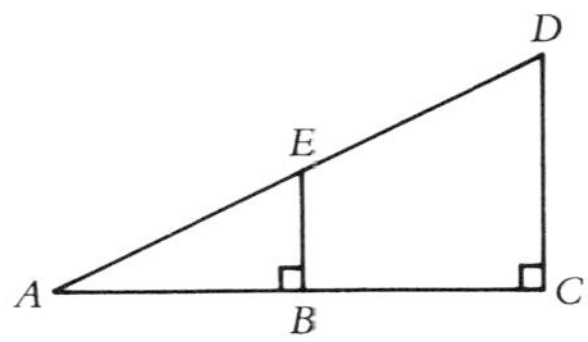

Figure 11.4.1

This latter proportion can be described as: The ratio of the length of the leg opposite $\angle A$ to the length of the leg adjacent to $\angle A$ is the same in both triangles. When speaking of a leg of a right triangle, we are referring to those sides which are not the hypotenuse.

As might be suspected by our logical readers, this particular proportion is not dependent on our particular model in Fig. 11.4.1.

Now consider Fig. 11.4.2. It is easy to see that all of the right triangles, AB_1O_1, AB_2O_2, AB_3O_3, . . . , AB_nO_n are similar and thus the lengths of their corresponding sides are proportional. From

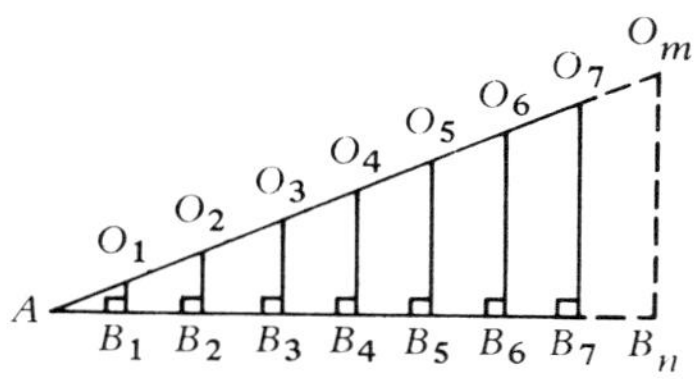

Figure 11.4.2

this fact we can derive the proportion:

$$\frac{B_1O_1}{AB_1} = \frac{B_2O_2}{AB_2} = \frac{B_3O_3}{AB_3} = \cdots = \frac{B_nO_n}{AB_n}$$

This particular proportion is completely dependent on the size of $\angle A$ and not the size of the right triangle. Consider Fig. 11.4.3. Since $\triangle ABE \sim \triangle ACD$, we have

$$\frac{BE}{AB} = \frac{CD}{AC}$$

Since $\triangle ABG \sim \triangle ACF$, we have

$$\frac{BG}{AB} = \frac{CF}{AC}$$

However,

$$\frac{BE}{AB} \neq \frac{BG}{AB}$$

This particular ratio is called the *tangent of* $\angle A$, which we will abbreviate as tan $\angle A$.

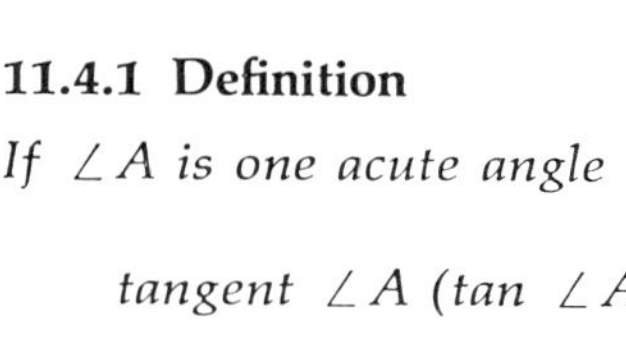

Figure 11.4.3

11.4.1 Definition

If $\angle A$ *is one acute angle of a right triangle*

$$\text{tangent } \angle A \text{ (tan } \angle A) = \frac{\text{Length of leg opposite } \angle A}{\text{Length of leg adjacent to } \angle A}$$

Note that the tangent of an angle depends only on the size of the angle: If $\angle A \cong \angle A'$, then tan $\angle A =$ tan $\angle A'$. Since $\measuredangle A$ and A' are congruent if and only if $m^\circ \angle A = m^\circ \angle A'$, it is customary to identify an angle with its measure. For instance, if

$$m^\circ \angle A = 49$$

we may write tan 49° instead of tan $\angle A$.

Example 1
See Fig. 11.4.4. In right triangle ABC, the sides are of lengths 5, 12, and 13. In this case

$$\tan \angle A = \frac{5}{12} \qquad \text{and} \qquad \tan \angle B = \frac{12}{5}$$

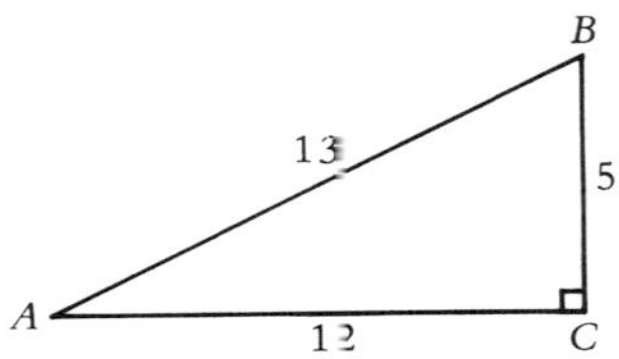

Figure 11.4.4

comment . . .

Note in this particular right triangle, the tangents of the two acute angles are reciprocals of each other. Is this true in a general right triangle?

Two other ratios associated with a right triangle depending only on the measure of the acute angles are the *sine* and *cosine*.

11.4.2 Definition

Let $\triangle ABC$ be a right triangle with right angle at C.

$$\textit{sine } \angle A \ (\textit{sin } \angle A) = \frac{\textit{Length of leg opposite } \angle A}{\textit{Length of hypotenuse}}$$

$$\textit{cosine } \angle A \ (\textit{cos } \angle A) = \frac{\textit{Length of leg adjacent to } \angle A}{\textit{Length of hypotenuse}}$$

Example 2
See Fig. 11.4.5.

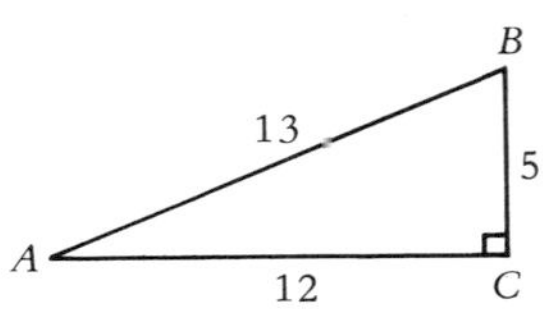

Figure 11.4.5
sin $\angle A = 5/13$ sin $\angle B = 12/13$
cos $\angle A = 12/13$ cos $\angle B = 5/13$

comment . . .

Note in this particular right triangle sin $\angle A = \cos \angle B$, sin $\angle B = \cos \angle A$. Or to put it another way,

$\sin \angle A = \cos (90^\circ - m^\circ \angle A)$ and $\sin \angle B = \cos (90^\circ - m^\circ \angle B)$

$\cos \angle A = \sin (90^\circ - m^\circ \angle A)$ and $\cos \angle B = \sin (90^\circ - m^\circ \angle B)$

Is this true in a general right triangle?

The tangent, sine, and cosine ratios are called *trigonometric ratios.* The word "trigonometric" comes from the Greek *trigonon* meaning triangle and *metron* meaning measure. The importance of these trigonometric ratios rests in the fact that if we know the ratio, say tan $\angle A = 5/12$, we can determine the measure of $\angle A$. On the other hand, if we know that the measure of $\angle A$ is, say, 30°, and that the side opposite $\angle A$ is 4, we can find the length of each of the remaining sides. (We already know the measure of the remaining acute angle, that being 60°. Can you explain why?)

To be able to solve the problem mentioned in the preceding paragraph, we need a table of trigonometric values. Such a table gives the tangent, sine, and cosine ratios of various angles from 0° to 90°. Depending on the accuracy of these tables, the included angles may be given in terms of degrees, minutes ($1/60$ of a degree), and seconds ($1/60$ of a minute); whereas, the ratios are usually accurate to four decimal places.

A portion of such a table might look as follows.

Degrees	*Tangent*	*Sine*	*Cosine*
30	.5774	.5000	.8660
31	.6009	.5150	.8572
32	.6249	.5299	.8480
33	.6494	.5446	.8387

Thus, tan 32° = .6249, sin 31° = .5150, cos 33° = .8387, and so on.

To approximate the value of the tangent, sine, and cosine of any acute angle, we draw, with the aid of a protractor, the particular angle on a piece of graph paper such as in Fig. 11.4.6.

Construct $\angle AOB$ such that $m^\circ \angle AOB = 30^\circ$. Using a ruler, we measure $\overline{AB}$ and discover $AB = 21/16$ to the nearest $1/16$ in. while $OA = 36/16$ to the nearest $1/16$ in. Thus, tan $\angle AOB = 21/36 \approx .58$. If we want a better approximation to tan $\angle AOB$ we

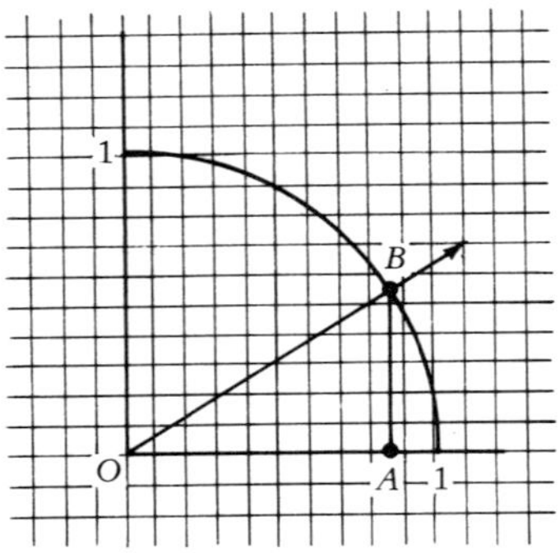

Figure 11.4.6

would need more accurate measuring devices. Measuring, we discover that $OB = 4\frac{1}{16}$ to the nearest $\frac{1}{16}$ in. Thus, sin $\angle AOB \approx$ .51 and cos $\angle AOB \approx .88$.

Many problems can be solved making use of trigonometric ratios. For instance:

1. A tree casts a shadow 80 ft long. The angle of elevation of the sun is 32°. Estimate the height of the tree.
2. A 10-ft ladder rests against a wall at a point 8 ft above the ground. Find the angle the ladder makes with the ground.

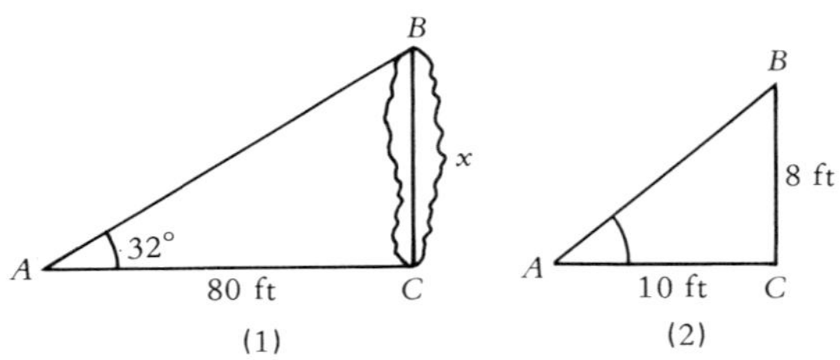

Figure 11.4.7

In Fig. 11.4.7, we represent each of these problems in terms of right triangles. In Problem 1, we use the tangent ratio to determine the height of the tree.

$$\tan 32° = \frac{\text{Height of tree}}{80} = \frac{x}{80}$$

$$.6249 = \frac{x}{80}$$

Thus,

$$x = 49.992 \text{ ft}$$

In Problem 2, we use the tangent ratio to determine $\angle A$.

$$\tan \angle A = \frac{8}{10} = .8000$$

Using a trigonometric table, we would find that this is the tangent ratio of an angle of approximately 39°.

It is not our purpose to investigate trigonometric ratios and the solution of triangles exhaustively. Instead, we introduced this brief discussion to suggest some of the power of results concerning similar triangles. It would not do to stop at this point and leave the reader with the impression that trigonometric ratios are only useful when one is dealing with right triangles. Quite the contrary; these ratios can be used to define functions of real numbers independent of triangles. These functions as a class, called *circular functions,* include the tangent functions, the sine functions, and the cosine functions. We will not investigate these functions in this book.

Exercise 11.4.1

1. Given triangle ABC with right angle at C as in Fig. 11.4.8.

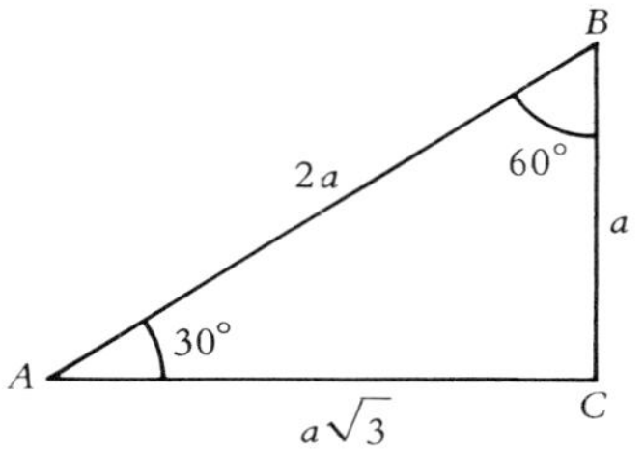

Figure 11.4.8

   Use it to find each of the following.
   a. sin 30° b. cos 60°
   c. tan 60° d. cos 30°
   e. tan 30° f. sin 60°
2. Use the graphing method of this section to approximate the tangent ratio, sine ratio, and cosine ratio of each of the following angles: 20°, 45°, 50°, 60°, 75°.
3. What can you say about the size of the trigonometric ratios for angles close to 0° and 90°? What values would you assign to each of the following ratios?
   a. sin 0° b. cos 0°
   c. tan 0° d. sin 90°
   e. cos 90° f. tan 90°

11.5 SUMMARY

1. Let O be a point and k be a positive real number. For any point A of a plane, the size transformation $S(O, k)$ maps A to the point A' so that O, A, and A' are collinear and $OA' = k \cdot OA$.

2. Size transformation preserves (a) collinearity, (b) betweenness, and (c) angle measure.
3. In a size transformation of magnitude k, the distance between images is k times the distance between their preimages.
4. Let α and β be any two figures in a plane. Then α and β are similar (written $\alpha \sim \beta$) if and only if there is a composite of reflections or size transformations mapping α to β.
5. A similarity is a mapping of the plane which sends each pair of points A and B into points A' and B', such that for some fixed positive real number k, $A'B' = k \cdot AB$.
6. (SAS Similarity) If two triangles have two sides of one proportional to two sides of the other and the included angles are congruent, then the triangles are similar.
7. (SSS Similarity) If two triangles have three sides of one proportional to the three sides of the other, then the triangles are similar.
8. (AA Similarity) If two triangles have two angles of one respectively congruent to the corresponding angles of the other, then the triangles are similar.
9. (Pythagorean Theorem) In any right triangle, the square of the length of the hypotenuse is equal to the sum of the squares of the lengths of the other two sides.

LINES IN COORDINATE GEOMETRY

12

12.1 PROLOGUE

Recall that in Chapters 2 and 5 it appeared that the graphs of sets such as $\{(x, y)|x + y = 2\}$, $\{(x, y)|2x + y = 5\}$, and $\{(x, y)|2x + 3y = 8\}$ were straight lines. We were tempted to conclude this because all the points that we graphed seemed to lie on a straight line. However, we were not able to completely justify our speculation for two reasons, both of which stem from the fact that our set of ordered pairs was infinite and we could only graph a very small finite subset of them.

For example, suppose we wish to graph $\{(x, y)|x + y = 2\}$. One subset of this could be $\{(0, 2), (2, 0), (1, 1)\}$. We graph these three points and we observe that they seem to be on a straight line. Then we enlarge our set by considering the following $\{(1\frac{1}{2}, \frac{1}{2}), (-5, 7)\}$. These also seem to be on the same straight line. Perhaps we should try some irrational numbers, so we try $(\pi, 2 - \pi), (-\sqrt{2}, 2 + \sqrt{2})\}$. The graph of all of these points is in Fig. 12.1.1 (only some of them are labeled). As far as we are able to observe, it seems that all the points lie on the same line. From this limited experiment, we are tempted to say that the graph of all the ordered pairs would be this straight line, or more formally,

$$\{(x, y)|x + y = 2\} = \{\text{all the points on the line joining } (0, 2) \text{ and } (2, 0)\}$$

As stated earlier, there are two important ideas we ought to worry about. First, perhaps $\{(x, y)|x + y = 2\}$ does not completely fill up the line. Second, maybe it *does* fill up the line, but there are also *other* points which do *not* fall on the line. Figures 12.1.2 and 12.1.3 help you understand these two possibilities.

In the first case, what we are really concerned with is does the set $\{(x, y)|x + y = 2\}$ contain *all* the points of the line (that is, are there any gaps)? In the second case, we are asking the

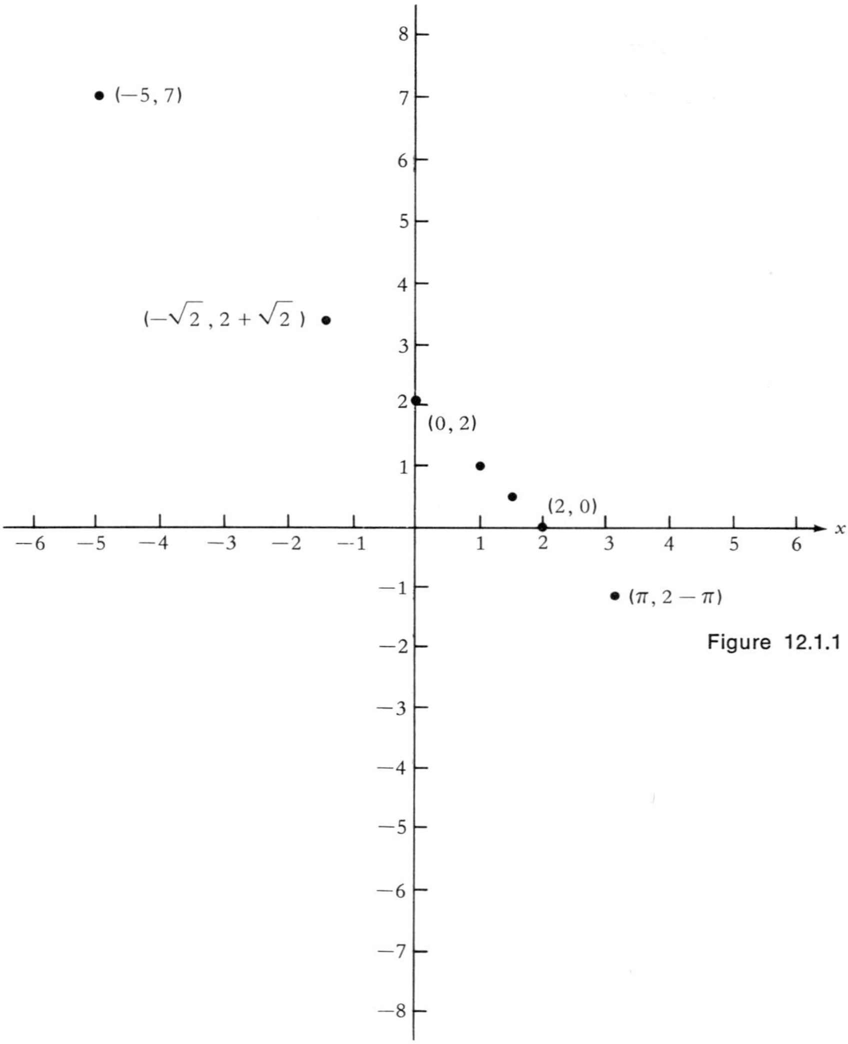

Figure 12.1.1

question does the set $\{(x, y)|x + y = 2\}$ contain *only* those points which are on the line (that is, are there additional points *not* on the line)? These two questions should be no surprise to you, for recall that in Chapter 5 we discussed what it meant to say that two sets A and B were equal; namely, if set A contained *all* the elements of B and if set A contained *only* those elements in B. Thus, in order to satisfactorily justify that $\{(x, y)|x + y = 2\} = \{$all points on the line joining (0, 2) and (2, 0)$\}$, we must consider both questions. A major section of this chapter

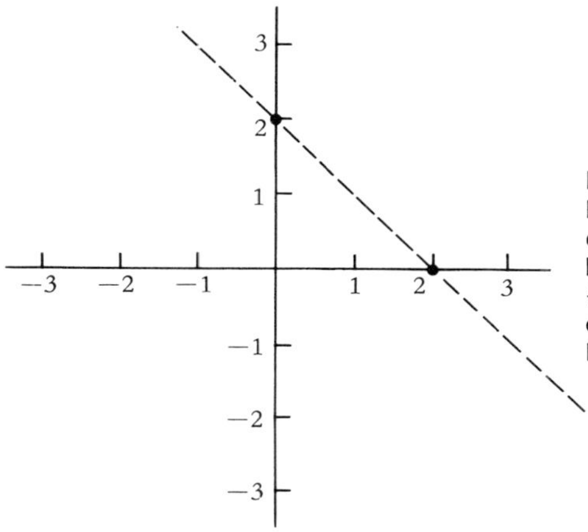

Figure 12.1.2
Perhaps $\{(x, y)|x + y = 2\}$ does not fill up the entire line. That is, perhaps $\{(x, y)|x + y = 2\}$ does not contain *all* the points of the line

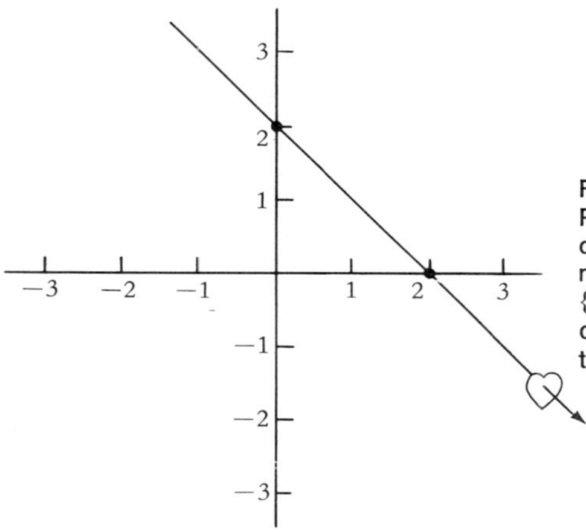

Figure 12.1.3
Perhaps $\{(x, y)|x + y = 2\}$ contains other points that are not on the line. That is, perhaps $\{(x, y)|x + y = 2\}$ does not contain *only* the points of the line

shall be devoted to this task, in fact we shall consider the more general idea that $\{(x, y)|ax + by = c, a, b \neq 0\} = \{$all points on the line joining $(0, c/b)$ and $(c/a, 0)\}$. Once we have accomplished this, we shall then be able to describe the coordinates of any line in the plane. To begin our investigation, you should attempt the next experiment.

Experiment 1

Step 1

Find a list of at least 10 representative ordered pairs which are members of the set $\{(x, y)|2x + 3y = 12\}$. At least two of these should be pairs of irrational numbers.

Step 2

Graph your set of ordered pairs. Do they lie on a line?

Step 3

Let (x_1, y_1) equal the first ordered pair of your list and (x_2, y_2) equal the second ordered pair of your list. Then calculate the value of

$$\frac{y_2 - y_1}{x_2 - x_1}$$

Step 4

Now let (x_1, y_1) equal the third ordered pair and (x_2, y_2) the fourth ordered pair of your list. Calculate

$$\frac{y_2 - y_1}{x_2 - x_1}$$

again. What do you observe?

Step 5

Continue calculating

$$\frac{y_2 - y_1}{x_2 - x_1}$$

using the fifth and sixth, the seventh and eighth, and the ninth and tenth ordered pairs. What do you observe each time? If (x_1, y_1) and (x_2, y_2) were any two ordered pairs in the set, what would you expect to get for

$$\frac{y_2 - y_1}{x_2 - x_1}?$$

Step 6

Repeat your experiment for $\{(x, y)|4x - 5y = 20\}$.

Step 7

Also, repeat your experiment for $\{(x, y)|y = 2x^2 + 1\}$.

12.2 SLOPES OF LINES

If you performed the above experiment successfully, you should have observed that for a line it appears that: If (x_1, y_1) and (x_2, y_2) are any two points on the line and (x_3, y_3) and (x_4, y_4) are any two other points on the line, then

$$\frac{y_2 - y_1}{x_2 - x_1} = \frac{y_4 - y_3}{x_4 - x_3}$$

comment . . .

The ratios

$$\frac{y_2 - y_1}{x_2 - x_1} \quad \text{and} \quad \frac{y_4 - y_3}{x_4 - x_3}$$

are only defined as long as the line is not vertical. (Why?) For this reason, this case will automatically be excluded from consideration in this chapter. Thus, whenever we speak of a line or the slope of a line it should be clear that the line is not vertical.

This idea is quite obvious for horizontal lines, for then if we look at Fig. 12.2.1 we have the fact that $y_1 = y_2 = y_3 = y_4$, since all the y-coordinates of the ordered pairs of a horizontal line are equal. Thus, $y_2 - y_1 = y_4 - y_3 = 0$ and therefore,

$$0 = \frac{y_2 - y_1}{x_2 - x_1} = \frac{y_4 - y_3}{x_4 - x_3}$$

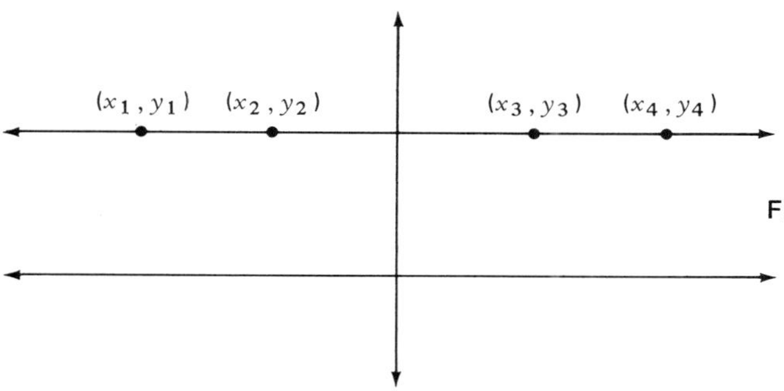

Figure 12.2.1

Now, to consider the more general case, when the line is neither horizontal nor vertical, look at Fig. 12.2.2. As the diagram suggests, suppose we draw a vertical line through P and a horizontal line through Q so that they intersect at a point T. Can you see that the coordinates of T are (x_1, y_2)? Likewise, draw a vertical line through R and a horizontal one through S to intersect at U. Its coordinates are (x_3, y_4). Now we have two right triangles PTQ and RUS. Since $\overleftrightarrow{PT}$ and $\overleftrightarrow{RU}$ are vertical lines, we have $\overleftrightarrow{PT} \| \overleftrightarrow{RU}$. Similarly, we have $\overleftrightarrow{TQ} \| \overleftrightarrow{US}$. Thus, in $\triangle PTQ$ and $\triangle RUS$, $\angle P \cong \angle R$ and $\angle Q \cong \angle S$. (Why?) Therefore, we know that

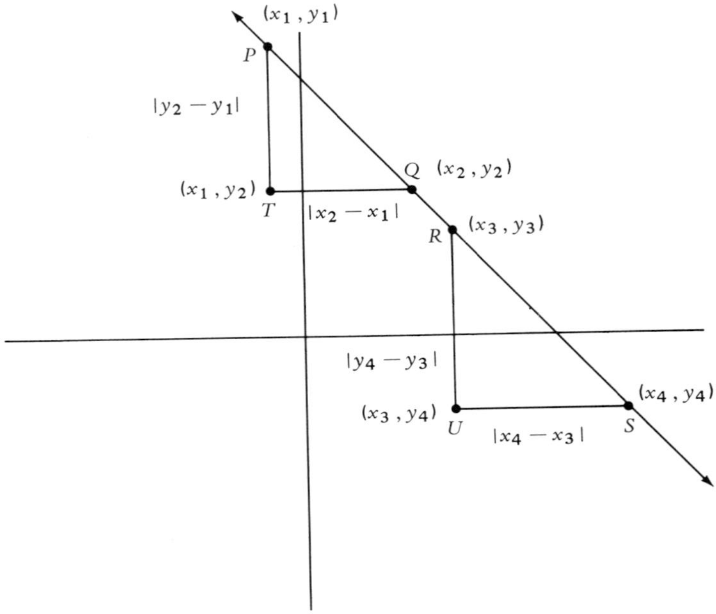

Figure 12.2.2

$\triangle PTQ \sim \triangle RUS$. Now, what do we know about similar triangles? We know that corresponding sides are proportional. In particular,

$$\frac{PT}{TQ} = \frac{RU}{US}$$

but since $PT = |y_2 - y_1|$, $TQ = |x_2 - x_1|$, $RU = |y_4 - y_3|$, and $US = |x_4 - x_3|$, we get, by substitution,

$$\frac{|y_2 - y_1|}{|x_2 - x_1|} = \frac{|y_4 - y_3|}{|x_4 - x_3|} \tag{12.1}$$

Now this is not quite what we want; in our investigation we did not work with absolute values. To resolve this problem, we must consider two families of lines, those which extend through the first and third quadrants, and those which extend through the second and fourth quadrants. See Figs. 12.2.3 and 12.2.4.

As you can see, the line we drew in Fig. 12.2.2 extends through the second and fourth quadrants. So, for the time being, we will restrict our discussion to this family of lines. You will be asked in an exercise to consider a similar argument for the other family of lines.

Now, if the line extends through the second and fourth quadrants, we can see that

$$y_4 < y_3 < y_2 < y_1$$

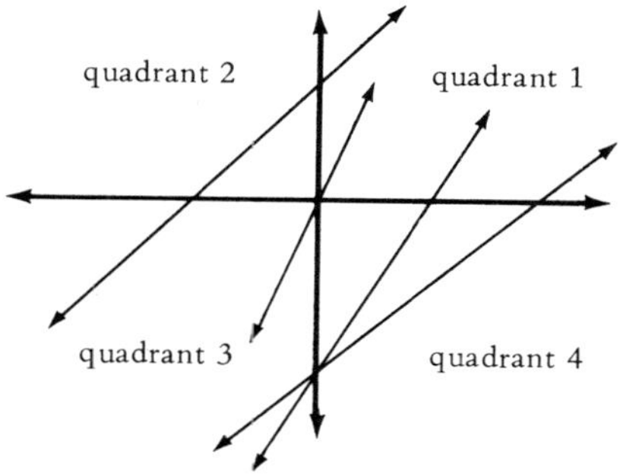

Figure 12.2.3
The family of lines which extend through the first and third quadrants

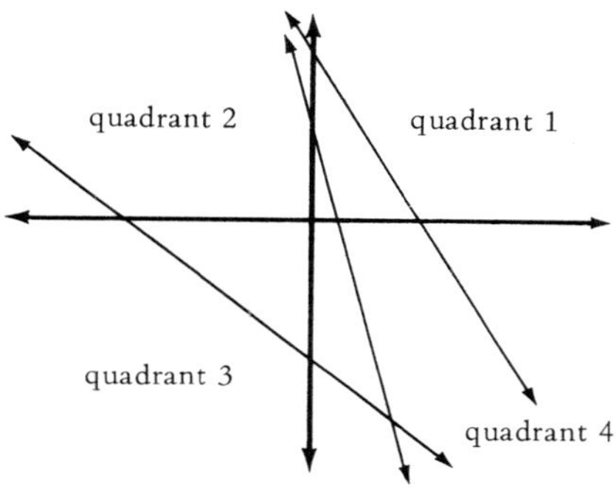

Figure 12.2.4
The family of lines which extend through the second and fourth quadrants

Thus,

$$y_4 - y_3 < 0$$

and

$$y_2 - y_1 < 0$$

That is, $y_4 - y_3$ and $y_2 - y_1$ are both negative numbers. Thus, $|y_2 - y_1| = -(y_2 - y_1)$ and $|y_4 - y_3| = -(y_4 - y_3)$. (Why?)

Also, $x_1 < x_2 < x_3 < x_4$ and, therefore, $x_2 - x_1$ and $x_4 - x_3$ are both positive. Thus, we know that $|x_2 - x_1| = x_2 - x_1$ and $|x_4 - x_3| = x_4 - x_3$. (Why?) Therefore, substituting into Eq. (12.1) we can eliminate the absolute values and get

$$\frac{-(y_2 - y_1)}{(x_2 - x_1)} = \frac{-(y_4 - y_3)}{(x_4 - x_3)}$$

Multiplying both sides by -1 gives us

$$\frac{y_2 - y_1}{x_2 - x_1} = \frac{y_4 - y_3}{x_4 - x_3}$$

12.2.1 Definition

If (x_1, y_1) and (x_2, y_2) are any two distinct points of a line, the ratio $(y_2 - y_1)/(x_2 - x_1)$ is called the slope *of the line.*

Since we saw earlier that $y_2 - y_1$ and $y_4 - y_3$ were negative while $x_2 - x_1$ and $x_4 - x_3$ were positive, we can see that all the lines which extend through the second and fourth quadrants have a negative slope. Also, you should be able to convince yourself that all the lines which extend through the first and third quadrants have positive slope. Sometimes this ratio $(y_2 - y_1)/(x_2 - x_1)$ is expressed as $\Delta y/\Delta x$ which is thought of as

$$\frac{\text{A change in } y}{\text{The corresponding change in } x} \tag{12.2}$$

One of the exercises will indicate why we need the word "corresponding" in the denominator. Also, this ratio can be thought of as the ratio of the vertical change (rise) to the horizontal change (run), as one proceeds from one point to the next. For this reason, the slope is often expressed as Rise/Run.

Example 1

Calculate the slope of the line which contains the points (2, 3) and (−1, 7) and determine which family the line belongs to. If we let $(x_1, y_1) = (2, 3)$ and $(x_2, y_2) = (-1, 7)$, then

$$\frac{y_2 - y_1}{x_2 - x_1} = \frac{7 - 3}{-1 - 2} = \frac{4}{-3} = -\frac{4}{3}$$

Thus, the slope is $-4/3$. Since it is negative, the line extends through the second and fourth quadrants.

Notice that if we set $(x_1, y_1) = (-1, 7)$ and $(x_2, y_2) = (2, 3)$, then

$$\frac{y_2 - y_1}{x_2 - x_1} = \frac{3 - 7}{2 - (-1)} = \frac{-4}{3} = -\frac{4}{3}$$

Thus, the choice of which point to call (x_1, y_1) and which (x_2, y_2) is purely arbitrary.

Example 2

A certain line has a slope of $-3/4$ and the point (1, 2) is on this line. Find the coordinates of three other points on the line. Suppose $(x_1, y_1) = (1, 2)$. Let (x_2, y_2) be any other point on the line. Then

$$\frac{y_2 - y_1}{x_2 - x_1} = \frac{y_2 - 2}{x_2 - 1} = -\frac{3}{4} = \frac{-3}{4}$$

Then, cross multiplying, we get

$$4(y_2 - 2) = -3(x_2 - 1)$$

or

$$4y_2 - 8 = -3x_2 + 3$$

or

$$4y_2 + 3x_2 = 11$$

So (x_2, y_2) is any ordered pair which satisfies the equation $4y_2 + 3x_2 = 11$. Thus, $(0, 11/4)$ is one, $(11/3, 0)$ is another, and $(5, -1)$ is a third.

An alternate approach is to say that since

$$\frac{y_2 - 2}{x_2 - 1} = \frac{-3}{4}$$

we can say that if $y_2 - 2 = -3k$, then $x_2 - 1 = 4k$ and $k \varepsilon R$, $k \neq 0$. Therefore, $y_2 = 2 - 3k$ and $x_2 = 1 + 4k$. Thus, we can obtain ordered pairs on the line of the form $(1 + 4k, 2 - 3k)$ for any nonzero k. Some typical points would be for $k = 1, -1$, and 2, giving us $(5, -1)$, $(-3, 5)$, and $(9, -4)$, respectively.

Exercise 12.2.1

1. Find the slopes of the lines which go through the following points. Then indicate whether they go through the first and third quadrants or the second and fourth quadrants.
   a. $(0, 0)$ and $(1, 2)$
   b. $(-1, -7)$ and $(2, -3)$
   c. $(-3, 4)$ and $(-8, -9)$
   d. $(5, 8)$ and $(-2, 4)$
   e. $(12, -3)$ and $(-7, -5)$
   f. $(3, \pi)$ and $(7, 4 + \pi)$
   g. $(4, -\sqrt{2})$ and $(5, -3\sqrt{2})$
2. a. Find three other points which are on the line with slope of $2/7$ and which contain the point $(-1, 4)$.
   b. Sketch a graph of the line.
3. a. Refer to Experiment 1 of this chapter. Using the same ordered pairs you selected for Step 1, calculate the ratio of $(y_1 - y_2)/(x_1 - x_2)$. Do you get the same results? Do the same for Steps 2–5 of the experiment. Does it make any difference which ordered pair you select for (x_1, y_1) and (x_2, y_2)?
   b. Use the same ordered pairs as above to calculate the ratio $(y_2 - y_1)/(x_1 - x_2)$. What do you observe?
   c. Explain what you think is meant by the phrase "the corresponding change in x" in Eq. (12.2).
4. a. Sketch and label a diagram similar to the diagram in Fig. 12.2.2, only make the line go through the first and third quadrant.
   b. Starting with the fact that

$$\frac{|y_2 - y_1|}{|x_2 - x_1|} = \frac{|y_4 - y_3|}{|x_4 - x_3|}$$

show that

$$\frac{y_2 - y_1}{x_2 - x_1} = \frac{y_4 - y_3}{x_4 - x_3}$$

c. Show that

$$\frac{y_2 - y_1}{x_2 - x_1}$$

is a positive number and hence that all lines in this family have positive slope.

Experiment 2

Step 1

Suppose l_1 is a line joining (2, 3) and (7, 5) and l_2 is a line joining (3, 1) and (13,5).
a. Calculate the slopes of l_1 and l_2.
b. Graph the two lines. What do you observe?

Step 2

Repeat Step 1 where l_1 joins (−6, 6) and (6, 2), and l_2 joins (−2, −1) and (−8, 1).

Step 3

Repeat Step 1 where l_1 joins (−3, −4) and (0, 0) and l_2 joins (6, 8), and (9, 12).

Step 4

a. From your observations in Steps 1, 2, and 3, make a conjecture about lines which have equal slopes. Put your statement in If—then form.
b. What is the converse of your conjecture? Do you think it is true? Can you test it with examples?

12.3 SLOPES AND PARALLEL LINES

If you did Experiment 2, you probably conjectured that if two lines have the same slope, then they are either parallel or they coincide. Let us examine a diagram of this situation and see if we can explain why this is true (Fig. 12.3.1).

Suppose l_1 and l_2 are two lines with equal slope. Now, of course, either l_1 and l_2 are the same line (that is, they coincide) or they are different. It is perfectly possible for them to be the same

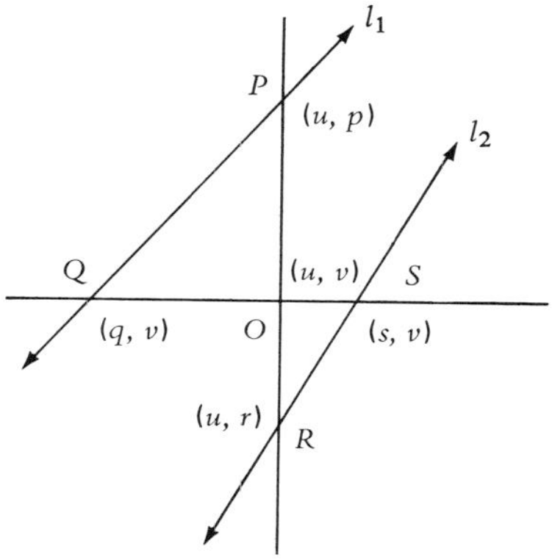

Figure 12.3.1

line for they would then have the same slope. So, we must show that if they do not coincide, then they are parallel. What does it mean to say that two lines do not coincide? Obviously, they will either intersect or be parallel. So let's consider Fig. 12.3.1 and see if we can understand why they must be parallel.

comment . . .

This discussion is valid as long as the lines are not horizontal. This case will be considered in an exercise.

Let point $O(u, v)$ be a point which is not on l_1 or l_2.

Now, since l_1 and l_2 are lines which are not vertical or horizontal, we know they must intersect every pair of vertical and horizontal lines; in particular, $\{(x, y)|x = u\}$ and $\{(x, y)|y = v\}$. Let $P(u, p)$ and $R(u, r)$ be the points of intersection of $\{(x, y)|x = u\}$ with l_1 and l_2, respectively. Then let $Q(q, v)$ and $S(s, v)$ be the points where $\{(x, y)|y = v\}$ intersects l_1 and l_2, respectively. Then we have two right triangles $\triangle POQ$ and $\triangle ROS$. Also, we have $PO = |p - v|$, $OQ = |u - q|$, $RO = |r - v|$, and $OS = |u - s|$. Now, what is the slope of l_1? Using points P and Q we get $(p - v)/(u - q)$. Likewise, using points R and S we find that the slope of l_2 is $(r - v)/(u - s)$. Now, since l_1 and l_2 have equal slopes we can say that $(p - v)/(u - q) = (r - v)/(u - s)$. Thus, we can also say $|p - v|/|u - q| = |r - v|/|u - s|$. Thus, we have two right triangles with the legs of one proportional to the corresponding legs of the other. So, from our chapter on similarity, we see that $\triangle POQ \sim \triangle ROS$. (Why?) Thus, $\angle P \cong \angle R$. (Why?) This leads us to conclude that $l_1 \| l_2$. (Why?) Thus, we have justified a very significant principle: If two lines have equal slope, then they either coincide or they are parallel.

In Experiment 2, you are also asked to consider the converse of this statement which is: If two lines are parallel or coincide, then they have equal slopes. This statement is obvious if the lines coincide. If they are parallel, the statement is still true and you will be asked to justify it in an exercise.

Since we have a statement and its converse which are both true, we are able to summarize them as the following theorem.

12.3.1 Theorem

Two nonvertical lines are parallel or coincide if and only if they have equal slopes.

The reader is reminded that this statement really has two parts:

1. If two nonvertical lines are parallel or coincide, then they have equal slopes.
2. If two nonvertical lines have equal slopes, then they are parallel or they coincide.

comment . . .

We have thus defined an equivalence relation on the family of all nonvertical lines in a coordinate plane. That is, lines l_1 and l_2 are equivalent if they have the same slope. This equivalency implies that the two lines are either parallel or coincident.

Example 1

Determine if the line joining (1, 3) and (2, 10) intersects, is coincident with, or is parallel to the line joining (4, −1) and (5, 6).

The slope of the first line is $(3 - 10)/(1 - 2) = -7/-1 = 7$ and the slope of the second line is $(6 - (-1))/(5 - 4) = 7/1 = 7$. Therefore, by Theorem 12.3.1, we know the lines are either parallel or coincident. (Which part?) Now we can determine whether the lines are parallel or coincident by determining the slope of the line joining a point of the first line with a point of the second. The slope of the line joining (1, 3) and (4, −1) is $(3 - (-1))/(1 - 4) = 4/-3 = -4/3$. Therefore, we know that (1, 3), (2, 10), and (4, −1) are not on the same line. (Why?) Therefore, the two lines are not coincident; they are parallel.

Exercise 12.3.1

1. Determine whether the lines which join the following pairs of points intersect, are coincident, or are parallel to each other.
   a. (2, 7) and (5, 12); (−1, −8) and (−4, −13)
   b. (3, −1) and (10, 7); (7, 4) and (0, 12)

c. $(-12, 15)$ and $(-10, 6)$; $(-16, 33)$ and $(-6, -12)$
d. $(2, -\sqrt{3})$ and $(-1, -2\sqrt{3})$; $(\sqrt{3}, \sqrt{3})$ and $(0, -1 + \sqrt{3})$
e. $(20\pi, -4\pi)$ and $(9\pi, 6\pi)$; $(3, 4)$ and $(13, 15)$
f. $\left(\sqrt{7}, \frac{-5\sqrt{7} + 8}{4}\right)$ and $\left(\pi, \frac{-5\pi + 8}{4}\right)$; $(4, -3)$ and $(-8, 8)$

2. Show that the points $(3, 1)$, $(6, 3)$, $(10, 0)$, and $(7, -2)$ are the vertices of a parallelogram.
3. Throughout this chapter we have been confining our discussion to slopes of lines which are neither vertical nor horizontal. Try to extend Theorem 12.3.1 to include horizontal lines.
4. Suppose in Fig. 12.3.2 $\overleftrightarrow{CD}$ and $\overleftrightarrow{EF}$ are two parallel lines and A is a point such that $A \notin \overleftrightarrow{CD}$ and $A \notin \overleftrightarrow{EF}$. Without using Theorem 12.3.1, justify the following statements:
   a. $\triangle CAD \sim \triangle FAE$.
   b. Show that $\overleftrightarrow{CD}$ and $\overleftrightarrow{FE}$ have equal slopes.
   c. Which part of Theorem 12.3.1 has this justified?
   d. Draw a new sketch of Fig. 12.3.2 where point A and $\overleftrightarrow{FE}$ are on opposite sides of $\overleftrightarrow{CD}$. Find a pair of similar right triangles which demonstrate that the slopes of $\overleftrightarrow{CD}$ and $\overleftrightarrow{EF}$ are equal.

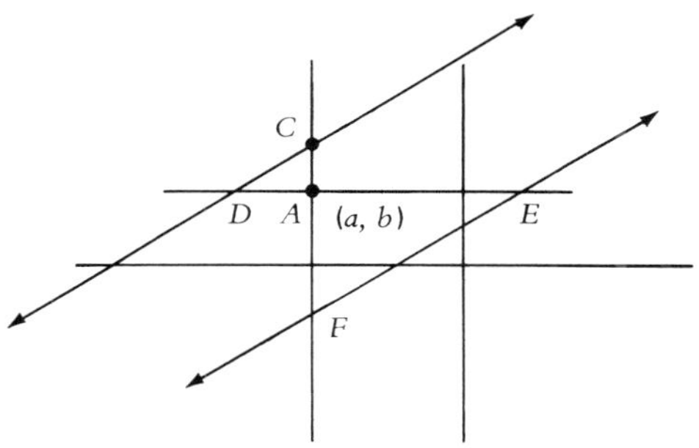

Figure 12.3.2

12.4 DESCRIBING THE COORDINATES OF LINES IN A PLANE

Now that we have discussed the concept of the slope of a line, we are finally ready to convince ourselves logically of what we have intuitively believed for some time; namely, $\{(x, y) | ax + by = c, a \neq 0, b \neq 0\}$ is equal to {all points of the line joining the points $(0, c/b)$ and $(c/a, 0)\}$. To facilitate our discussion, let $P = \{(x, y) | ax + by = c, a \neq 0, b \neq 0\}$ and $L = \{$all points of the line joining the points $(0, c/b)$ and $(c/a, 0)\}$. Now, from earlier encounters with equality of sets, we realize that we must first show that P contains all the points which are in L and, second, that P contains only those points which are in L.

In order to show that P contains all the points of L, let us first observe that obviously $(0, c/b) \in P$ and $(c/a, 0) \in P$. (Why?) So, let us suppose that $R(r, s)$ is any other point in L. That is, $R \neq (0, c/b)$ and $R \neq (c/a, 0)$ but $R \in L$. Let us see if we can show that $R \in P$. Refer to Fig. 12.4.1 and do the exercises that are indicated.

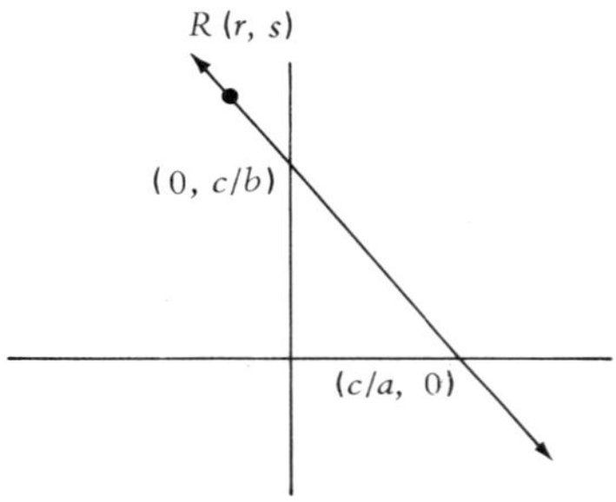

Figure 12.4.1

Exercise 12.4.1

1. a. Calculate the slope of L using the points $(0, c/b)$ and $(c/a, 0)$ and simplify it.
   b. Calculate the slope of L using $R(r, s)$ and $(0, c/b)$.
   c. Since R, $(0, c/b)$, and $(c/a, 0)$ are all $\in L$, what do you know about the slopes calculated in parts a and b? Which part of Theorem 12.3.1 justifies this?
   d. Use the results of parts a, b, and c to show that $ar + bs = c$.

   From the preceding work, you have demonstrated that if $R(r, s) \in L$, then $ar + bs = c$. But this then means that $R \in P$. Therefore, we have shown that P contains all the points of L.
2. Now let us see if we can show that P contains *only* those points which are in L. So suppose $U(u, v)$ is any point in P. Can we show that U must be in L? Try to answer the next set of questions.
   a. Calculate the slope of the line joining (u, v) and $(0, c/b)$.
   b. Since $(u, v) \in P$ we know that $au + bv = c$. (Why?) Use this to show that $v = (c - au)/b$.
   c. Substitute your result of part b for v in the slope you found in part a.
   d. What do the results of parts a, b, and c tell you about

L and the line joining (u, v)? Which part of Theorem 12.3.1 tells you this?

From the preceding work, you should have been able to show that L and the line joining (r, v) and $(0, c/b)$ are either parallel or coincident. Since $(0, c/b) \in L$ (why?), we can conclude that the two lines are coincident and, therefore, $(u, v) \in L$. Thus, we have seen that if $U(u, v)$ is any point of P, it must be in L. That is, P contains only those points which are in L.

Since we have shown that P contains all the points and only the points of L, we know that $P = L$. Actually, there is a flaw in this argument, for it is based on the assumption that the points $(0, c/b)$ and $(c/a, 0)$ are distinct points. If they are not distinct, then, of course, they do not determine a unique line. This case shall be considered in an exercise. As a result of this discussion, we are able to state a second fundamental idea.

12.4.1 Theorem

If $P = \{(x, y) | ax + by = c, a \neq 0, b \neq 0\}$, then $P = \{$all points of the line which joins any pair of distinct points in $P\}$, or more simply, $P = \overleftrightarrow{AB}$, where $A \neq B$ and $A, B \in P$.

As a direct result of this theorem, it has become acceptable to refer to the equation $ax + by = c$, where $a \neq 0$, and $b \neq 0$ as a line, even though it has been pointed out earlier that actually the equation $ax + by = c$ defines a condition which may be used to describe either a binary relation, in which case it would define a line, or a ternary relation, in which case it would define a plane. This seldom poses any problem, however, since it is usually clear from the context which dimensions we are operating in. Thus, from now on we shall also use this convention. Furthermore, since $ax + by = c$ defines a straight line, any equation of this form is called a *linear equation*. At this point there are some additional useful ideas we should consider.

The first is a direct result of Theorem 12.4.1; namely, the slope of the line $ax + by = c$ is $-a/b$. This idea can be very useful, for it enables us to determine the slope of a line when we know its equation without having to determine two points of the line.

We also get the following well-known idea: If the equation $ax + by = c$ is expressed in the form $y = mx + q$, then m is the slope of the line and $(0, q)$ is the *y-intercept*.

It is easy to see why this is so, for if $ax + by = c$, then $by = -ax + c$ and $y = (-a/b)\ x + c/b$. This is of the form $y = mx + q$ with $m = -a/b$ and $q = c/b$. Thus, we see that m is the slope and $(0, q)$ is its y-intercept. When a linear equation is written in the form $y = mx + q$, it is said to be in the *slope-intercept form.*

Example 1

Use the slope-intercept form of the equation $7x - 5y = 10$ to graph the line it represents.

We want to write $7x - 5y = 10$ in the form $y = mx + q$. Thus, we get $-5y = -7x + 10$ or $y = (-7/-5)x + (10/-5) = (7/5)x - 2$. Thus, the slope is $m = 7/5$ and the y-intercept is the point $(0, -2)$. To graph the line, we simply locate the y-intercept and then "count off" a slope of $7/5$, thus locating a second point on the line. The line joining these two points is the line of the equation $7x - 5y = 10$ (Fig. 12.4.2).

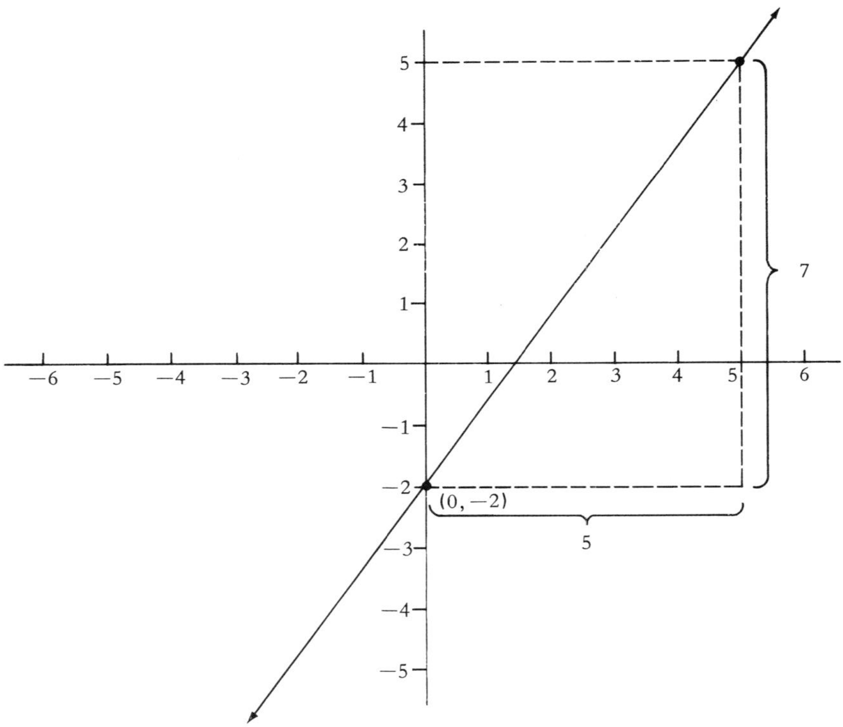

Figure 12.4.2

Example 2

Determine the equation of the line which has a slope of $\frac{3}{5}$ and which goes through the point $(10, 2)$. Since the line has slope $\frac{3}{5}$, the slope-intercept form is $y = mx + q$ with $m = \frac{3}{5}$. Thus, $y = \frac{3}{5}x + q$. Now $(10, 2)$ is on the line. Thus, $(10, 2)$ "satisfies" the equation $y = \frac{3}{5}x + q$. That is, $(10, 2) \varepsilon \{(x, y)\ y = \frac{3}{5}x + q\}$. So $2 = \frac{3}{5}(10) + q$ or $2 = 6 + q$ or $q = -4$. Therefore, the equation is $y = \frac{3}{5}x - 4$. If this is written in the form $ax + by = c$, it becomes $-3x + 5y = -20$.

Example 3

Determine the equation of the line which joins $(2, -3)$ and $(-5, 6)$. The solution is essentially the same as the previous example, except that we must first calculate the slope from the two given points. Thus,

$$m = \frac{-3 - 6}{2 - (-5)} = \frac{-9}{7} = -\frac{9}{7}$$

Thus, the slope-intercept form is $y = (-\frac{9}{7})x + q$. Since $(2, -3) \varepsilon \{(x, y) | y = (-\frac{9}{7})x + q\}$, we get $-3 = -\frac{9}{7}(2) + q$ or $-3 = -\frac{18}{7} + q$ or $q -\frac{3}{7}$.

Thus, we get the equation $y = (-\frac{9}{7})x - \frac{3}{7}$ or, multiplying by 7 and rearranging, we get $9x + 7y = -3$.

Exercise 12.4.2

1. Find the slopes and y-intercepts of the following linear equations and sketch their graphs.
   a. $2x + 3y = 6$
   b. $5x - 2y = 8$
   c. $3y + 7x = 15$
   d. $-9y - 4x = 12$
   e. $x = \frac{1}{3}y + 10$
2. Find the equations of the following lines.
   a. slope is 5 and y-intercept is $(0, 8)$
   b. slope is 5 and x-intercept is $(8, 0)$
   c. slope is $-\frac{2}{3}$ and $(2, 3)$ is a point on it
   d. slope is $\frac{5}{7}$ and $(14, 4)$ is a point on it
3. Find the equation of the lines which join the following pairs of points.
   a. $(5, 2)$ and $(7, -8)$
   b. $(3, -8)$ and $(-5, 6)$
   c. $(4, 5)$ and $(13, 7)$
   d. $(-11, -4)$ and $(-12, -13)$
   e. $(\pi, 3)$ and $(3 + \pi, 3 - \pi)$
4. a. It was indicated that the argument supporting Theorem 12.4.1 was only valid as long as $(0, c/b)$ and $(c/a, 0)$ were distinct points. Show that if $P(0, c/b)$ and $Q(c/a, 0)$ are not distinct, then $c = 0$ and thus the line goes through the origin.
   b. Show that the point $(1, -a/b) \varepsilon \{(x, y) | ax + by = 0, a \neq 0, b \neq 0\}$.
   c. Show that if $R(r, s) \varepsilon \overleftrightarrow{PQ}$, then $R \varepsilon \{(x, y) | ax + by = 0, a \neq 0, b \neq 0\}$.
   d. Show that if $U(u, v) \varepsilon \{(x, y) | ax + by = 0, a \neq 0, b \neq 0\}$, then $U \varepsilon \overleftrightarrow{PQ}$.

Experiment 3

Step 1

Suppose l_1 is a line joining (2, 3) and (7, 5) and l_2 joins (3, 1) and (5, −4).

a. Calculate the slopes of l_1 and l_2.
b. Graph the two lines. What do you observe?

Step 2

Repeat Step 1 where l_1 joins (−3, 5) and (−6, 6) and l_2 joins (−2, −1) and (−1, 2).

Step 3

Repeat Step 1 where l_1 joins (−3, −4) and (0, 0) and l_2 joins (10, −8) and (6, −5).

Step 4

a. From your observations in Steps 1, 2, and 3, can you make a conjecture? Put your statement in if—then form.
b. What is the converse of your conjecture? Do you think it is true?

12.5 SLOPES AND PERPENDICULAR LINES

In Experiment 3, you probably discovered that in contrast to parallel lines, the slopes of perpendicular lines, although not equal, do seem to be related to each other in a rather unique way. For example, you found that the slope of one line was $2/5$ and the slope of a line perpendicular to it was $-5/2$. In other examples, the slopes of the two lines were $-1/3$ and 3; and $4/3$ and $-3/4$. Notice that in each case if we multiply the two numbers together we get $(2/5)(-5/2) = (-1/3)(3) = (4/3)(-3/4) = -1$. When two numbers are related in this way, we say that each number is the negative reciprocal of the other.

12.5.1 Definition

Two numbers are negative reciprocals *of each other if and only if their product is negative one.*

Thus, $-13/5$ and $5/13$, $1/\sqrt{2}$ and $-\sqrt{2}$, $-a/b$ and b/a are additional examples of pairs of numbers which are negative reciprocals of each other.

Let us see if we can show that if two lines are perpendicular, then their slopes are negative reciprocals of each other. Suppose, in Fig. 12.5.1, $l_1 \perp l_2$ and they intersect at the point $R(r, s)$.

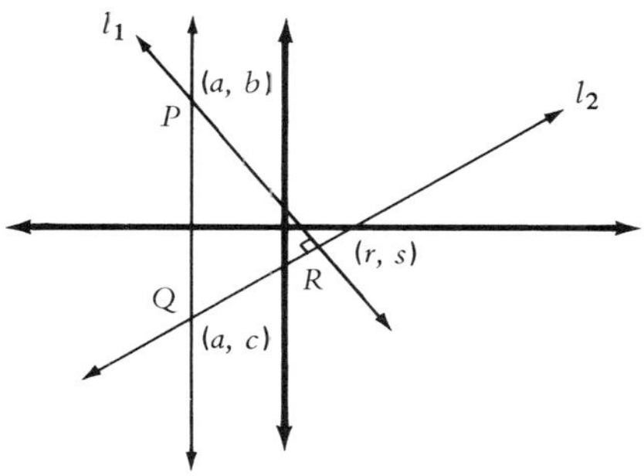

Figure 12.5.1

Since neither l_1 nor l_2 are vertical, there exists a vertical line, say $\{(x, y)|x = a\}$, where $a \neq r$ and intersects l_1 at $P(a, b)$, l_2 at $Q(a, c)$.

The slope of l_1 is $(b - s)/(a - r)$ and the slope of l_2 is $(c - s)/(a - r)$. Now we want to show that they are negative reciprocals; that is, we want to show that

$$\frac{(b-s)}{(a-r)} \cdot \frac{(c-s)}{(a-r)} = \frac{(b-s)(c-s)}{(a-r)^2} = -1$$

Since $\triangle PRQ$ is a right triangle, we have $(PQ)^2 = (PR)^2 + (QR)^2$, where $PQ = |b - c|$, $PR = \sqrt{(a-r)^2 + (b-s)^2}$, and $QR = \sqrt{(a-r)^2 + (c-s)^2}$.

Thus,

$$\begin{aligned}(b-c)^2 &= [(a-r)^2 + (b-s)^2] + [(a-r)^2 + (c-s)^2] \\ &= 2(a-r)^2 + (b-s)^2 + (c-s)^2\end{aligned}$$

or

$$b^2 - 2bc + c^2 = 2(a-r)^2 + b^2 - 2bs + s^2 + c^2 - 2cs + s^2$$

or

$$-2bc = 2(a-r)^2 + 2s^2 - 2bs - 2cs$$

Then

$$-2bc + 2bs + 2cs - 2s^2 = 2(a-r)^2$$

or

$$-2(b-s)(c-s) = 2(a-r)^2$$

and therefore

$$\frac{(b-s)(c-s)}{(a-r)^2} = -1$$

This shows that the slopes of l_1 and l_2 are negative reciprocals. It can also be shown that the converse of this idea is also true and you will be asked to consider it in an exercise. Thus, we have another very useful theorem.

12.5.2 Theorem

Two lines are perpendicular if and only if their slopes are negative reciprocals.

This, of course, implies two statements:

1. If two lines are perpendicular, then their slopes are negative reciprocals.
2. If the slopes of two lines are negative reciprocals, then the lines are perpendicular.

comment . . .

Theorem 12.5.2 is referring to lines which are not vertical or horizontal.

Example 1

Find the slope of a line which is perpendicular to a line which joins the points (5, 6) and (8, 11).

The slope of the line joining (5, 6) and (8, 11) is

$$\frac{6-11}{5-8} = \frac{-5}{-3} = \frac{5}{3}$$

Therefore, the slope of the line perpendicular to it is $-3/5$.

Example 2

Find the equation of the line which goes through (7, 1) and is perpendicular to the line whose equation is $7x + 4y = 2$.

We can see that the slope of the line $7x + 4y = 2$ is $-7/4$. Therefore, the line through it has a slope of $4/7$. Thus, $y = 4/7x + q$. Since it must go through the point (7, 1), we get $1 = 4/7(7) + q$ and $q = -3$. Thus, the equation is $y = 4/7x - 3$ or $-4x + 7y = -21$.

Exercise 12.5.1

1. Find the slope of a line which is perpendicular to a line whose slope is:

   a. $3/4$ b. $5/9$
   c. $-2/3$ d. 4
   e. -3 f. $1\frac{1}{4}$
   g. $.1$ h. $1/\sqrt{7}$
   i. $-\pi/2$

2. Find the slope of a line which is perpendicular to a line which joins the following pairs of points:
   a. $(1, 4)$ and $(3, -7)$ b. $(-2, -3)$ and $(0, 3)$
   c. $(1, 7)$ and $(-2, 3)$ d. $(\frac{1}{2}, 3)$ and $(-\frac{1}{4}, \frac{2}{3})$
   e. $(\sqrt{2}, \pi)$ and $(-2, 3)$ f. (x_1, y_1) and (x_2, y_2)
3. Determine by slopes which of the following are vertices of right triangles.
   a. $(2, 2), (4, 1), (4, 6)$ b. $(2, 5), (-4, 3), (-3, 0)$
   c. $(1, 1), (4, 4), (7, 2)$ d. $(1, 1), (3, 0), (0, -4)$
4. In each problem, find the equation of the line which is perpendicular to the line of the equation given and goes through the point $(-2, 6)$.
   a. $3x + 2y = 8$ b. $-4x + 5y = 10$
   c. $5x - 7y = 15$ d. $12x + 4y = 11$
   e. $\sqrt{11}x + \frac{2}{3}y = 11$
5. In Fig. 12.5.2, suppose l_1 and l_2 are two lines which intersect at $R(r, s)$. $P(a, b)$ and $Q(a, c)$ are the points of intersection of the line $\{(x, y) | x = a\}$ with l_1 and l_2, respectively.
   a. What is the slope of l_1? l_2?
   b. Suppose the slopes are negative reciprocals. Show that

   $$(bc - bs - cs + s^2) = -(a - r)^2$$

   c. Find $(PR)^2$, $(QR)^2$, $(PQ)^2$.
   d. Use the results of parts b and c to show that $(PQ)^2 = (PR)^2 + (QR)^2$.
   e. What does part d tell you about $\triangle PRQ$?
   f. Which part of Theorem 12.5.2 have you justified?

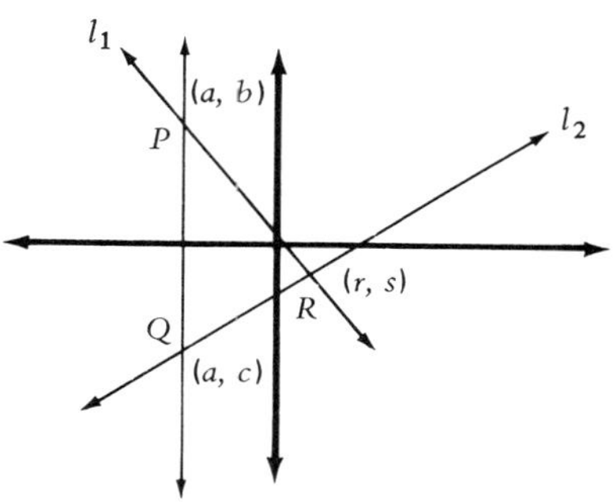

Figure 12.5.2

12.6 THE MIDPOINT OF A LINE SEGMENT

Suppose that $P(x_1, y_1)$ and $Q(x_2, y_2)$ are the endpoints of a line segment. What are the coordinates of M, the *midpoint* of this segment? Let us first consider the possibility that P and Q are both on a vertical line in Fig. 12.6.1. Without loss of generality, let us assume that $y_2 > y_1$. Then $PQ = |y_2 - y_1| = y_2 - y_1$. Suppose the coordinates of M, the midpoint of $\overleftrightarrow{QP}$, are (x_1, m). Then $MP = |m - y_1| = m - y_1$. Since M is the midpoint of $\overline{PQ}$, we get $m - y_1 = \frac{1}{2}(y_2 - y_1) = \frac{1}{2}y_2 - \frac{1}{2}y_1$ or $m = \frac{1}{2}y_2 + \frac{1}{2}y_1 = \frac{1}{2}(y_1 + y_2)$. Thus, the coordinates of M are $(x_1, \frac{1}{2}(y_1 + y_2))$. Similarly, it can be shown that if P and Q are on a horizontal line, then the midpoint is $(\frac{1}{2}(x_1 + x_2), y_1)$.

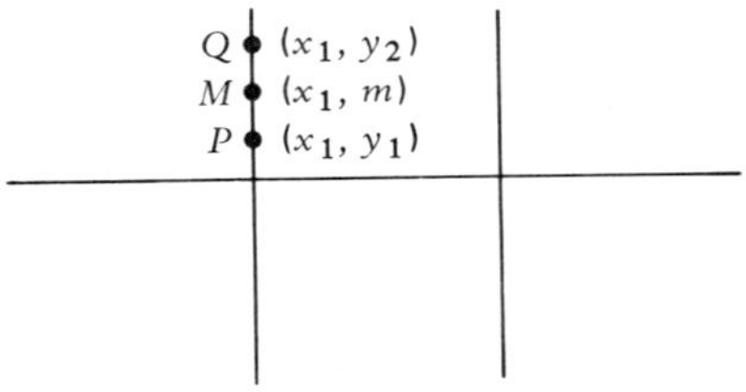

Figure 12.6.1

Now suppose $\overleftrightarrow{PQ}$ is neither vertical nor horizontal. (See Fig. 12.6.2.) Pass horizontal lines through Q, M, and P (why possible?) intersecting some vertical line at $R(a, y_2)$, $S(a, m_2)$, and $T(a, y_1)$, respectively. Now, since M is the midpoint of $\overline{PQ}$, S must be the midpoint of $\overline{RT}$. (Why?) Thus, the coordinates of S are $(a, \frac{1}{2}(y_1 + y_2))$; that is, $m_2 = \frac{1}{2}(y_1 + y_2)$. In a similar way, by drawing vertical lines through Q, M, and P it can be shown that $m_1 = \frac{1}{2}(x_1 + x_2)$. Thus, the coordinate of the midpoint M must be $(\frac{1}{2}(x_1 + x_2), \frac{1}{2}(y_1 + y_2))$.

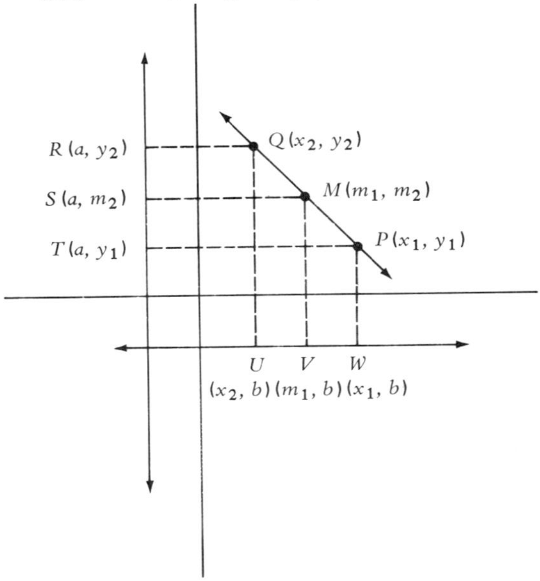

Figure 12.6.2

12.6.1 Theorem

The coordinates of the midpoint of the segment joining (x_1, y_1) and (x_2, y_2) are $(\frac{1}{2}(x_1 + x_2), \frac{1}{2}(y_1 + y_2))$.

Example 1

Find the coordinates of the midpoint of the line segment whose endpoints are (7, 4) and (−3, 2). Let (m_1, m_2) be the coordinates of the midpoint. Then $m_1 = \frac{1}{2}(7 + -3) = \frac{1}{2}(4) = 2$; and $m_2 = \frac{1}{2}(4 + 2) = 3$. Thus, (2, 3) is the midpoint.

Exercise 12.6.1

1. Find the coordinates of the midpoints of the segments which have the following endpoints:
   a. (5, 7) and (−3, 8) b. (−1, −6) and (−4, −3)
   c. (6, 7) and (13, −20) d. $(\pi, 3)$ and $(\sqrt{2}, 4)$
2. A triangle has $P(2, 5)$, $Q(3, -6)$, and $R(8, -9)$ as vertices.
   a. Find the coordinates of S, the midpoint of $\overline{PQ}$.
   b. Find the coordinates of T, the midpoint of $\overline{QR}$.
   c. Show that $\overleftrightarrow{ST} \parallel \overleftrightarrow{PR}$.
   d. Show that $ST = \frac{1}{2}PR$.

12.7 SUMMARY

1. Two distinct nonvertical lines are parallel if and only if their slopes are equal.
2. Two nonvertical lines are perpendicular if and only if their slopes are negative reciprocals.
3. The equation $ax + by = c$ represents a line, provided a and b are not both zero.
4. The midpoint of the line segment which joins (x_1, y_1) and (x_2, y_2) is $(\frac{1}{2}(x_1 + x_2), \frac{1}{2}(y_1 + y_2))$.

AREA REVISITED

13

13.1 PROLOGUE

In Chapter 7 we talked about the concept of area. We saw that the problem of calculating areas of figures was primarily one of designing clever and efficient ways to count the area units. We considered the rectangle and saw that the most efficient way to determine its area was to multiply the number of square units in each row by the number of rows, thus suggesting the familiar formula $A = l \times w$. Since we will be basing further area principles on this formula, we shall refer to it as the following.

13.1.1 Postulate

The area of a rectangle *is equal to the product of its length times its width.* $A = l \times w$.

We then proceeded to the parallelogram and asked if it would be possible to find a rectangle which was equal in area to the parallelogram. Figure 13.1.1 was suggested. It seemed apparent

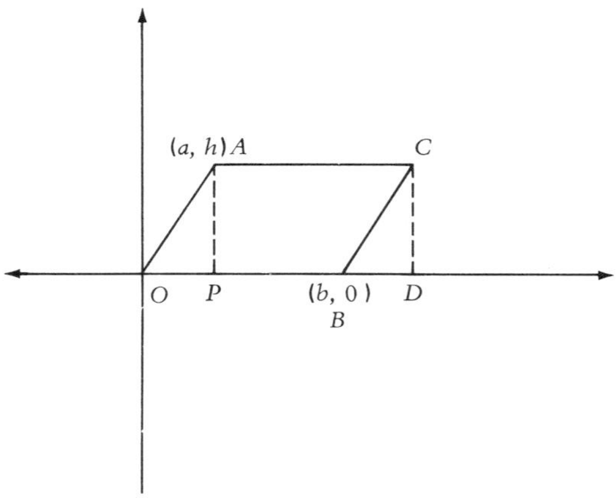

Figure 13.1.1

that if the area of $\triangle AOP$ were equal to the area of $\triangle CDB$, then the area of rectangle $APDC$ would equal the area of parallelogram $AOBC$. The problem at the time was that we had difficulty in convincing ourselves that the two triangles were equal in area. In fact, this was one of the reasons why we investigated the concept of congruent triangles. In this chapter we shall be able to resolve this problem and we shall be able to determine area formulas for other common geometric figures such as the triangle, trapezoid, regular polygon, and circle. Not only will we discuss how to calculate the areas of these figures, but we will conduct experiments which will enable us to discover how various area formulas are derived from others.

in our opinion . . .

We realize that our readers are probably already familiar with many of the area formulas and for that reason the "discovery" is liable to be artificial. Nevertheless, we believe that such experiments should still be performed. First, it is possible that our readers initially memorized the formulas as isolated, unrelated facts. If this is the case, these experiments will provide our readers with additional insights into area, giving them a rational basis for the formulas. Second, these experiments may serve as models for experiments which our readers, as teachers, can set up for their students. There is a movement in the schools to conduct mathematics laboratories where students are encouraged to use manipulative materials to help them investigate a mathematical principle. Although area and volume are not the only topics which are suitable for the laboratory approach, it is a good place to start, since the procedures are simple and the materials inexpensive.

Experiment 1

Purpose

To discover the formula for the area of a parallelogram.

Preconcepts

The student should be able to:

1. determine the area of a rectangle
2. distinguish between figures which are parallelograms and those which are not
3. identify the bases of a given parallelogram
4. construct, sketch, or draw altitudes to a given set of bases of a given parallelogram

Materials

1. heavy paper or cardboard
2. protractor
3. ruler
4. scissors

Procedure

1. On your paper or cardboard, carefully draw a parallelogram which has the dimensions of Fig. 13.1.2.

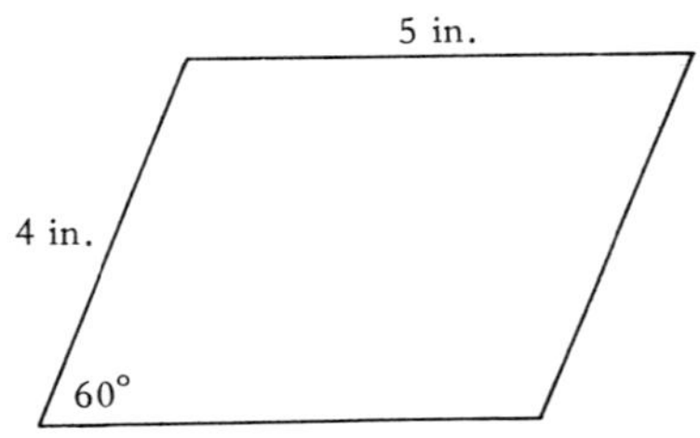

Figure 13.1.2

2. Cut out this figure.
3. Using this figure as a pattern, trace out at least three more duplicates.
4. Take one of your parallelograms and see if you can cut it into two pieces and rearrange the pieces so that they form a rectangle. Figure 13.1.1 may help you.
5. Take one of your other parallelograms and see if you can cut along a different line but still end up with a rectangle of the same size and shape as the first one.
6. With your figure still arranged as a rectangle, what dimensions of the rectangle do you need to measure in order to calculate its area?

   ____
7. Use your ruler to measure these dimensions and record your answers below.
   Dimension of rectangle: ______________________________
8. Calculate the area of the rectangle and record your answer.
   Area of rectangle: ______________________________
9. What do you believe is the area of the parallelogram? ________
   ______________________________ Why? ______________________________
   ____
10. Put your rectangle back into the shape of the parallelogram. What part of the parallelogram is equal to the length of the rectangle? ______________________________
    What part of the parallelogram is equal to the width of the rectangle? ______________________________
11. Take a third copy of the parallelogram and see if you can

divide the parallelogram into two pieces and rearrange them to form a rectangle which is *not* congruent to the first rectangle you made.

12. Measure the dimensions of this rectangle and determine its area. Record your measurement.
    Dimensions: ____________________
    Area: ____________________
13. Compare the results of Step 12 with those of Step 8.
    What do you observe? ____________________
14. State what you believe to be the area principle you developed: ____________________
15. What do you think is the formula which summarizes this principle? ____________________
16. Test your conjecture by drawing and cutting out at least one other parallelogram of a different size and shape.

13.2 THE AREA OF A PARALLELOGRAM

The previous experiment provided us with an intuitive development of the area principle for a parallelogram. It now remains for us to formalize the concrete procedures into an abstract argument.

Let us now return to our parallelogram diagram (Fig. 13.2.1). Recall that we dropped a perpendicular from A to $\overleftrightarrow{OB}$, intersecting $\overleftrightarrow{OB}$ at P. If the coordinates of A are (a, h), the coordinates

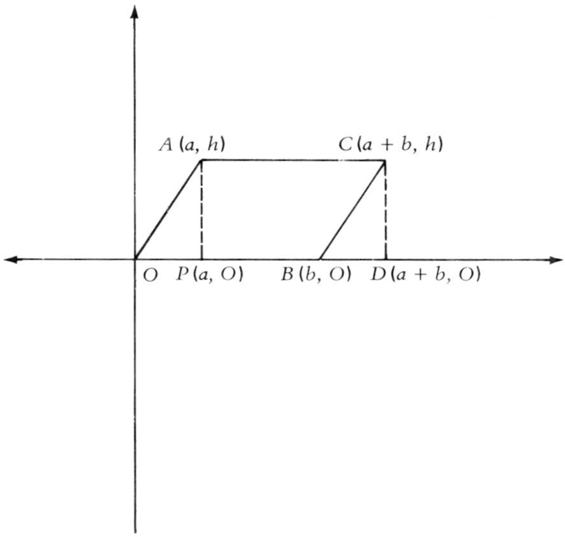

Figure 13.2.1

of P are $(a, 0)$. Similarly, we dropped a perpendicular from C to $\overleftrightarrow{OB}$, intersecting it at D. Convince yourself that the coordinates of C are $(a + b, h)$ and of D are $(a + b, 0)$. Now we can show $\triangle APO$ and $\triangle CDB$ are congruent, for $\angle O \cong \angle B$ (why?), $\angle P \cong \angle D$. (Why?) Also, we can see that $OP = |a|$ and $BD = |a + b - b| = |a|$, so $\overline{OP} \cong \overline{BD}$. Therefore, $\triangle APO \cong \triangle CDB$. (Why?) Thus, the area of $\triangle APO =$ the area of $\triangle CDB$, and the area of parallelogram $AOBC$ must equal the area of rectangle $APDC$.

Now the area of the rectangle is $PD \times AP$, and, since $PD = |a + b - a| = |b|$ and $AP = |h|$, we have the area of rectangle $= |b| \cdot |h|$. Now observe that $OB = |b|$ as well. That is, the length of the rectangle is equal to the base of the parallelogram. Also, the width of the rectangle is the distance between the bases of the parallelogram (called the *altitude* or *height*).

Thus, the area of the parallelogram is

$$CB \times AB = |b| \cdot |h| = b \cdot h$$

Why can we eliminate the absolute value?

We are now able to suggest our second area principle.

13.2.1 Theorem

The area of a parallelogram *is equal to the product of its base times its height. $A = bh$.*

Exercise 13.2.1

1. Find the area of a parallelogram which has a base of 121 in. and an altitude of 53 in.

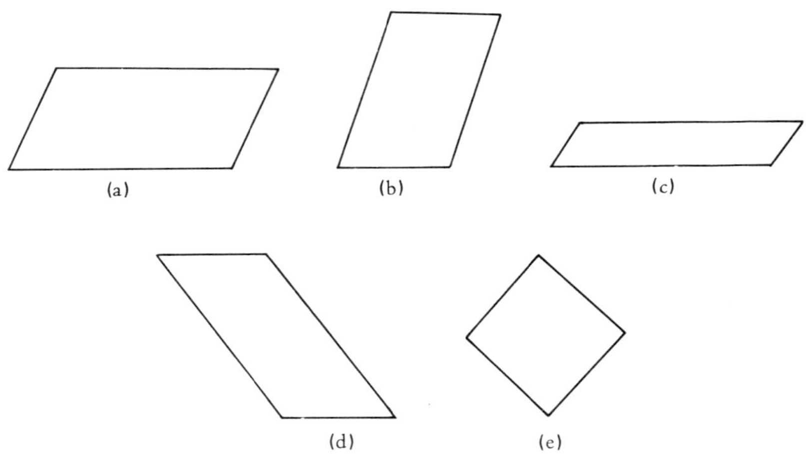

Figure 13.2.2

2. Find the area of a parallelogram whose vertices are (0, 0), (8, 0), (11, 4), and (3, 4).
3. Give an argument to show that a figure with vertices of (0, 0), $(b, 0)$, $(a + b, h)$, and (a, h) is always a parallelogram, providing that all the points are distinct.
4. Suppose $ABCD$ is a parallelogram and $m°(\angle B) = 45$, $AB = 4$ in., and $BC = 10$ in. Find its area.
5. Use your ruler to measure the necessary dimensions for calculating the area of each of the diagrams in Fig. 13.2.2 and determine their areas.

Experiment 2

Purpose

To discover the formula for the area of a triangle.

Preconcepts

The student should be able to:
1. determine the area of a parallelogram
2. construct, sketch, or draw an altitude to a given base of a given triangle

Materials

1. heavy paper or cardboard
2. protractor
3. ruler
4. scissors

Procedure

1. Draw any triangle on your paper or cardboard and cut it out.
2. Use your triangle as a pattern and cut out a duplicate.
3. Take one of your two triangles and see if you can cut it into two pieces and then rearrange them so that they form a parallelogram. Describe where you made the cut.
   ____
4. Using your ruler, measure the base and corresponding altitude of the parallelogram. Record your answers below.
   Base of parallelogram: ______________________
   Altitude of parallelogram: ______________________
5. Calculate the area of the parallelogram.
   ____
6. Is this also the area of the triangle? ______________
   Why? ______________________________
7. Place your parallelogram on top of the second triangle. What part of the triangle is equal to the base of the parallelogram?
   ____

8. Measure the length of the altitude of the triangle which is drawn to the base of the triangle which equals the base of the parallelogram. Record the measurement here: ______
9. How does the altitude of the parallelogram compare with the altitude of the triangle?

______
10. If you wanted to determine the area of the triangle without cutting it into pieces, what parts of the triangle would you measure? ______

______
11. How would you use these measurements to calculate the area of the triangle? ______

______
12. What do you think is the area principle you have discovered?

______
13. What do you think is the area formula which represents this principle? ______
14. Test your conjecture by repeating the above steps on at least one other triangle.

13.3 THE AREA OF A TRIANGLE

The previous experiment should have provided us with an intuitive development of how to determine the area of a triangle. A formal development can parallel our concrete procedure.

Suppose in Fig. 13.3.1 $Q(a, h)$, $O(0, 0)$, and $P(b, 0)$ are the vertices of a triangle. The results of your experiment may have suggested that we should draw a line through U, the midpoint of $\overline{QP}$, parallel to $\overleftrightarrow{OP}$, intersecting altitude $\overline{QR}$ at M and side $\overline{OQ}$

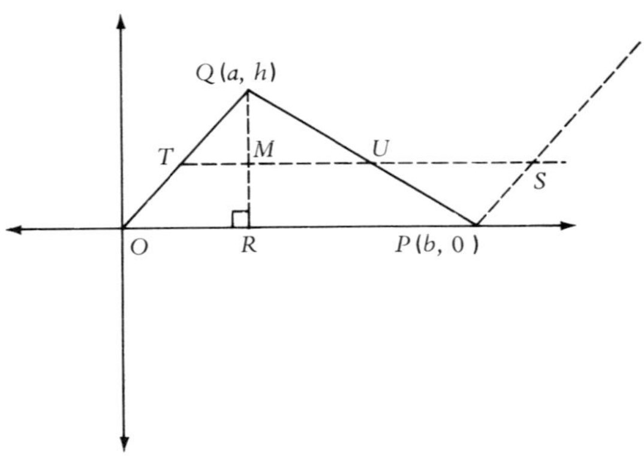

Figure 13.3.1

at T. Draw a line through P parallel to $\overleftrightarrow{OQ}$. Let the point S be the intersection of this line with $\overleftrightarrow{TM}$.

Now, first you should be able to convince yourself that $TOPS$ is a parallelogram. Second, you should be able to see that $\triangle QTU \cong \triangle PSU$ and that consequently the area of $\triangle QTU =$ the area of $\triangle PSU$. (This will be left as an exercise.)

Thus, we have that

Area of $\triangle QOP$ equals the area of parallelogram $TOPS$.
But the area of parallelogram $TOPS = OP \times RM$.
But OP is the base of QOP and RM is ½ the altitude of $\triangle QOP$. (Why?)

Thus, we get

$$\text{Area of } \triangle QOP = OP \times \tfrac{1}{2}QR = \tfrac{1}{2} \times OP \times QR$$

We are able to formulate the following principle.

13.3.1 Theorem

The area of a triangle *is equal to one-half the product of its base and altitude.* $A = \tfrac{1}{2}bh$.

Exercise 13.3.1

1. Find the area of a triangle which has a base of 32 in. and a height of 15 in.
2. Find the area of a triangle which has (0, 0), (4, 3), and (6, 0) as vertices.
3. a. Show that a triangle with $(0, 0)$, $(\tfrac{1}{2}b, h)$, and $(b, 0)$ as vertices is isosceles.
   b. Suppose the length of one of its legs is d. Express h as a function of d.
   c. Create a formula for finding the area of an isosceles triangle in terms of d and b.
   d. Show that the triangle with vertices (4, 5), (12, 6), and (5, 2) is isosceles.
   e. Use your formula created in part c to find the area of this triangle.
   f. Use your formula to show that the area of an equilateral triangle of side s is $A = s^2\sqrt{3}/4$.
4. Figure 13.3.2 is the diagram used for our development of the area of a triangle.
   a. Find the coordinates of M, T, U, R, and S.
   b. Show that $TOPS$ is a parallelogram.
   c. Show that $\triangle QTU \cong \triangle PSU$.

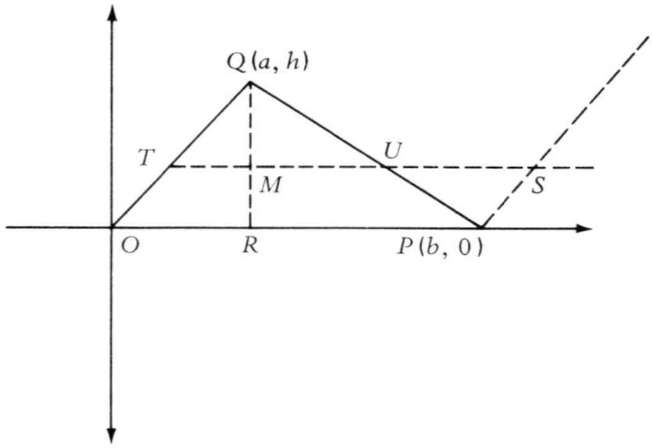

Figure 13.3.2

5. Use your ruler to measure the necessary dimensions for calculating the area of each of the diagrams in Fig. 13.3.3 and determine the areas.

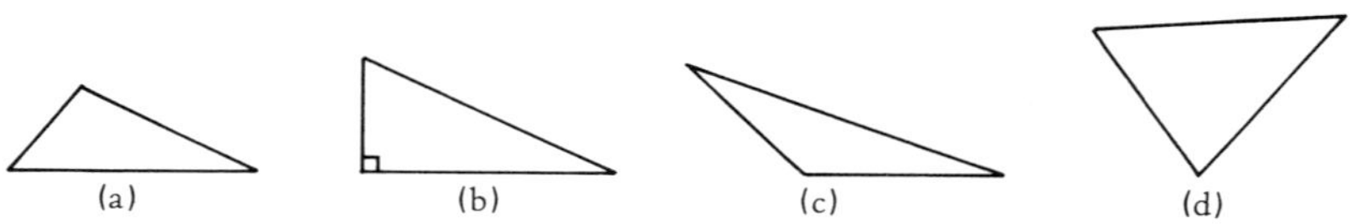

Figure 13.3.3

Experiment 3

Purpose

To discover the formula for the area of a trapezoid.

Preconcepts

The student should be able to:
1. determine the area of a triangle and a parallelogram
2. distinguish between those quadrilaterals which are trapezoids and those which are not
3. identify the bases, legs, and altitude of a given trapezoid

Materials

1. cardboard
2. protractor
3. ruler
4. scissors

Procedure

1. Draw any nonisosceles trapezoid and cut it out.
2. Label its vertices A, B, C, and D in order in a counterclockwise direction, with $\overline{AB}$ being one base and $\overline{CD}$ the other base.

3. Use your ruler to locate M, the midpoint of $\overline{EC}$. Draw in $\overline{AM}$.
4. Cut the trapezoid into two pieces along $\overline{AM}$.
5. See if you can arrange the two pieces so that they form a single triangle.
6. This triangle has three altitudes. One of them is equal to the altitude of the trapezoid. Measure the length of that altitude. Altitude: ______
7. Measure the base of the triangle which corresponds to the altitude you measured. ______
8. Use these measurements to calculate the area of the triangle.

   ______
9. Is this also the area of the trapezoid? ______
   Why? ______
10. We already recognized that the altitude of the triangle is equal to the altitude of the trapezoid. Is the base of the triangle equal to either one of the bases of the trapezoid?

    ______

    How is the base of the triangle related to the bases of the trapezoid? Keep rearranging your pieces back and forth from a triangle until you can see this relationship.

    ______
11. If you did not want to cut the trapezoid, what measurements would you make so that you could determine its area? ______

    ______
12. How would you use these measurements to calculate the area? ______
13. What do you think is the area principle you have discovered?

    ______
14. What do you think is the area formula which summarizes this principle? ______
15. Test your conjecture by repeating the above steps on at least one other trapezoid.

13.4 THE AREA OF A TRAPEZOID

The preceding experiment indicated one way of formally developing the formula for the area of a trapezoid. Our method shall parallel that experiment. In Fig. 13.4.1, $OPQR$ is a trapezoid with vertices $(0, 0)$, $(b, 0)$, (c, h), and (a, h), respectively, with $|c - a| \neq |b|$. (Why?) Let M be the midpoint of $\overline{QP}$ and draw $\overleftrightarrow{RM}$ interesecting $\overleftrightarrow{OP}$ at S. With little difficulty, you should be able to show $\triangle RQM \cong \triangle SPM$. Therefore, the area of trapezoid $OPQR$ is equal to the area of $\triangle ROS$. Now, area of $\triangle ROS = \frac{1}{2} \times OS \times h$. (Why?) But $OS = OP + PS$ and $PS = QR$. So the

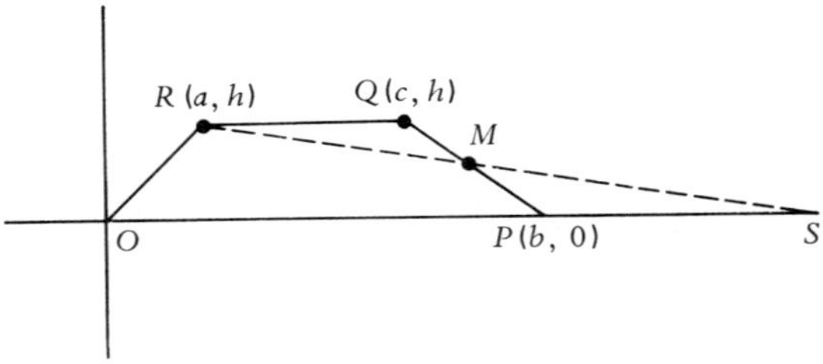

Figure 13.4.1

area of the trapezoid is

$$OPQR = Area\ of\ \triangle ROS = \tfrac{1}{2} \times (OP + PS) \times h$$
$$= \tfrac{1}{2} \times h \times (OP + PS)$$

But OP is the measure of one base and PS is the measure of the other. Therefore, we are able to formulate the following principle.

13.4.1 Theorem

The area of a trapezoid *is equal to one-half the product of its altitude and the sum of its bases.* $A = \tfrac{1}{2}h(b_1 + b_2)$.

Exercise 13.4.1

1. Find the area of a trapezoid which has one base of 112 in., another of 313 in., and an altitude of 17 in.
2. Find the area of a trapezoid which has (0, 0), (7, 0), (6, 8), and (3, 8) as vertices.
3. Explain why the following is true:
   A quadrilateral with $(0, 0)$, $(b, 0)$, (c, h), and (a, h) as vertices is a trapezoid if and only if b, c, h, and a are not zero and $c - a \neq b$.
4. Refer to Fig. 13.4.1.
   a. Calculate the coordinates of M.
   b. Calculate the coordinates of S.
   c. Show that $PS = RQ$.
5. Suppose $ABCD$ is a trapezoid with A, B, C, and D having coordinates of (a, h), $(0, 0)$, $(b_2, 0)$, and $(b_2 - a, h)$, respectively. Also, $AB = s$ and $AD = b_1$ and $b_1 < b_2$ (Fig. 13.4.2).
   a. Use the distance formula to show that $s = DC$, thus showing that $ABCD$ is an isosceles trapezoid.

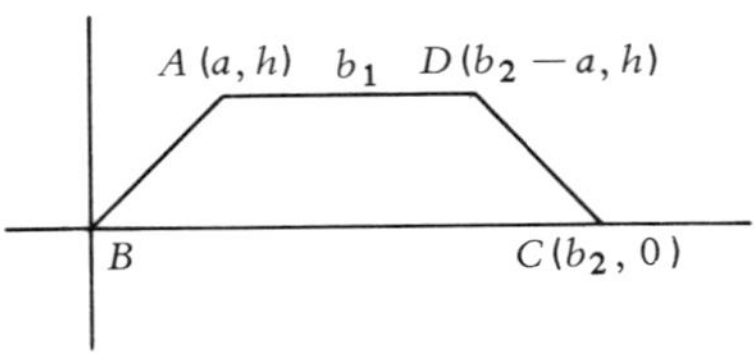

Figure 13.4.2

b. Show that $a = (b_2 - b_1)/2$.
c. Draw a perpendicular through A intersecting $\overleftrightarrow{BC}$ at E. What are the coordinates of E?
d. Express h in terms of s, b_1 and b_2.
e. Find a formula for the area of an isosceles trapezoid in terms of b_1, b_2, and s.

6. Use your ruler to measure the necessary dimensions for calculating the area of each of the diagrams in Fig. 13.4.3 and determine the areas.

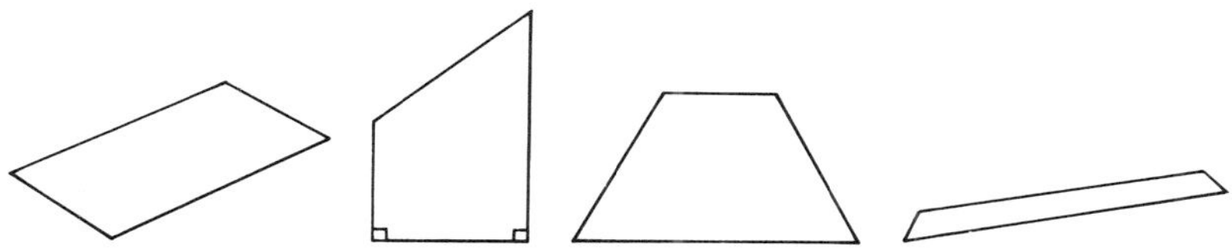

(a) (b) (c) (d)

Figure 13.4.3

13.5 THE AREA OF POLYGONS

The problem of calculating areas of other polygons is rendered quite simple since we are able to dissect them into triangles, as indicated in Fig. 13.5.1. We simply have to determine the areas of each of the triangles and then add them up.

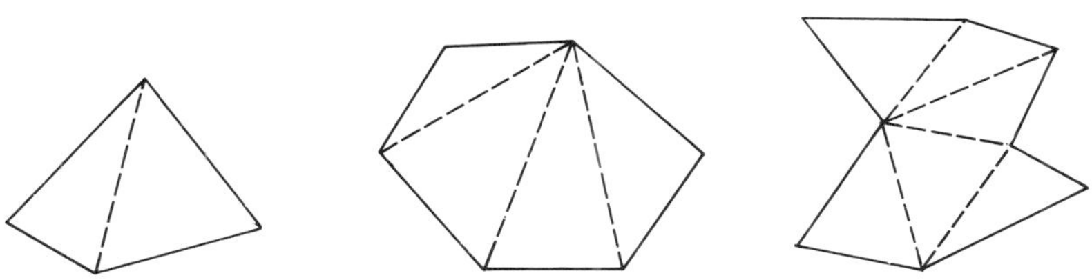

Figure 13.5.1

If, however, we limit ourselves to *regular polygons* (that is, polygons that are equilateral and equiangular), then we are able to formulate a rather simple formula for calculating the area. Before we proceed with this, it is essential that we review some basic properties of regular polygons. First, although there are some names for the more common polygons (such as quadrilateral, pentagon, hexagon, heptagon, and so on), we shall simply refer to any polygon of n sides as an *n-gon*. Second, it is a basic property that for every regular n-gon, the vertices all lie on the same circle. We say that this circle is *circumscribed* about the n-gon. The center of this circle shall be referred to as the *center of the regular n-gon*. Furthermore, the midpoints of all the sides of the n-gon lie on another circle which has the same center as the circumscribed circle. This circle is called the *inscribed circle*. The radius of the circumscribed circle is called the *radius of the regular n-gon*, and the radius of the inscribed circle is called the *apothem* of the n-gon (see Fig. 13.5.2).

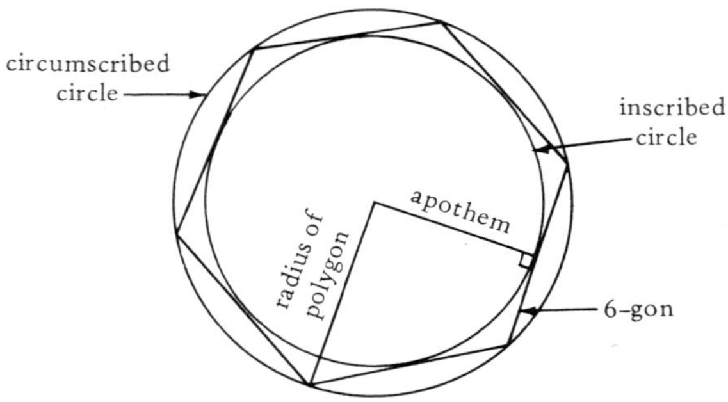

Figure 13.5.2

It should be clear that the apothem, when drawn to the midpoint of any side, is perpendicular to the side.

Now let us consider any regular n-gon. Suppose we call its vertices $V_1, V_2, V_3, \ldots, V_n$ (Fig. 13.5.3).

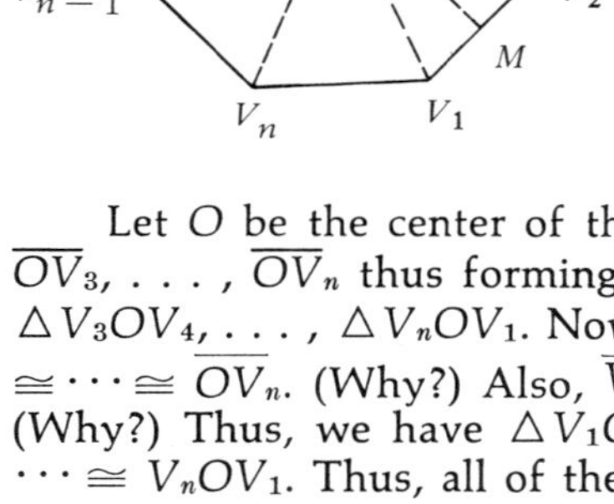

Figure 13.5.3

Let O be the center of the n-gon. Then draw in $\overline{OV_1}$, $\overline{OV_2}$, $\overline{OV_3}, \ldots, \overline{OV_n}$ thus forming n triangles, $\triangle V_1OV_2$, $\triangle V_2OV_3$, $\triangle V_3OV_4, \ldots, \triangle V_nOV_1$. Now clearly $\overline{OV_1} \cong \overline{OV_2} \cong \overline{OV_3} \cong \cdots \cong \overline{OV_n}$. (Why?) Also, $\overline{V_1V_2} \cong \overline{V_2V_3} \cong \cdots \cong \overline{V_nV_1}$ (Why?) Thus, we have $\triangle V_1OV_2 \cong \triangle V_2OV_3 \cong \triangle V_3OV_4 \cong \cdots \cong V_nOV_1$. Thus, all of the n triangles are equal in area. Thus, we only have to find the area of one triangle, say $\triangle V_1OV_2$, and then multiply that by n.

To find the area of $\triangle V_1OV_2$ we need the altitude from O to

$\overline{V_1V_2}$. But the altitude strikes the base at its midpoint (Why?) But then this altitude is the apothem of the n-gon (call it a). Thus, the area of $\triangle V_1OV_2 = \frac{1}{2}a \times V_1V_2$ or if we let $s = V_1V_2$, the area of $\triangle V_1OV_2 = \frac{1}{2}as$. Therefore, the area of the n-gon is $n \times (\frac{1}{2}as) = \frac{1}{2}a(ns)$. Now n is the number of sides and s is the length of each side, so ns is the perimeter of the n-gon (say p). Thus, we get our next area principle.

13.5.1 Theorem

The area of any regular n-gon *is equal to one-half the product of the apothem and the perimeter.* $A = \frac{1}{2}ap$.

Exercise 13.5.1

1. Find the area of a regular 12-gon which has an apothem of 37.3 in. and one side of 20 in.
2. Find the area of a regular hexagon which has a side of 18 in. [*Hint:* Join the center of the hexagon to two consecutive vertices, thus forming an isosceles triangle. Then find the measure of the base angles.]
3. The equilateral triangle is a regular 3-gon. Use the area formula for an n-gon to develop a formula for the area of the triangle in terms of s, the length of its side. [*Hint:* Observe that the medians of any triangle all concur at a point whose distance from the base is $\frac{1}{3}$ the length of the median.]
4. Show that the apothem drawn to the midpoint of a side of a regular polygon is perpendicular to the side.

Experiment 4

Purpose

To discover the formula for the area of a circle.

Preconcepts

The student should be able to:

1. determine the area of a parallelogram
2. identify the altitude to the base of a given parallelogram
3. determine the circumference of a circle
4. describe what is meant by the circumference of a circle
5. identify the radius of a circle

Materials

1. A model of a circle which is cut into two semicircles. Each semicircle is divided into an equal number of wedges, joined together by a leather strap around the circumference.
2. Ruler

Procedure

1. See if you can rearrange the pieces so that they form a figure which approximates a parallelogram.
2. Measure the dimensions you need in order to determine the area of the parallelogram. ________.
3. Calculate the area of the parallelogram. ________
4. Do you think this will be close to the area of the circle? ________ Why would it not be the exact area of the circle? ________________
5. Look at the base of the parallelogram. What does it become as you put it back into a circle? ________________
6. How do you calculate this length, without measuring it? ________________
7. Look at the altitude of the parallelogram. What does it become as you reform this circle? ________________
8. If you did not want to cut the circle into pieces, what *single* measurement could you make in order to calculate the area of the circle? ________
9. How would you use this measurement to determine the area of the circle? ________________
10. What do you think is the area principle you have discovered? ________________
11. What do you think is the formula for the area of the circle? ________________

13.6 THE AREA OF A CIRCLE

We will look at the development of the area of a circle in two different ways. First, let us parallel the preceding experiment. With the model, you should have been able to approximate a parallelogram-like shape by arranging them as in Fig. 13.6.1. Now, actually, this is not a parallelogram. So, in this respect this

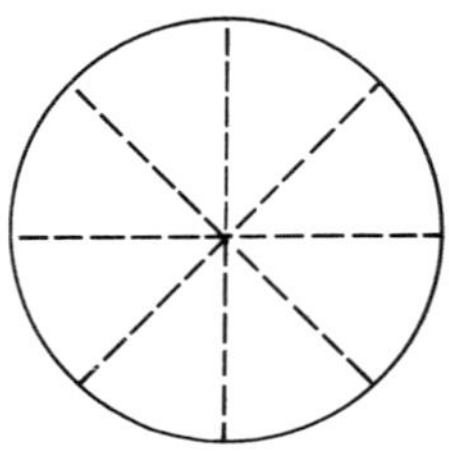
(a)

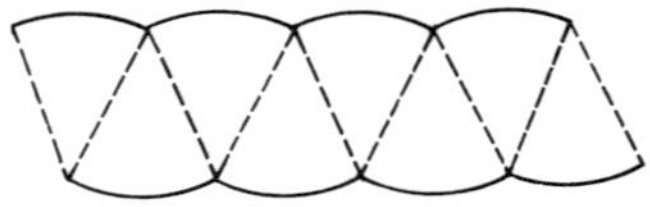
(b)

Figure 13.6.1

approach differs from our work with parallelograms, triangles, and trapezoids for in those cases they actually formed those figures which they were rearranged to form. However, let us examine this parallelogram-like shape more closely. In Figs. 13.6.2, 13.6.3, and 13.6.4, you will see what happens to this shape as we increase the number of wedges.

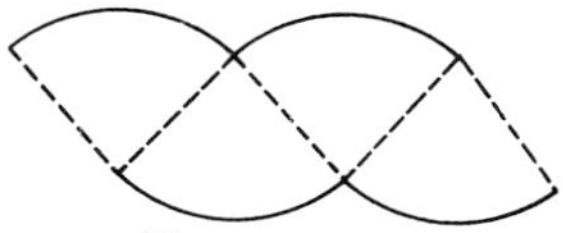

Figure 13.6.2
4 wedges

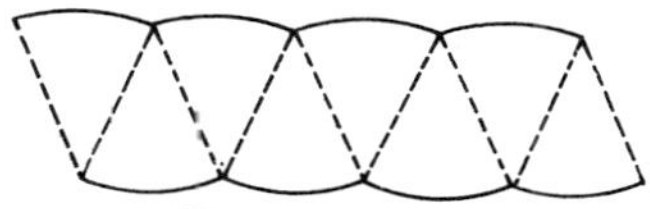

Figure 13.6.3
8 wedges

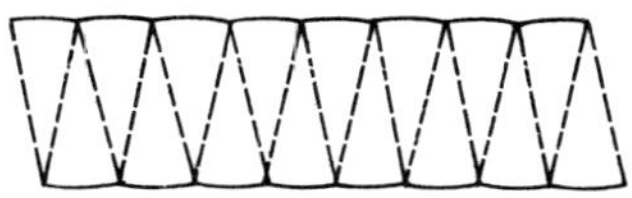

Figure 13.6.4
16 wedges

The reason why the figure is not really a parallelogram is because its bases are not straight. But observe that as the number of wedges increases, the bases begin to "straighten" and the figure begins to look more and more like a parallelogram. Actually, what we are considering here is a very important idea in mathematics—the idea of a limit. In mathematics, we might think of this idea in the following way: If n is the number of wedges, then as n gets bigger and bigger, the shape approaches the shape of a parallelogram. In fact, we might consider writing:

$$\lim_{n \to \infty} \text{Parallelogram-like shape} = \text{Parallelogram}$$

which means as the number of wedges increases, the parallelogram-like shape approaches the parallelogram as a limit.

Thus, let us think of this shape as a parallelogram. Then, under this assumption, we have that the area of the parallelogram equals bh, where b is the base of the parallelogram and h is its height. Now if you watch closely as you transform the parallelogram back into the circle, you should see that the base becomes half of the circumference of the circle and the height is the radius of the circle (Fig. 13.6.5).

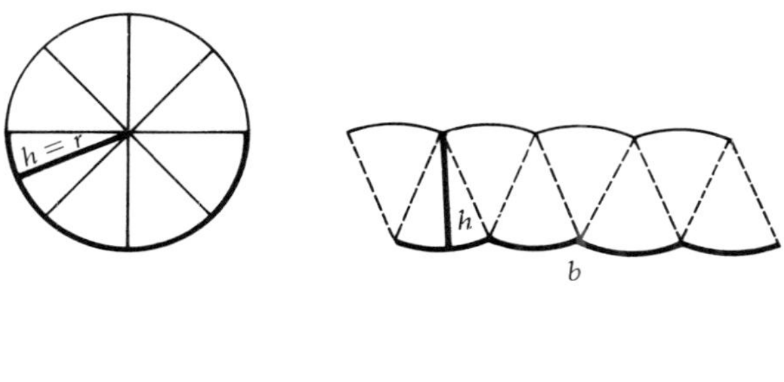

Figure 13.6.5

So, if the area of the circle equals bh, by substituting, we get

$$A = (\tfrac{1}{2}C)(r)$$

From an earlier chapter, we discovered that $C = 2\pi r$, so our area now is

$$\begin{aligned} A &= [\tfrac{1}{2}(2\pi r)]r \\ &= \pi r^2 \end{aligned}$$

which is the formula for the area of a circle. Thus, we get the following principle.

13.6.1 Theorem

The area of a circle *is the product of half the circumference and the radius or it is the product of pi and the square of the radius,* $A = \pi r^2$.

Let us consider the area from a different viewpoint. Look at the next series of regular n-gons (Figs. 13.6.6–13.6.8). What do you observe? For each of the regular n-gons, the radius remained fixed, but the number of sides increased. As this occurred, what happened to the shape of the n-gon? The shape looks more and more like a circle. Using our earlier notation, we might say

$$\lim_{n \to \infty} n\text{-gon} = \text{Circle}$$

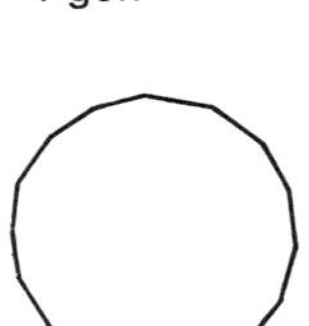

Figure 13.6.6
4-gon

Figure 13.6.7
8-gon

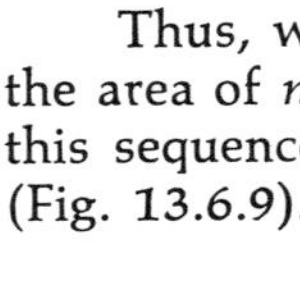

Figure 13.6.8
16-gon

Thus, we can investigate the area of a circle by considering the area of n-gons of fixed radius as n increases. Let us reconsider this sequence of regular n-gons all inscribed in the same circle (Fig. 13.6.9).

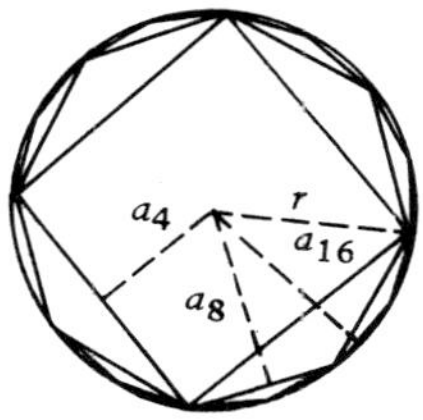

Figure 13.6.9

Now it appears as if the area of each n-gon is getting closer and closer to the area of the circle. If we let A_n be the area of the n-gon (that is, A_4 is the area of the square, A_8 of the octagon, and so on) and A be the area of the circle, then it looks as if $\lim_{n \to \infty} A_n = A$. That is, the limit of the area of the n-gons as n increases is the area of the circle. Now, for a regular n-gon we know that $A_n = \frac{1}{2}a_n p_n$, where a_n is the apothem of the n-gon and p_n is the perimeter of the n-gon. Now look at the apothems for the 4-, 8-, and 16-gons in the diagram. Do you see that as n increases the apothem gets closer and closer to the radius in length? That is, $\lim_{n \to \infty} a_n = r$. Likewise, consider the sequence of perimeters of the n-gons as n increases. What does p_n approach? It seems to be getting closer and closer to the circumference of the circle. That is, $\lim_{n \to \infty} p_n = C$. Now, since $A_n = \frac{1}{2}a_n p_n$, if we substitute these limiting values, we get $A = \frac{1}{2}(r)(C) = \frac{1}{2}(r)(2\pi r) = \pi r^2$, or expressed as limits, we get

$$A = \lim_{n \to \infty} A_n = \lim_{n \to \infty} \tfrac{1}{2}a_n p_n = \tfrac{1}{2}(r)(2\pi r) = \pi r^2$$

which agrees with our earlier result.

in our opinion . . .

For those who might be inclined to question our introduction of the limit notation in this section, we believe that this is completely consistent with our effort to teach this material in a spiral fashion. It is not possible to treat this topic honestly without some acknowledgment of the limit concept. We believe that this is only one of several places in the elementary curriculum in which the limit concept naturally arises. The teacher should be aware of this and there is no harm in actually using the language and notation to accompany the intuitive development. We further believe that the fact that there are topics of this nature in the curriculum gives support to the suggestion that an adequately prepared elementary teacher should be introduced to some of the basic ideas of calculus. If that is the case, then the language and notation used here will provide them with an intuitive basis for more formal study later.

Exercise 13.6.1

In all of the following problems, use 3.14 as an approximation for π.

1. Calculate the area of a circle with a radius of 3 in.
2. Calculate the area of a circle with a diameter of 10 in.
3. Calculate the area of a circle which has a circumference of 37.68 ft.
4. The area of a circle is 153.86 sq in. What is its radius?
5. Calculate the area of a circle whose radius is equal to the ratio of its circumference to the diameter.

13.7 SURFACE AREA OF SOLIDS

Now that we have considered the areas of some of the most common two-dimensional shapes, it seems quite natural to consider the area of three-dimensional shapes. As we shall soon see, we have practically developed all the necessary formulas.

However, before we proceed much further, the reader should take special care to distinguish between upper and lower case letters in the remainder of this chapter. The term "base" can become confusing as we begin to discuss the base of a prism or pyramid because, if this base is a triangle or trapezoid or parallelogram, then *this* base will have a base. Likewise, the prism or pyramid has a height and so does the base of the prism or pyramid have a height. Thus, from now on, we shall *capitalize* the words "base" and "height" when they refer to the Base and Height of the prism or pyramid, and we shall use *lower case letters* when they refer to the base and height of a polygon such as a triangle, trapezoid, or parallelogram. Although they may sound strange, the phrases "the base of the Base" and "the height of the Base" should convey precise ideas. Likewise, for any prism or pyramid B shall represent the area of the Base, H the Height of the prism or pyramid, b for the length of the base of the Base and h for the height of the Base.

First, let us consider the right prism. There are actually two different areas we might consider; the surface area of the sides only (the rectangles) and the surface area of the sides and the two Bases. The former will be referred to as the *lateral surface area* and the latter as the *total surface area*. In all cases, total surface area equals the sum of the lateral surface area and twice the area of one of its Bases. Thus, if S is the total surface area, L is the lateral surface area, and B is the area of one of the Bases, then $S = L + 2B$.

Since for all prisms the Base is some kind of polygon, there is no problem to calculate B. Thus, we shall concentrate on calculating L, the lateral surface area. By definition, each side of a prism is a rectangle, so to calculate L we simply calculate the area of each rectangle and then add these areas up. Thus, if the measure of the sides of the Base are $s_1, s_2, s_3, \ldots, s_n$ and the Height of the prism is H, we have $L = Hs_1 + Hs_2 + Hs_3 + \cdots + Hs_n = H(s_1 + s_2 + s_3 + \cdots + s_n)$ but, since $s_1 + s_2 + s_3 + \cdots + s_n = p$ (the perimeter of the Base), we have $L = Hp$ (Fig. 13.7.1). Now picture a prism which has many lateral faces and a regular n-gon for Bases. What does it almost look like? It looks very nearly like a cylinder, doesn't it? Then we might guess that as the number of faces increases, the surface area of the prisms approaches the surface area of the cylinder.

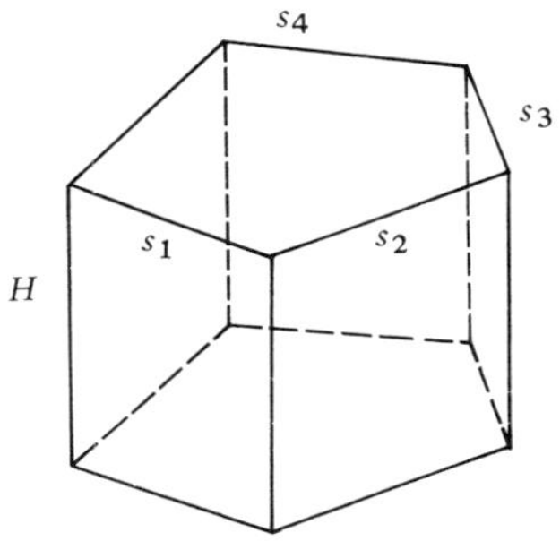

Figure 13.7.1

That is, we might surmise that if L_n is the lateral surface area of a right regular n-gonal prism of fixed Height and radius, and if L_c is the lateral surface area of the cylinder,

$$\lim_{n \to \infty} L_n = L_c$$

If this is true, then, as before, we see that the perimeter of the Base of the prism gets closer to the circumference of the cylinder. Thus, we might guess that

$$L_c = HC = H(2\pi r) = 2\pi rH$$

Thus, we get another principle.

13.7.1 Theorem

a. The lateral surface area of a prism *is equal to the product of the height of the prism and the perimeter of the base of the prism. $L = Hp$.*

b. The lateral surface area of a right circular cylinder *is equal to the product of the height of the cylinder and the circumference of the base. $L_c = HC = 2\pi rH$.*

Now, for the right pyramid, the problem is quite similar. The lateral surface area will again be the sum of the areas of the lateral faces, except that now the lateral faces are triangles (Fig.

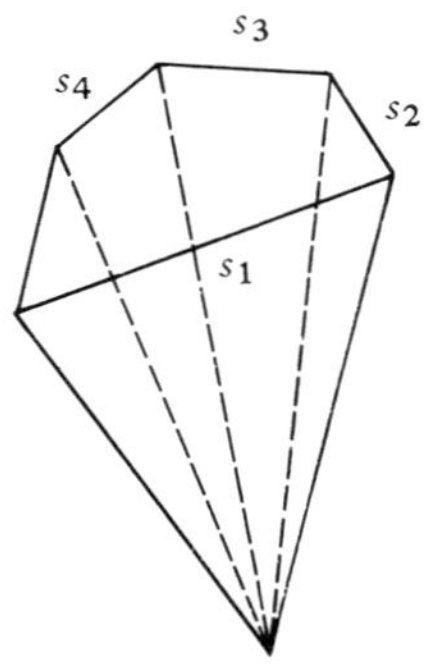

Figure 13.7.2

13.7.2). Thus, if $s_1, s_2, s_3, \ldots, s_n$ are the measures of the sides of the Base, then $L = \frac{1}{2}ls_1 + \frac{1}{2}ls_2 + \frac{1}{2}ls_3 + \cdots + \frac{1}{2}ls_n$, where l is the height of each lateral face. Now, be careful; l is not the Height of the pyramid—it is the height of each lateral face. This is called the *slant height* (Fig. 13.7.3).

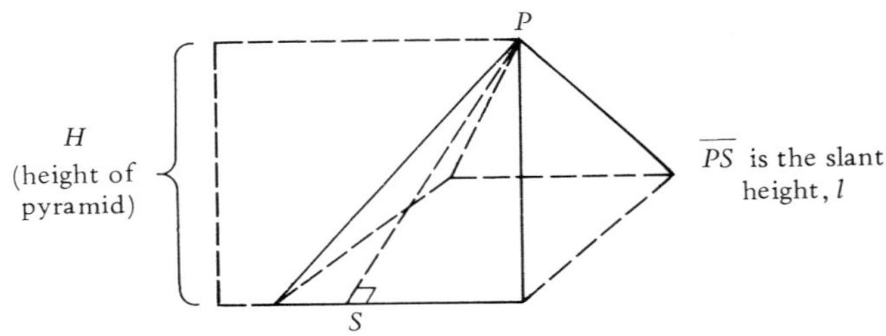

Figure 13.7.3
H: height of pyramid; *PS:* slant height *l*

Which is greater: H or l? Now we can rewrite

$$L \quad \text{as} \quad \tfrac{1}{2}l(s_1 + s_2 + s_3 + \cdots + s_n) = \tfrac{1}{2}lp$$

where p is the perimeter of the Base.

Now, if you picture a right regular n-gonal pyramid of fixed Height and radius, it should be clear that the pyramid looks more and more like a right circular cone as n increases. That is, if L_n is the lateral surface area of a right regular n-gonal pyramid of fixed Height and radius, and L is the lateral surface area of the cone,

$$\lim_{n \to \infty} L_n = L_c$$

and thus the formula for the lateral surface area of a cone becomes $L_c = \frac{1}{2}lC$ or $L_c = \frac{1}{2}l(2\pi r) = \pi rl$, where l is the slant height of the cone. This gives us the following principle.

13.7.2 Theorem

a. *The* lateral surface area of a right pyramid *is one-half the product of the slant height and the perimeter of the base.* $L = \frac{1}{2}lp$.

b. *The* lateral surface area of a right circular cone *is one-half the product of the slant height and the circumference of the base.* $L_c = \frac{1}{2}lC = \pi rl$.

Experiment 5

The last solid we shall consider is the sphere. In an experiment in an earlier chapter, you were exposed to one method of determining the surface area of a sphere. If you were successful in winding the string around the circle and the hemisphere, you may now have an idea for the formula. However, this experiment was quite difficult to perform and it may not have been successful. There is another experiment which is considerably easier. Place a spherically shaped object such as a basketball on a table and place the palms of your hands on opposite sides of the sphere with your fingers outstretched and your fingertips touching the table top. Now, roll the sphere one complete revolution. Consider the area which was swept out on the table between your hands.

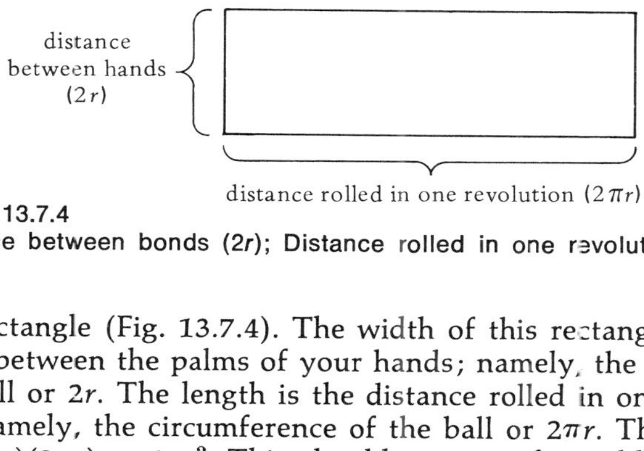

Figure 13.7.4
Distance between bonds ($2r$); Distance rolled in one revolution ($2\pi r$)

It is a rectangle (Fig. 13.7.4). The width of this rectangle is the distance between the palms of your hands; namely, the diameter of the ball or $2r$. The length is the distance rolled in one revolution; namely, the circumference of the ball or $2\pi r$. Thus, the area is $(2r)(2\pi r) = 4\pi r^2$. This should compare favorably with the string experiment for, if you did it correctly, the string covering the hemisphere should have been twice the length of the string covering the circle. Since the area of the circle is πr^2, the area of the hemisphere is $2\pi r^2$ and the surface area of the whole sphere would be $4\pi r^2$. This circle is often referred to as a *great circle* of the sphere. Thus, the surface area of a sphere is frequently given as the area of four great circles. Thus, we have our final area principle.

13.7.3 Theorem

The surface area of a sphere *is equal to the area of four great circles:* $A = 4\pi r^2$.

Exercise 13.7.1

In each of the following problems, by perimeter we shall mean the sum of the measures of all the edges of the figure and by the surface area we shall mean the sum of the areas of all the faces.

1. Calculate the perimeter and surface area of each of the four rectangular solids whose dimensions are listed in Table 13.7.1 and complete the table.

Table 13.7.1

Length (in.)	*Width (in.)*	*Height (in.)*	*Perimeter (in.)*	*Surface area (sq. in.)*
8	6	1		
10	3	2		
30	4	½		
9	4	2		

2. Draw scale drawings of the four rectangular solids, using the same scale for each. On the basis of your drawings, how would you compare the volumes of the four solids? Which would you estimate has the smallest interior? Which the largest? Looking at your results of Exercise 1, do you think that the surface area and perimeter of a rectangular solid determine the volume of the solid?
3. Determine the dimensions of a rectangular solid different from those above which has a perimeter of 60 in.
4. Determine the dimensions of a rectangular solid different from those above which has a surface area of 124 sq in.
5. What would be the dimensions of a cube whose perimeter is 60 in.? What is its surface area?
6. What would be the dimensions of a cube whose surface area is 124 sq in.? What is its perimeter? Which cube (this one or the one in Problem 5) do you think has greater volume?
7. Determine the length of each edge of an equilateral tetrahedron (a tetrahedron in which all faces are congruent equilateral triangles) whose perimeter is 60 in. What is its surface area?
8. Determine the length of an edge of an equilateral tetrahedron whose surface area is 124 sq in. What is its perimeter? Which tetrahedron do you believe to have the greater volume —this one or the one in Problem 7?
9. Write an algebraic expression for each perimeter and surface

area of the following figures (use only the variables given in the problems).

a. a rectangular solid whose length, width, and Height are a, b, and c, respectively
b. a cube whose edge is s
c. an equilateral tetrahedron whose edge is s
d. a right triangular prism of Height H whose Bases are equilateral triangles with each side s
e. a right triangular prism of Height H whose Bases are right triangles with one leg a and the other leg b
f. a right triangular prism of Height H whose Bases are isosceles triangles with sides l, l, and b
g. a right regular octagonal prism of Height H whose Bases have sides of s and an apothem of a

10. Suppose you are going to make a square piece of cardboard into a rectangular solid by cutting four squares out of the corners and folding the edges up, as indicated in Fig. 13.7.5. If the dimensions of the large square are s, and the dimensions of each little square are h, write algebraic expressions for the perimeter and surface area of the resulting rectangular solid (including the top even though it does not have one).

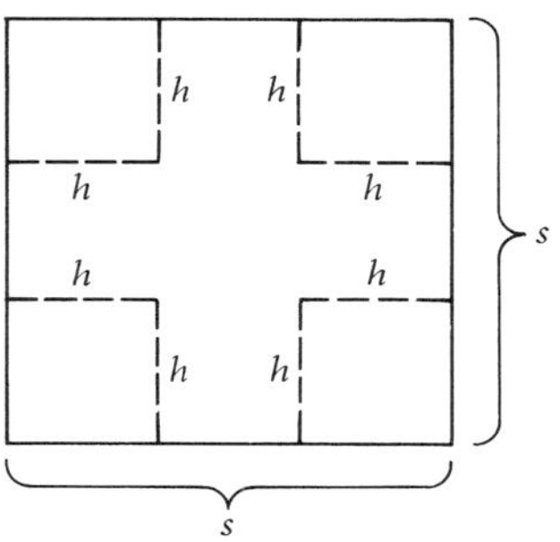

Figure 13.7.5

11. Calculate the surface area of a sphere whose radius is 20 in.
12. A certain silo is constructed of a cylinder and its roof is a hemisphere. The total height of the silo is 40 ft and the radius of the hemisphere is 8 ft (Fig. 13.7.6). How many gallons of

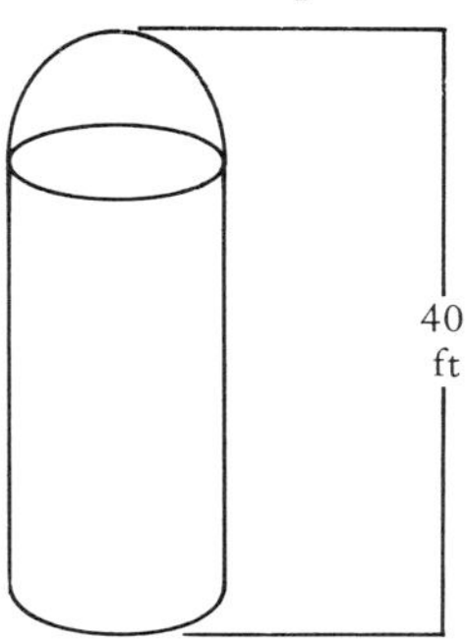

Figure 13.7.6

paint will be required to paint the silo, including the roof, if one gallon covers 450 sq ft?

13. Calculate the slant height of a right circular cone if its Height is 12 in. and the radius of its Base is 5 in.
14. Calculate the surface area of the cone in Problem 13.

13.8 SUMMARY

1. The area of a rectangle is $A = lw$.
2. The area of a parallelogram is $A = bh$.
3. The area of a triangle is $A = \frac{1}{2}bh$.
4. The area of a trapezoid is $A = \frac{1}{2}h(b_1 + b_2)$.
5. The area of regular polygon is $A = \frac{1}{2}ap$.
6. The area of a circle is $A = \pi r^2$.
7. The lateral surface area of a right prism is $L = Hp$.
8. The lateral surface area of a right circular cylinder is $L_c = HC = 2\pi rH$.
9. The lateral surface area of a right pyramid is $L = \frac{1}{2}lp$.
10. The lateral surface area of a right circular cone is $L_c = \frac{1}{2}lC = \pi rl$.
11. The surface area of a sphere is $A = 4\pi r^2$.

VOLUME

14

14.1 PROLOGUE

The problem of measuring volume is simply a natural extension of our previous discussion of measure. When we discussed linear measure, we noticed that it was a correspondence between one-dimensional sets of points such as line segments and the nonnegative real numbers. When we attempted to apply this measure to two-dimensional figures, we encountered the concepts of perimeter and circumference. But through some experimentation, we discovered that this was not an adequate way to compare sizes of geometric figures in two dimensions for it was very possible to have two polygons with the same size perimeter, yet the interior of one could be considerably larger than the interior of the other. Thus, we were led to area measure which is a correspondence between the interiors of plane segments and the nonnegative real numbers. In the last set of exercises, we extended our attention to three-dimensional figures and we saw that here again it was possible to apply linear and area measure to such figures, but they were not sufficient to distinguish between the sizes of the three-dimensional figures with respect to the size of their interiors.

14.2 VOLUME AND THE UNIT OF MEASURE

Intuitively, we can think of the volume of a figure being the amount of space contained inside a figure or how much the figure can "hold." What type of unit can we use to determine volume? Perhaps we should first consider what characteristics this unit ought to have. If we think back and consider the other units of measure we encountered, we are reminded that in each case the unit possessed the property we were measuring; that is, a unit of linear measure had length, a unit of area had area, and a unit for measuring an angle was itself an angle. Similarly, then, a unit for measuring volume should itself have volume. There was another characteristic of units as well, one which did not become too apparent until we studied area and that was that several units of the same size should fit together compactly. Thus, we use squares for units of area since squares can be placed side by side without leaving spaces. Similarly, linear units can be placed end to end and degree units can be placed adjacent to each other. Thus, our unit of volume should possess the following properties:

1. It should possess volume, and
2. several of the units of volume should fit together compactly leaving no spaces between.

Exercise 14.2.1

1. Turn back to Problem 1 of Exercise 7.3.1 of your book and decide which of the items listed there would make suitable units of volume.
2. Which of the following would be used as units of volume? Are there circumstances under which you might change your answer?
   a. a grain of salt
   b. a box of salt
   c. a marble
   d. a beer bottle
   e. a molecule of water
   f. a pail of water
   g. an ice cube
   h. a balloon filled with air

For the time being, our consideration of volume shall be restricted to determining a unit of fixed shape. Under this restriction, we seek a shape which fits together nicely. Since we use the square as the primary unit of measure of area, it is no surprise that we use the three-dimensional counterpart, the cube, as the unit of measure for volume.

Thus, we have such volume units as cubic inch, cubic centimeter, cubic yard, cubic kilometer, and so on, each of which is a cube with its edge 1 in., 1 cm, 1 yd, and 1 km, respectively. Thus, when we ask for the volume of a three-dimensional figure, we are asking for the number of cubes of a certain size it would take to fill the space in the interior of the figure.

At this point perhaps we should point out that the word volume is used in two different ways. If you pondered the points made at the beginning of this discussion, you probably noticed that the shape of the unit is not always as vital as it was in considering area. This is because we have liquids and gases (and solids to some extent) which retain a constant volume at a constant temperature and pressure yet can readily change their shape. In such cases, the shape of the unit is not important. Some people make the distinction between these two concepts of volume by referring to this latter case as *capacity* and only using volume for referring to the number of cubic units used. In the United States there are two major kinds of capacity, *liquid measure* and *dry measure*. Thus, the word volume is used in a general sense to refer to any measure of space occupied by a three-dimensional object and it is also used in a more specific sense to indicate the number of cubic units contained in a figure as opposed to the capacity of a figure which indicates the amount of liquid capable of being contained in a three-dimensional object.

In most cases, it does not become necessary to differentiate between the two meanings of the word volume, but the distinction was pointed out here because in many of the experiments of this chapter, the reader is invited to compare capacities of containers and to then deduce from this, identical comparisons of their volumes (as in the restricted sense). Thus, underlying much of the development of this chapter is the tacit assumption that figures of equal capacities have equal volumes (in the restricted sense) and conversely.

Exercise 14.2.2

1. Look up the definition of 1 liter. This is the basic metric unit for capacity. What is its equivalent in cubic centimeters? cubic inches?
2. Look up other units of capacity in the United States' system. Is the quart for liquid measure the same as the quart for dry measure?

14.3 THE VOLUME OF RECTANGULAR SOLIDS AND OTHER RIGHT PRISMS

If we realize that determining the volume of a rectangular solid is simply a problem of counting the number of cubic units in its interior, then it is not at all difficult to see that multiplying the length times the width times the Height will give us the correct result. Consider Fig. 14.3.1. If we think of the solid being completely filled by the cubic units, then we can see that there are several "layers" of cubes. If we can determine how many cubic units in each layer, then it is easy to determine how many cubes there are altogether. So, let's examine just one layer (Fig. 14.3.2). If the length is l units, then l cubes will fit along the length, and if the width is w units, then w cubes will fit along the width. Thus, there will be lw cubes in the bottom layer. Each layer above it will have the same number of cubes—lw also. If

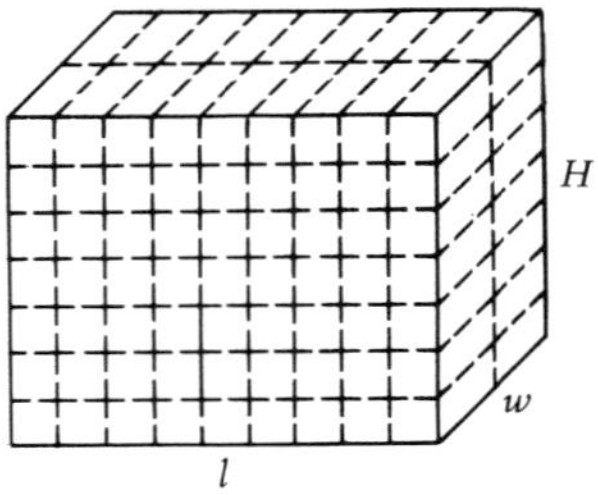

Figure 14.3.1

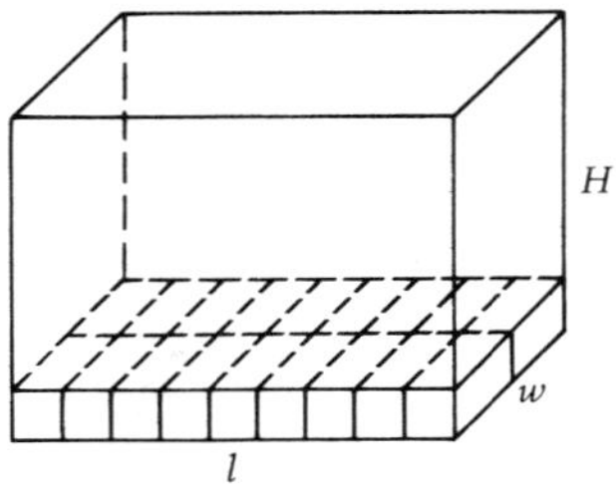

Figure 14.3.2

the Height is H units, then H layers of cubes will fill the solid. Thus, there are H layers of cubes with lw cubes in each layer; lwH is the total number of cubes.

A slightly different but very useful way to consider this problem is to recognize that the number of cubes which covers the Base of the solid in one layer will equal the area of the Base of the solid. (Why?) The volume would be equal to the area of the Base times the Height of the solid. Let us continue with this idea and consider other prisms (Figs. 14.3.3–14.3.5). If you look carefully at each prism, you will see that in each case the problem is essentially the same. The prism consists of layer upon layer of cubes. The number of cubes in each layer is the same as the number of cubes it takes to cover the Base. But the number of cubes it takes to cover the Base is precisely the same

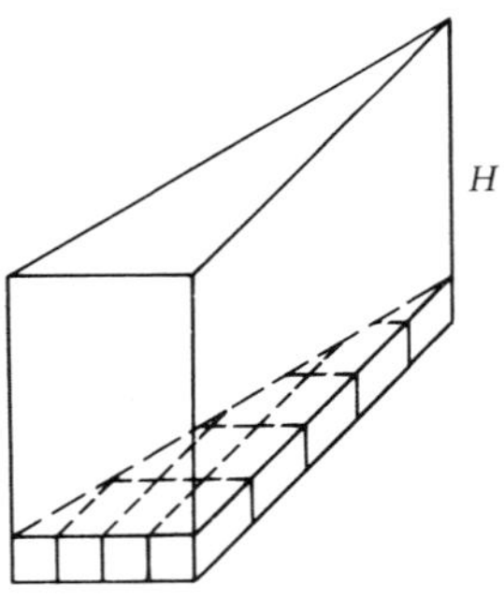

Figure 14.3.3
Triangular prism

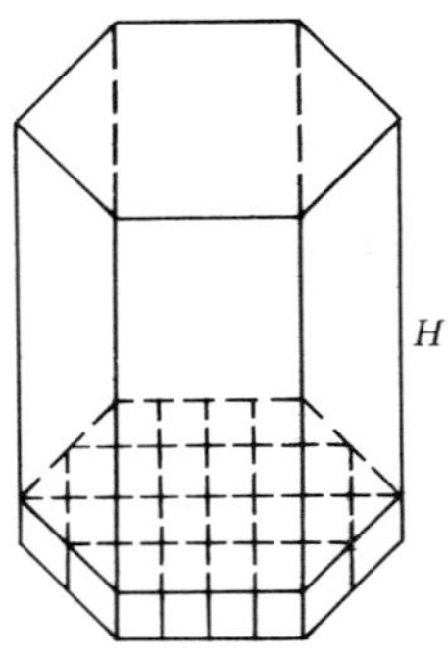

Figure 14.3.4
Hexagonal prism

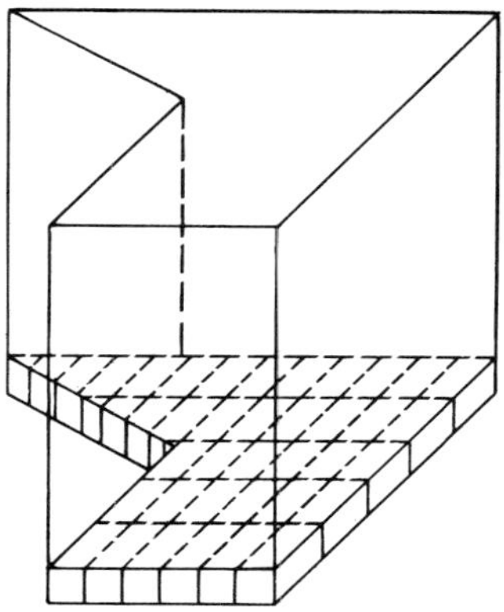

Figure 14.3.5
Any prism

as the number of squares it takes to cover the Base, which is the area of the Base—call it B. Thus, if the prism is H units high, then there will be H layers of cubes with B cubes in each layer. Thus, altogether there will be BH cubes in the prism.

Thus, we have our first important volume principle.

14.3.1 Postulate

The volume of any right prism *is equal to the product of the area of the Base of the prism times the Height of the prism.* $V = B \times H$.

Exercise 14.3.1

1. Write the algebraic expressions for the volumes of the following prisms using the notation discussed earlier.
   a. a triangular prism
   b. a trapezoidal prism
   c. a prism whose Bases are parallelograms
   d. a cube with edges of s
   e. a hexagonal prism whose Bases are regular hexagons with s for a side of the Base and a for the apothem of the base
2. Calculate the volumes of each prism below.
   a. a cube where $s = 6$ in.
   b. a triangular prism where $b = 3$ ft, $h = 8$ ft, and $H = 4$ ft
   c. a triangular prism where the Base is a right triangle, with one leg 12 cm, the hypotenuse 13 cm, and $H = 20$ cm
   d. a trapezoidal prism where $b_1 = 3$ yd, $b_2 = 2$ ft, $h = \frac{1}{2}$ ft, and $H = 3$ in.
   e. a regular hexagonal prism where $s = 4$ in. and $H = 6$ in.
3. How would you determine the volume of a circular cylinder? Justify your answer using a carefully worded explanation similar to that in the previous section.

Experiment 1

Step 1

Write an algebraic expression for the volume of the rectangular solid obtained from cutting four squares of sides h from the corners of a square of side s and folding up the edges. (See Fig. 14.3.6.)

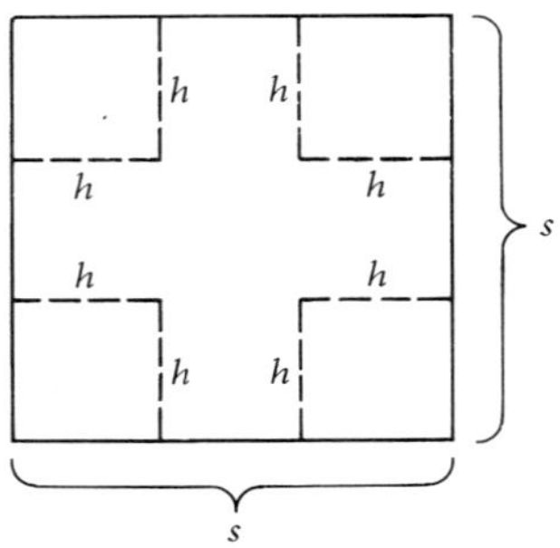

Figure 14.3.6

Step 2

Use your expression above to calculate the volume of the resulting solid for the different values of h and s in Tables 14.3.1–14.3.7. You may want to do the work in teams. Also feel free to use any available desk calculator to aid you.

Step 3

Graph each table on a separate axis, using the vertical axis for V and the horizontal axis for h. From your graphs, estimate for each value of s the value of h for which the volume is a maximum: Record your estimates in tabular form (Table 14.3.8).

Table 14.3.1

$s = 6$ in.

h	0	$\frac{1}{4}$	$\frac{1}{2}$	$\frac{3}{4}$	1	$1\frac{1}{4}$	$1\frac{1}{2}$	$1\frac{3}{4}$	2	$2\frac{1}{2}$	3
V											

Table 14.3.2

$s = 9$ in.

h	0	$\frac{1}{2}$	1	$1\frac{1}{2}$	2	$2\frac{1}{2}$	3	$3\frac{1}{2}$	4	$4\frac{1}{2}$
V										

Table 14.3.3

$s = 12$ in.

h	0	½	1	1½	2	2½	3	3½	4	4½	5	5½	6
V													

Table 14.3.4

$s = 15$ in.

h	0	1	2	3	4	5	6	7	7½
V									

Table 14.3.5

$s = 18$ in.

h	0	1	2	3	4	5	6	7	8	9
V										

Table 14.3.6

$s = 24$ in.

h	0	1	2	3	4	5	6	7	8	9	10	11	12
V													

Table 14.3.7

$s = 50$ in.

h	0	2	4	6	8	10	12	14	16	18	20	22	24
V													

Table 14.3.8

s	6	9	12	15	18	24	50
h							
max. V							

Step 4

From your results in Step 3, make a conjecture for the value of h at which the volume will be a maximum for any value s. Express h and the maximum volume as functions of s.

14.4 VOLUME OF A PYRAMID

The majority of this section will depend on a series of experiments which you are asked to perform. As part of your materials, you will need to have available at least three pairs of prisms and pyramids of various shapes and sizes. More specifically, for each pyramid you construct, you will need a prism whose Height is the same as the Height of the pyramid and whose Base is equal in area to the Base of the pyramid. Thus, a brief description of how to construct these figures will follow. We suggest that you make the figures out of heavy construction paper or cardboard and that you make them large enough to work with in your experiments.

We suggest that you construct your pyramids first and then your prisms, as it will be easier to adjust the Height of each prism to that of the corresponding pyramid. Although it is not necessary, it is easier to construct right pyramids with regular polygons as Bases. In such a case, the triangular faces are congruent isosceles triangles. Perhaps the next sequence of steps will be useful. We shall construct a regular octagonal right pyramid. The procedure would be similar for others.

1. Construct the regular octagon of a suitable size. To do this, you have to determine the size of certain central angles. See Fig. 14.4.1. If O is the center of the regular octagon, then connecting O to each of the vertices $A_1, A_2, \ldots, A_8$ will form eight congruent angles. Thus, the measure of each angle is $360°/8 = 45°$.

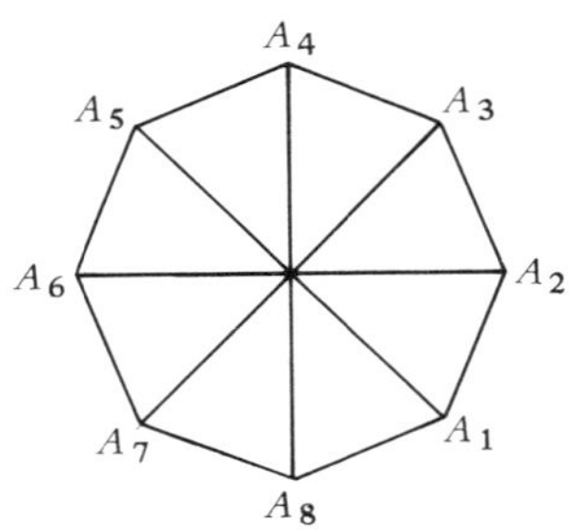

Figure 14.4.1

2. Pick a point O on your paper and draw any line through it. Then at O measure eight adjacent angles each of 45° (Fig. 14.4.2).

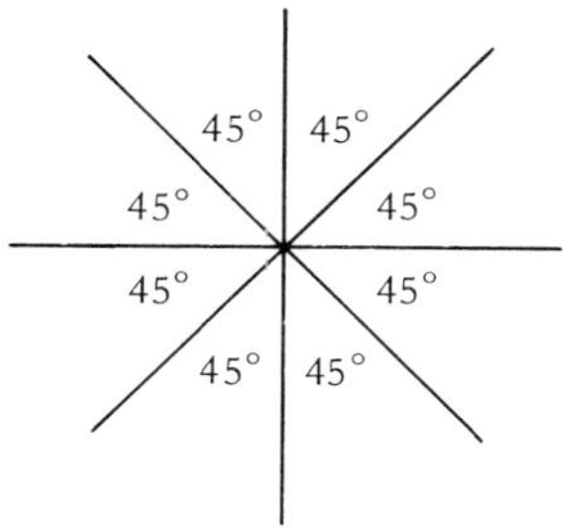

Figure 14.4.2

3. Now all you have to do is decide how big you want your octagon and locate one vertex on each ray. So select a suitable radius on your compass and using O as center, mark off eight arcs, one cutting each ray. Then connect the vertices (Fig. 14.4.3). If you cut out the polygon you can use it as a pattern for the Bases of the pyramid and prism.

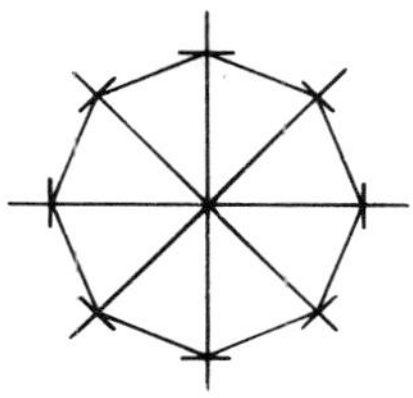

Figure 14.4.3

4. Now to complete your pyramid. Trace the Base on your paper or cardboard. Leave yourself lots of room (that is, place your Base in the center of the paper). Now set your compass at a convenient radius (this will be the length of the legs of the isosceles triangles) and swing an arc using a vertex of the polygon as the center. Use a consecutive vertex as the center; swing a second arc. Connect the two vertices with the point determined by the two arcs. You now have constructed one of the faces of your pyramid, say $\triangle ABC$.
5. Now proceed to construct seven other isosceles triangles congruent to $\triangle ABC$, with each vertex at C and with one leg coincident with a leg of the preceding triangle. One efficient way to do this is to draw an arc with C as the center and AC as the radius. Then, use your compass set to a radius of AB and mark off seven more segments. (See Fig. 14.4.4.)

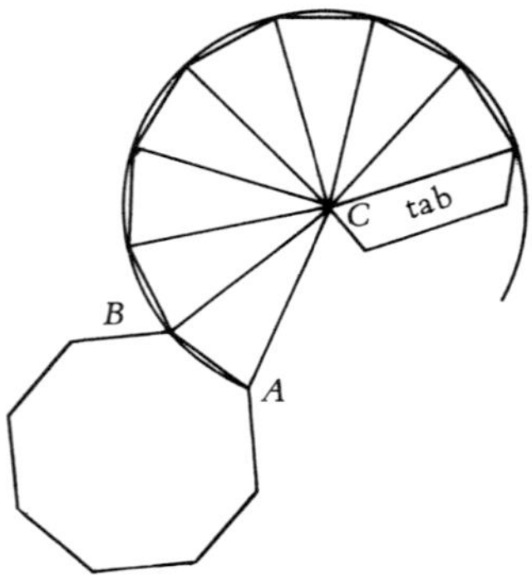

Figure 14.4.4

6. Now your pyramids are to be open at the bottom so you will want to leave the Base hinged so you can open it up. Otherwise, you could draw in the tabs for fastening together. So just draw in a tab on the leg of the last triangle constructed. Fold along the appropriate lines and you have your pyramid. You can construct other pyramids in a similar way and this method will work for any regular polygon as the Base.
7. Once you have assembled your pyramid, rest it on its Base and determine as accurately as you can its Height; that is, the distance from the apex to the plane of its Base. This distance will also be the Height of your prism.
8. Now the construction of the prism is easy. First, using the same n-gonal pattern as before, trace another polygon. Then draw a line which coincides with one of the sides of the polygon. With your compass set at a radius equal to a side of the polygon, and using the two vertices as centers, mark off seven more segments (Fig. 14.4.5).

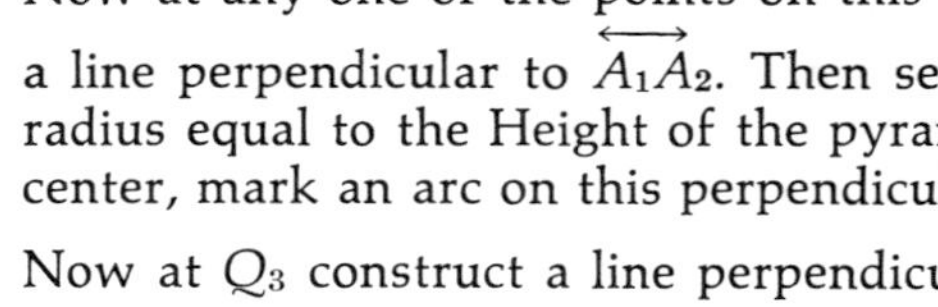

Figure 14.4.5

9. Now at any one of the points on this line (say A_3) construct a line perpendicular to $\overleftrightarrow{A_1A_2}$. Then set your compass to a radius equal to the Height of the pyramid and using A_3 as the center, mark an arc on this perpendicular. Call the point Q_3.
10. Now at Q_3 construct a line perpendicular to $\overleftrightarrow{A_3Q_3}$.
11. Now mark off $Q_1, Q_2, Q_3, \ldots, Q_9$ so that $\overline{Q_1Q_2} \cong \overline{Q_2Q_3} \cong \overline{Q_3Q_4} \cong \cdots \cong \overline{A_1A_2}$.
12. Draw $A_1Q_1, A_2Q_2, A_3Q_3, \ldots$ (Fig. 14.4.6).

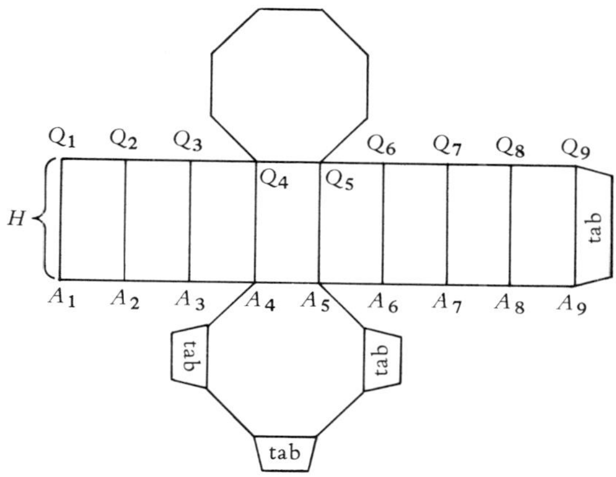

Figure 14.4.6

13. Draw in some suitable tabs for one of the Bases only, as you want to leave the other Base free to open up.
14. Fold along the suitable lines and you now have a prism whose Bases are congruent to the Base of your pyramid and whose Height is equal to the Height of the pyramid. Make other pairs of pyramids and prisms until you have at least three different sets. You are now ready to perform your experiment.

Experiment 2

Purpose

To determine the volume of a pyramid.

Preconcepts

The student should be able to:

1. identify Bases and Heights of pyramids and prisms
2. determine the volume of any right prism

Materials

1. At least three pairs of prisms and pyramids, where for each pair the Bases of the prism are congruent to the Base of the pyramid and the Height of the prism is equal to the Height of the pyramid.
2. Some pourable material, such as sugar, salt, sand, rice, barley, and so on.

Procedure

1. Take your first pair of prism and pyramid. Set them in front of you. Check to make sure that the Bases of the two figures are congruent. Also, check to see if the Heights are equal.

2. Now, the purpose of the experiment is to determine the volume of a pyramid. To do this, you will compare its volume with the volume of a prism. So, first look carefully at the prism and pyramid in front of you. Make a conjecture about how their volumes compare. Record your guess here.

   ____

3. Test your conjecture by filling the pyramid up with the available material and then pouring it into the prism. Continue this process until the prism is full. How many times did you have to fill the pyramid in order to fill the prism? The prism's volume is how many times bigger than the pyramid's volume? Did the experimental results coincide with your original guess? Record your findings on your lab worksheet.
4. Repeat Steps 1–3 for your second and third pair of prisms and pyramids. Record your results on your lab worksheet.
5. Complete the rest of your lab worksheet.
6. Prepare a lab report in which you explain what conclusions you have drawn and why you arrived at these conclusions. Cite any difficulties you encountered in your experiment. Indicate what aspects of your experiment may have contributed to errors in your results.

Lab Worksheet

1. Experiment Purpose ______________________________
2. Materials
   a. Indicate the kind of prisms and pyramids you used. (Example: one triangular prism and pyramid, one square prism and pyramid, and one octagonal prism and pyramid.)

      ____

      ____

      ____

   b. Indicate the material you used to fill your prisms and pyramids.

      ____

3. Investigation

Table 14.4.1

Type used	*Initial guess*	*Experimental results*
First pair	I think the volume of the prism is about ____ times as big as the volume of the pyramid.	I found the volume of the prism to be about ____ times as large as the volume of the pyramid.

Table 14.4.1

Type used	*Initial guess*	*Experimental results*
Second pair	I think the volume of the prism is about ____ times as big as the volume of the pyramid.	I found the volume of the prism to be about ____ times as large as the volume of the pyramid.
Third pair	I think the volume of the prism is about ____ times as big as the volume of the pyramid.	I found the volume of the prism to be about ____ times as large as the volume of the pyramid.

4. On the basis of your experiment, what do you conclude about the volumes of prisms and pyramids?

____

____

5. Under what conditions will your conclusions in Question 4 be valid?

____

____

6. What is the formula for the volume of a prism?

____

7. State what you think is the volume principle for pyramids.

____

____

What do you think the volume formula is?

____

____

8. Use whatever instruments you need to determine as accurately as you can the volumes (in cubic units) of the three pyramids you used in your experiment. Show all necessary work and indicate what measurements you made.

Exercise 14.4.1

1. Write algebraic expressions for the volumes of each of the following pyramids:
   a. a square pyramid of Height H and the sides of the Base s
   b. a regular tetrahedron of Height H and each edge s
   c. any tetrahedron of Height H and the base b and height h
   d. a trapezoidal pyramid of Height H, bases b_1 and b_2 and height h
   e. a pyramid of Height H, with its Base a parallelogram with base b and height h

f. a regular hexagonal pyramid of Height H with the edges of its Base s
g. a pyramid of Height H whose Base is a regular n-gon with each side s and its apothem a

2. Find the volume of the following pyramids:
   a. a triangular pyramid with $H = 2$ in., $b = 5$ in., and $h = 3$ in.
   b. a trapezoidal pyramid with $H = 2$ mm, $b_1 = 5$ cm, $b_2 = 7$ cm, and $h = 9$ mm
   c. Look up the dimensions of the Great Pyramid of Cheops and determine its volume. Also, calculate the ratio of the area of its Base to its Height
   d. the regular octagonal pyramid whose Base is in Fig. 14.4.5 and whose Height is 4 in.
3. A prism and pyramid have congruent Bases, but the ratio of their volumes is $2/7$. What is the ratio of their Heights?
4. Design a laboratory experiment for discovering the formula for the volume of a cone. What do you expect the formula will be? Why do you think so?

14.5 THE VOLUME OF A SPHERE

By now you should have become very adept at discovering principles on your own. See if you can perform the next experiment and discover the formula for the volume of a sphere.

Experiment 3

Purpose

To discover the formula for the volume of a sphere.

Preconcepts

The student should be able to:
1. identify the Base and Height of a pyramid
2. calculate the volume of a pyramid
3. factor polynomials using the distributive property
4. calculate the surface area of a sphere

Materials

1. a large apple, as nearly spherical as possible
2. a knife

Procedure

1. Take your apple and cut it in half so that each half approximates a hemisphere. Then cut each half in half resulting in

quarters, and cut the quarters in half. Thus, you should have eight wedges of apple (Fig. 14.5.1).

(a) (b)

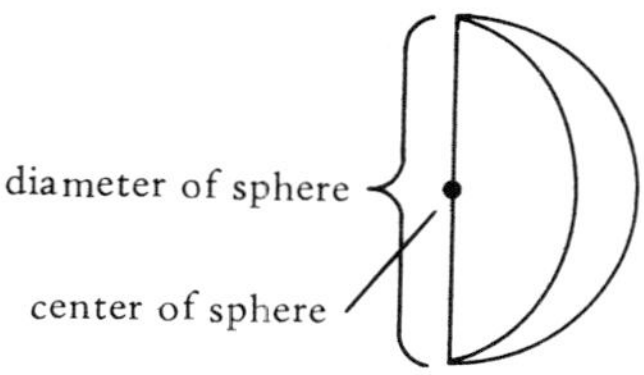

(c)

Figure 14.5.1
(a) Sphere;
(b) Top view of sphere cut into 8 wedges;
(c) Side view of 1 wedge

2. At this point it is important to notice that the straight edge of each wedge is the diameter of the sphere. If you do not see this, put some wedges together to form a portion of the sphere. Since each wedge contains a diameter, the center of the sphere must be on each wedge. Can you locate it?

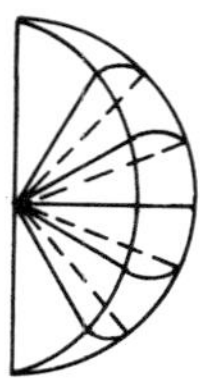

Figure 14.5.2

3. Now select one wedge and proceed to cut it into smaller pieces making cuts as shown in Fig. 14.5.2. Notice that each cut goes through the center of the sphere. Thus, your wedge is now cut into six pieces and should look like Fig. 14.5.3.

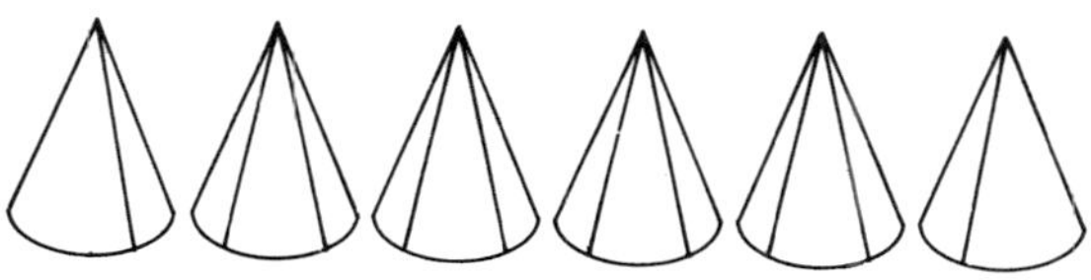

Figure 14.5.3

4. Now examine each small piece; what geometric shape does each one approximate? ____________
5. Why must we be careful to say that each piece *approximates* this geometric shape rather than that it *is* the shape?

____

____

6. How could we alter our experiment so that each little piece would look even more like this geometric shape? ____________

____

7. Let us treat these pieces as if they were the geometric shapes suggested above. Can you classify them into two categories? Name these two categories. ____________
8. Now, if you were to cut each wedge as you did this one, do you see that you would have completely dissected the sphere into many more of these geometric shapes? Pick up one of the geometric shapes. What is its Height? If you are not sure, attempt to put several of them back together to form the sphere. Observe the relationship between the Height of this shape and the sphere. What do you observe?

____

____

9. Similarly, how is the Base of this shape related to the sphere?

____

10. Now pretending that the sphere is completely dissected into n of these shapes, suppose we let the volume of each shape be represented by $V_1, V_2, V_3, \ldots, V_n$, respectively. Then the volume of the sphere V would be equal to the sum of the volumes of these shapes; that is, $V = V_1 + V_2 + V_3 + \cdots + V_n$. If we let B_1 represent the area of the Base of the shape whose volume is V_1, and B_2 be the area of the Base of the shape whose volume is $V_2, \ldots$ what would B_n represent?

____

11. How can you represent the Height of each shape?

____

12. Express each volume below by an algebraic expression:
    a. $V_1 =$ ____________
    b. $V_2 =$ ____________
    c. $V_3 =$ ____________

    •
    •
    •

    d. $V_n =$ ____________
13. Now rewrite the sentence below, replacing $V_1, V_2, V_3, \ldots, V_n$ by the corresponding algebraic expression

    $V = V_1 + V_2 + V_3 + \cdots + V_n$

    $V =$ ____________

14. Use the generalized distributive property to write the expression above.
$V =$ ____________________ × (________________________)
15. Look carefully at the expression in the parentheses above. How is that expression related to the sphere?
____
16. Replace the quantity in the parentheses by a suitable equivalent algebraic formula.
$V =$ ____________________ × (________________________)
17. Simplify your answer. What resulting formula do you arrive at? $V =$ ________________________________
18. State the geometric principle represented by the above formula. ________________________________
____

Exercise 14.5.1

1. Calculate the surface area and volume of the following spheres with the following radii:
   a. $r = 2$ in.
   b. $r = 7$ cm
   c. $r = 121$ yd
2. The volume to the nearest hundredth of a sphere is 678.24 cu in. What is the radius of the sphere?
3. Gasoline is stored in a spherical tank whose diameter is 20 ft. How many gallons does the tank hold?
4. A silo is made from a cylinder topped by a hemisphere for a roof.
   a. Calculate the surface area of the entire silo.
   b. Calculate the volume of the silo. The diameter of the cylinder is 12 ft, and the Height, from the ground to the top of the roof, is 30 ft (Fig. 14.5.4).

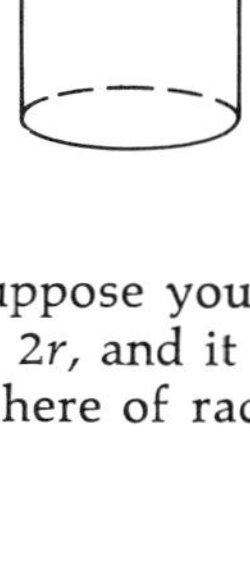

Figure 14.5.4

5. Suppose you have a cylinder which has a radius of r, a Height of $2r$, and it is full of water. Now suppose you slowly lower a sphere of radius r into the cylinder until it touches the bottom

of the cylinder. Of course, this causes water to overflow. After the water has stopped overflowing, you remove the sphere.

a. What is the volume of the water which remained in the cylinder?
b. What is the ratio of the remaining water to the volume of the cylinder?
c. Design an experiment which would use the above idea to discover the formula for the volume of a sphere. (See Fig. 14.5.5.)

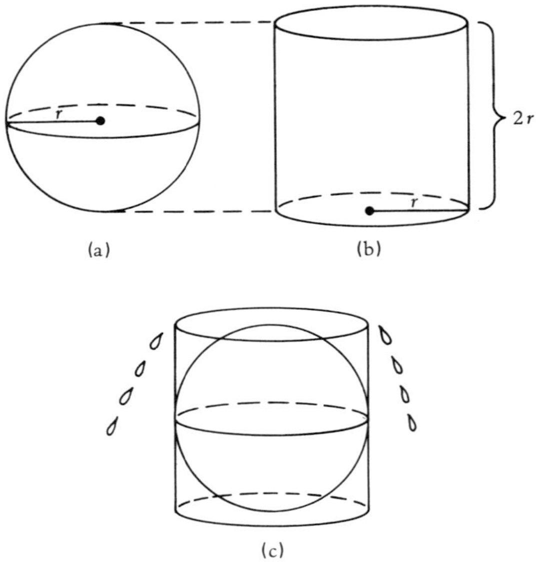

Figure 14.5.5
(a) Sphere;
(b) Cylinder

comment . . .

Since many of the above formulas were discovered by the student through lab experiments, we will leave it up to him to complete this Chapter Summary.

14.6 SUMMARY

1. The volume formula for any right prism is $V = B \times H$.
2. The volume formula for any right circular cylinder is $V =$
3. The volume formula for any right pyramid is $V =$
4. The volume formula for any right circular cone is $V =$
5. The volume formula for a sphere is $V =$

PROOF

15

15.1 PROLOGUE

The reader may have noticed that so far in this book the authors have conspicuously avoided the use of the word "proof." This omission was intentional, primarily because the authors are of the opinion that frequently the word "proof" in a mathematical context has a rather distasteful stigma attached to it in the minds of many readers. There frequently is an implication of terror surrounding the word, and as soon as a student is asked to "prove" something, his mind closes instantly like a trap and he thinks to himself, "I've always had trouble proving things in the past; I know I can't prove anything now and I'll never be able to prove anything in the future; the situation is utterly hopeless." If, indeed, you happen to be one of these individuals, you ought to realize several things. First, you are certainly not alone in your feelings; there are many people who react the same way to one degree or another. Second, your lack of confidence in your ability to do proofs is probably due, to a large extent, to external factors over which you had very little control—such as the way in which the concept of mathematical proof has been presented to you in your mathematics classes and through your mathematics textbooks. Third, you *are* capable of proving things successfully and, indeed, you have been for many years.

in our opinion . . .

Let us examine these last two ideas more carefully. We believe that all too frequently a student leaves a high school geometry class with a rather rigid concept of what a proof really is. Perhaps he thinks of a sheet of paper with a vertical line down through the middle with statements listed on the left and reasons listed on the right. He may even believe that statement 1 is always "the given" and reason 1 is always the word "given."

In other words, students frequently believe that the *format* is the most essential ingredient of a correct mathematical proof, and indeed, that a proof *is* this form. Because of this, the student frequently becomes preoccupied with format and exerts most of his effort in trying to come up with the first step, the second step, and so on without really understanding what it is his proof is really supposed to do. He believes that in his head there should suddenly appear an "instant proof' in all its majesty, perfection, and form. Why shouldn't he believe this? What do we authors frequently do in our textbooks? We say to the reader we shall now prove theorem such-and-such. Then what follows? A beautiful polished proof set down in a perfect form. Are we not in effect saying that this is what proof is all about? Are we not *implying* that no activities transpire during the time interval from the moment we first wish to prove a theorem to the moment we present our finished product? If, indeed, by proof we meant that you were to perform this kind of an activity, then there is small wonder why you are unable to do so. In fact, seldom, if ever, would anyone be able to do a proof under such circumstances. The fact is, however, that this is a complete misconception of what constitutes a proof.

15.2 WHAT IS A PROOF?

In simplest terms, a proof is an argument or an explanation which supports or justifies a particular idea. This is why we claim that, indeed, you have been capable of proving ideas for many years. Who has not been engaged in an argument? Who has not successfully convinced someone that they are right? From the moment we were able to speak (and probably even before that) we have attempted to convince our parents, our brothers and sisters, our friends, and our teachers to believe in our point of view. In the most general sense, such arguments, discussions, debates, or whatever you choose to call them, are attempts at proving things.

Throughout our experience we have doubtless encountered a variety of methods of proof such as fighting, protesting, cajoling, flattering, seducing, and illustrating. What becomes very clear is that not all methods succeed in convincing all people and, furthermore, a method which might succeed with an individual at one particular stage of his development may no longer succeed at a later stage. Whereas spanking an infant may prove to him why he should obey his parents, there comes a time when such a proof is no longer effective and he demands a more convincing type of argument. For a very young child who asks why the sky is blue, the answer "because it is" may satisfy him. Yet, later he regards such a response as senseless and demands more information. Let us consider another example in a mathematical setting. Consider the concept that the sum of the measures of the angles of a triangle is a straight line. A first grader may be convinced of the truth of this statement by virtue of the fact that his teacher told him it was true. When he reaches the second grade, he may begin to realize that teachers are human beings and, therefore, are fallible. So he says to the teacher, "How do I know you're right?" The teacher gives him a textbook and says, "It says so in this book." Thus, the second grader is convinced. But when this concept is encountered in the third grade the student asks, "How do we know the book is right?" This time the teacher says, "Let us test the idea. Make a triangle. Cut off the vertices of the triangle and place them next to one another like Fig. 15.2.1. See, the outside edges fall on a straight line." Thus, the third grader is convinced.

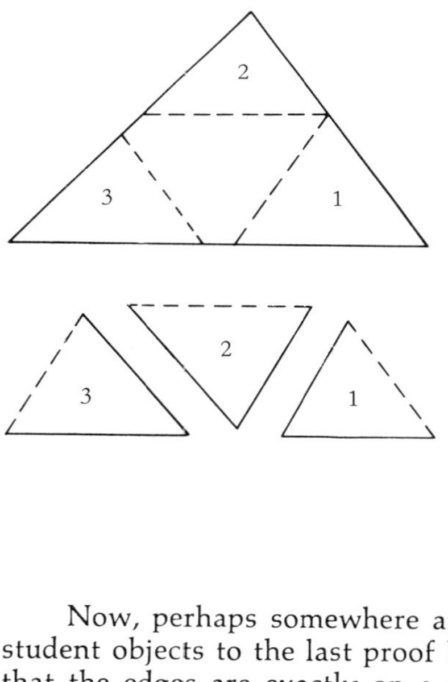

Figure 15.2.1

Now, perhaps somewhere around the fifth or sixth grade the student objects to the last proof by declaring, "How do we know that the edges are exactly on a straight line?" This time the teacher suggests that he measure the angles with a protractor and add them up. But now for the first time this attempt at a proof may serve to strengthen the student's doubts, for although he gets 180° for two of the triangles, he gets only 179° for a third one and he says, "Aha! It doesn't always work." But the clever teacher is ready for this and explains that the discrepancy was due to the fact that the protractor used was only accurate to one degree and, therefore, the 179° resulted from round-off error; and if, indeed, the student had a more accurate instrument, he would verify that the sum is 180°. So, now the student is convinced once again.

Now, sometime in the seventh or eighth grade, he questions again, "We could never check every possible triangle with a protractor; how do we know there isn't one somewhere that we haven't measured where this isn't true?" Now, fortunately for the teacher, the student has studied parallel lines and so the teacher uses the following explanation:

"If we pass a line through one of the vertices of the triangle so that it is parallel to the opposite side, three angles are formed at this point (Fig. 15.2.2). Call them ∡1, 2, and 3. Now, from the picture we can see that these three angles are adjacent and form a straight angle. Angles 1 and A are alternate interior angles and so are ∡3 and B. Therefore, we know that $\angle 1$ is congruent to $\angle A$ and $\angle 3$ is congruent to $\angle B$. Also, $\angle 2$ and $\angle C$ refer to the same angle. So, since the sum of ∡1, 2, and 3 is a straight angle, the sum of ∡A, B, and C must also form a straight angle."

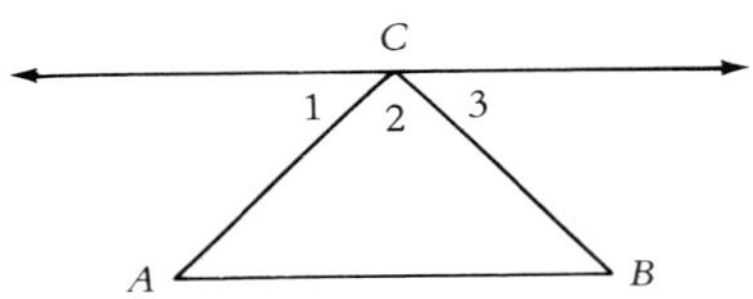

Figure 15.2.2

"Oh, I see," says the student, and, once again, he is convinced.

Now, by the time he is a tenth grader, a strange phenomenon occurs. The student hasn't even thought about the idea. It is the teacher who raises the question, "Do you think we can prove this idea?" The student protests, "Why do we need to prove that? We all know it is true. I've known it for years."

"I know you have heard the statement many times before," says the teacher, "but up until now, nobody has *proved* it to you. Now I can prove it to you and really convince you that it is true."

"But I already *am* convinced it is true," says the student.

"You should never be convinced something is true until someone proves it to you," warns the teacher.

"But if I am convinced it is true, then someone must have already proved it to me," argues the student.

"But nobody ever wrote 'given' and 'to prove' and listed steps and reasons, did they?" asks the teacher, who is getting quite annoyed.

"No," sighs the student, "nobody ever has."

"Then nobody has ever proved it to you," the teacher says triumphantly, "and that's what I'm going to do."

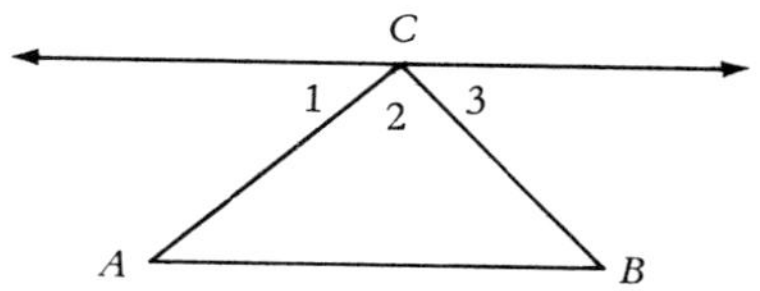

Figure 15.2.3

At this point, the teacher begins to put Fig. 15.2.3 and the following on the blackboard:

Statements	*Reasons*
1. ABC is a triangle.	1. Given.
2. Through point C construct a line parallel to $\overleftrightarrow{AB}$.	2. Through a point outside a line, one and only one line may be drawn parallel to the given line.
Etc.	Etc.

A number of years pass before our student once again returns to the same idea. He is now a college student, majoring in mathematics and he visits his college professor one day.

"Professor, I was thinking about that geometry theorem about the sum of the angles of a triangle being 180° last night and I'm not convinced that it is true."

"Oh, why is that?" asks the professor.

"Well," says the student, "I looked up the proof in my old high school notes and it seems to me that the whole proof is dependent on the fact that through a point outside a line one and only one parallel line can be drawn. I'm not convinced that that's a true statement. How do we know there are any parallels at all, or, for that matter, how do we know there is just *one* parallel, not many through the given point parallel to the given line? If I could be convinced that that idea is true, then I would be convinced about the property of the triangle."

The professor looked thoughtfully at the student and then replied, "I'm afraid I cannot convince you of the truth or falsity of that statement of parallelism. It must be an assumption which either you believe or you don't. Ever since Euclid first suggested this idea as a basic assumption, many mathematicians have attempted to prove it, but unsuccessfully. Finally, hundreds of years later, some mathematicians became convinced that it could not be proved and some of them demonstrated that. In fact, they showed that if you choose to believe that through a point not on a line there is more than one parallel, then all triangles have an angle sum less than 180°; and if you choose to believe that there are no parallel lines, all triangles have an angle sum greater than 180°."

"Do you mean to say that nobody really knows if the sum of the angles of a triangle is 180° or not?" asks the student.

"That's right," says the professor.

"Wow, what an idea!" exclaims the student. "That's incredible and to think that all of these years I was convinced that the sum was 180°!"

There are certainly many points to be drawn from this little story. First, it should be pointed out that the method used to convince the student at each stage of development was a type of proof, insofar as they all served to convince the student of the truth of a particular idea. Second, with the exception of the authoritarian methods used in the first two stages, all the types of proofs were appropriate for the developmental level of the student at each stage. Proof by measuring with a protractor would have been inappropriate for first grade students who couldn't use a protractor and couldn't add up the numbers; and it certainly would not have *convinced* them of the truth of the statement. Similarly, the logical argument given to the seventh or eighth grade student would not have convinced the fifth or sixth grade student if he was not yet aware of certain parallelism properties. Also, it certainly would have served no useful purpose to suggest to a third grader that whether or not the sum of the angles of a triangle is 180° is dependent on the basic assumptions one makes about parallel lines.

comment . . .

However, measurements, observations, and demonstrations can still be useful for an initial intuitive introduction to an idea one later proves.

A proof, then, whether in a mathematical setting or not, still is basically a convincing argument which supports or justifies an idea. What sets mathematical proof apart from other proofs you have done is that you are no longer allowed to prove by fighting, flattering, or seducing. Your methods are limited. Within high school or college contexts, even experimental techniques such as measuring, observing, and demonstrating are discouraged. You are encouraged to present your argument verbally, basing your arguments on previously encountered mathematical ideas. The form in which you present your argument is of little significance. The important point is that your argument is a clear, logically sound discourse which supports the conclusion asked for and uses previously encountered and accepted mathematical principles as the basis of the argument.

In the remainder of this chapter we will attempt to suggest

helpful steps that a person can take to help him complete mathematically correct proofs.

comment . . .

As we progress through the remainder of this section, we may notice a resemblance to the ideas suggested by G. Polya. Indeed, his thoughts have influenced the authors' developments, as they have hundreds of other mathematicians.

Exercise 15.2.1
Underlying any kind of proof there are certain assumptions that must be accepted in order for the proof to be convincing. Examine each proof discussed in the story about the student's encounter with the angle sum property of triangles, and list one or more assumptions upon which each argument was based.

15.3 PROOF—PHASE ONE: CONVINCING YOURSELF

Since a proof is first and foremost a convincing argument, you will have considerable difficulty in writing such an argument to convince someone else unless you yourself are first convinced of the truth of the statement. Thus, a most essential first step in preparing a proof is to convince yourself that the idea to be proved makes sense. Now, how do you do that? First you must understand the idea, that is, you must really know what it says, and then you must think about the idea and contemplate whether or not it seems reasonable.

How do you understand the problem? You have to read it, but you must do more than this. You have to analyze what you have read. You may have to read it several times. You may have to draw pictures or look at models. In your analysis of the statement, you may find it helpful to separate the hypothesis (what is given or what is assumed) from the conclusion (what you are to prove). In many instances, this is a rather easy task, particularly if the original statement is expressed in an *if— conditional* form; that is, of the form "If *A*, then *B*." When a statement you are to prove is stated in this form, the If— clause contains the hypothesis and the then— clause contains your conclusion.

Example 1
Consider the statement: If two sides of a triangle are congruent, then the angles opposite these sides are congruent.

The hypothesis is the If—clause, which is: Two sides of a

triangle are congruent. The conclusion is the then—clause: The angles opposite these sides are congruent.

Thus, we are given a triangle in which two sides are congruent and we are asked to prove that the angles opposite these sides are congruent.

Example 2

Consider the statement: A triangle is equilateral if it is equiangular.

Don't be fooled by this one. The hypothesis is the If—clause, which is the *second* clause in the statement: It is equiangular. The conclusion is the then—clause, which is the *first* part of the statement (even though the word "then" is missing): a triangle is equilateral.

Thus, we are given a triangle which is equiangular and we want to prove that it is equilateral.

This last example illustrates that If A, then B is equivalent to the statement B, if A. There are also other equivalent forms for the If—conditional statement, and they appear below.

The following statements are all equivalent to If A, then B.

B, if A A is a sufficient condition for B A, only if B B is a necessary condition for A	} alternate ways to state If A, then B.

If you find it difficult to determine the hypothesis and conclusion of a particular statement, you may find that your task can be simplified by first rewriting the statement so that it is an equivalent statement in If—then form.

Exercise 15.3.1

For each of the statements below:

a. indicate the hypothesis and conclusion of each statement
b. indicate if the statement is true or false.

1. A quadrilateral is a rhombus only if it is a parallelogram.
2. Two lines are parallel if, when cut by a transversal, the alternate interior angles are congruent.
3. That a triangle is isosceles is a necessary condition for it to be equilateral.
4. That a polygon is equilateral is a sufficient condition for it to be regular.
5. Two triangles are congruent only if there exists a one-to-one correspondence such that three angles of one triangle are congruent respectively to the corresponding angles of the other.

When we have a statement S in If—then form, there are

three transformations which may be performed on it, called the *converse, inverse,* and *contrapositive* of S. In general, they are formed in the following ways.

1. The converse of S: Interchange the hypothesis and conclusion of S.
2. The inverse of S: Contradict (negate) the hypothesis and and contradict the conclusion of S.
3. The contrapositive of S: Interchange the contradiction of the hypothesis with the contradiction of the conclusion of S.

These can be summarized as follows:

S	*Converse of* S
If A, then B	If B, then A
Inverse of S	*Contrapositive of* S
If not A, then not B	If not B, then not A

Example 3

Consider the following four statements:

S: If a figure is a triangle, then it is a polygon.
Converse of *S:* If a figure is a polygon, then it is a triangle.
Inverse of *S:* If a figure is not a triangle, then it is not a polygon.
Contrapositive of *S:* If a figure is not a polygon, then it is not a triangle.

If the reader studies the above example and other examples similar to it, he should become aware that the original statement S and its contrapositive are always either both true or both false. This means that S and its contrapositive are *equivalent statements.* Likewise, the inverse of S and the converse of S are equivalent statements. This becomes quite useful in proofs when occasionally we may find it easier tc prove the contrapositive of a statement rather than the statement itself.

In contrast, we should observe that a statement S is not equivalent to its converse. That is, it is possible to find examples of S where any one of the following is true.

1. S is true and the converse of S is true.
2. S is true and the converse of S is false.
3. S is false and the converse of S is true.
4. S is false and the converse of S is false.

This means that the truth values of S and its converse are independent of each other and, thus, they cannot be equivalent statements.

Likewise, S and its inverse are not equivalent.

Exercise 15.3.2

1. Give four statements in If—then form, one corresponding to each of the four conditions listed in the preceding paragraphs.
2. Form the converse, inverse, and contrapositive of each of the statements in Exercise 15.3.1. Then decide if each statement is true or false.

In the event that we encounter a statement S which is true, whose converse is also true, it sometimes becomes convenient to form the conjunction of the two statements. Thus, we could write *If A, then B and If B, then A*. There is, however, a shorter way to write such a statement.

Recall that *If A, then B* is equivalent to *B, if A*.

Recall, furthermore, that *If B, then A* is equivalent to *B, only if A*.

Thus, an alternate way to express the conjunction *If A, then B and if B, then A* is to write *B if and only if A*. Such a statement written in this form is called a *bi-conditional statement*. The essential idea to remember here is that any bi-conditional statement implies two nonequivalent statements: an if conditional statement and the converse of this statement. Thus, if we are asked to prove a bi-conditional statement, we are obligated to present two proofs, one for each of the two nonequivalent statements.

Example 4

1. Find two nonequivalent statements implied by the following bi-conditional statement.
2. Then decide if the bi-conditional statement is true or false.

> Two triangles are equilateral if and only if they are equiangular.

One way to separate the conjunction is as follows.

Statement 1: Two triangles are equilateral if they are equiangular.
Statement 2: Two triangles are equilateral only if they are equiangular.

These are now two nonequivalent statements. Frequently, however, these components are rewritten so that each is in an equivalent If—then form. Thus, they become

Statement 1: If two triangles are equiangular, then they are equilateral.
Statement 2: If two triangles are equilateral, then they are equiangular.

Now, to determine if the bi-conditional is true or false, we need to consider the truth value of each of the statements above. Since both statements are true, the bi-conditional is true.

Example 5

Do the same for the bi-conditional: A quadrilateral is a rectangle if and only if it is a square.

Statement 1: A quadrilateral is a rectangle if it is a square; or equivalently: If a quadrilateral is a square, then it is a rectangle.
Statement 2: A quadrilateral is a rectangle only if it is a square; or equivalently: If a quadrilateral is a rectangle, then it is a square.

Now, since Statement 1 is true, and Statement 2 is false, the bi-conditional is false. A bi-conditional is true only when both of its component statements are true.

Exercise 15.3.3

1. Form the converse of each statement below. Then use this statement and its converse to form the bi-conditional in if and only if form. Determine if the bi-conditional is true or false.
   a. If two lines are cut by a transversal so that the corresponding angles are congruent, then the lines are parallel.
   b. If two triangles are congruent, then they are similar.
   c. If set A equals set B, then set A is a subset of set B and set B is a subset of set A.
2. For each of the bi-conditional statements listed below:
   i. Write two nonequivalent statements implied by the statment.
   ii. Determine the truth value of the statement.
      a. Two sides of a triangle are congruent if and only if the angles opposite these sides are congruent.
      b. A quadrilateral is a parallelogram if and only if a diagonal divides it into two congruent triangles.
      c. The diagonals of a quadrilateral are perpendicular if and only if the quadrilateral is a rhombus.
      d. A polygon is regular if and only if it is equilateral and equiangular.

All of the preceding discussion was designed to give the reader a basis upon which he can study statements to gain some understanding of a statement he is asked to prove. As pointed out before, understanding the statement and becoming convinced of its truth is a necessary first step to completing the proof. Most people find it quite difficult to try to convince others that something is true unless they themselves also are convinced that it is true. Thus, it is often senseless to attempt to progress any further with a proof until this initial goal has been accomplished. In trying to convince yourself, get into the habit of asking yourself a series of questions about the problem. Below is a list of such questions you can ask yourself.

1. Can you write it in the If—then form?
2. What is the hypothesis?
3. What is the conclusion?
4. Is it possible to satisfy the hypothesis?
5. Does the conclusion make sense?
6. Does the hypothesis support the conclusion?
7. Can you draw a figure which satisfies the hypothesis?
8. Can you draw a figure which satisfies the hypothesis but which contradicts the conclusion? If not, why not?
9. Separate the various parts of the hypothesis. Can you write them down?
10. Separate the various parts of the conclusion. Can you prove one of the parts?

You may find that in attempting to convince yourself, you arrived instead at just the opposite conclusion, that is, you do not believe the statement is true. If this is the case, then your proof would involve presenting a counter-argument which disproves the statement.

However, at this point let us assume that we have finally convinced ourselves that, indeed, the statement is true. What is the next step?

15.4 PROOF—PHASE TWO: PLANNING AN ARGUMENT

Now that we have become convinced of the truth of the statement, we have to begin to plan an argument. How are we going to explain to someone else why the statement is true? What will we say? Upon what principles will we base our argument? Perhaps, during Phase One the answers to these questions have become obvious and we can use whatever arguments convinced us to convince others. Frequently, however, arguments which we use to convince ourselves are a mixture of objective ideas together with personal ideas influenced by our own insight and intuition. Also, it is one thing to have an idea internalized and quite a different thing to be able to communicate that idea accurately to others. This is often indicated by the statement, "I know why it is, but I just can't explain it." The purpose of this phase, then, is to plan an argument which retains any of the objective ideas we already had and replaces any personal notions which are intuitional or "feelings" rather than knowledge by objective ideas. Once again, the successful completion of this phase depends on your asking yourself additional questions and attempting to answer them. These answers may begin to present you with a plan. Some typical questions are listed below.

1. Have you seen this statement before, but in a slightly different form?
2. Do you know a principle which is related to this statement?
3. Do you know a theorem which could be useful?
4. Look at the conclusion! Can you think of a familiar statement which has the same conclusion?
5. Look at the hypothesis! Can you think of a familiar statement with the same hypothesis? Can you relate the conclusion of this statement to the conclusion of the statement you want to prove?
6. Can you think of a statement similar to this which has already been proved? Can you use it? Can you use its result? Can you use the method of proof? Can you introduce some auxiliary lines so that you can use this statement?
7. Could you restate the problem?
8. If you cannot prove the proposed statement, could you first prove a simpler but related statement?
9. Could you prove a more general statement? Could you prove a special case of the statement?
10. Have you used all of the conditions of the hypothesis?
11. Can you prove a part of the conclusion?
12. If you added conditions to the hypothesis, could you prove the conclusion? How can you establish these added conditions from the original conditions?

The whole purpose of asking yourself questions similar to those above is to assist you in formulating a series of intelligent and logical statements which indicate how the hypothesis of your statement is related to the conclusion. Once you have such a series of statements in mind, you are ready to go on to the next phase.

15.5 PROOF—PHASE THREE: PRESENTING YOUR ARGUMENT

This is the step in which we finally set our argument down on paper in some acceptable form. It may appear in a series of paragraphs, it may be a list of statements accompanied by reasons. The essential idea is that you present a sequence of statements which explain the connection between the hypothesis and the conclusion. Now, since the argument we present is supposed to be a convincing one, we must be sure that the statements we make will be acceptable to the people to whom our proof will be presented. The best way to assure ourselves of this is to accompany each of the statements with a supporting statement of a more general nature. Such statements are usually referred to as "reasons." These reasons can be of several types: definitions, postulates, previously proved theorems, or conditions which are in the hypothesis of the statement to be proved. One frequent difficulty with this is deciding upon what statement to give as a supporting reason. Frequently, the student thinks of several statements which he thinks might possibly support a given conclusion and he has difficulty deciding which one is most appropriate for his purposes. When the general statements are expressed in If—then form, there are some requirements which can help you make these decisions. These requirements are discussed below.

Suppose, in a proof, we wish to make a specific conclusion, C. We need to select a general statement S in If—then form. S will successfully support conclusion C providing the following three conditions are satisfied:

1. S is accepted as a true statement.
2. Conclusion C is a specific example of the conclusion stated in S.
3. All the conditions listed in the hypothesis of S must have been established in specific form prior to the use of statement S.

Example 1

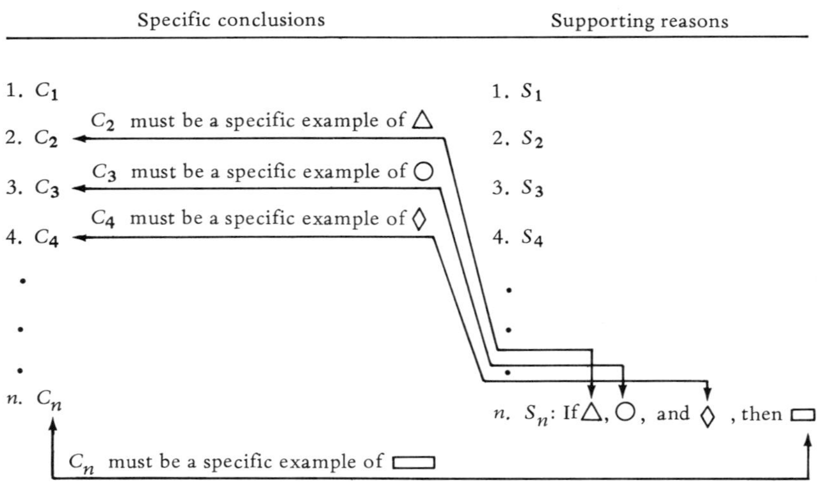

Example 2

Decide if reason 6 is suitable for supporting conclusion 6 in the following portion of a proof.

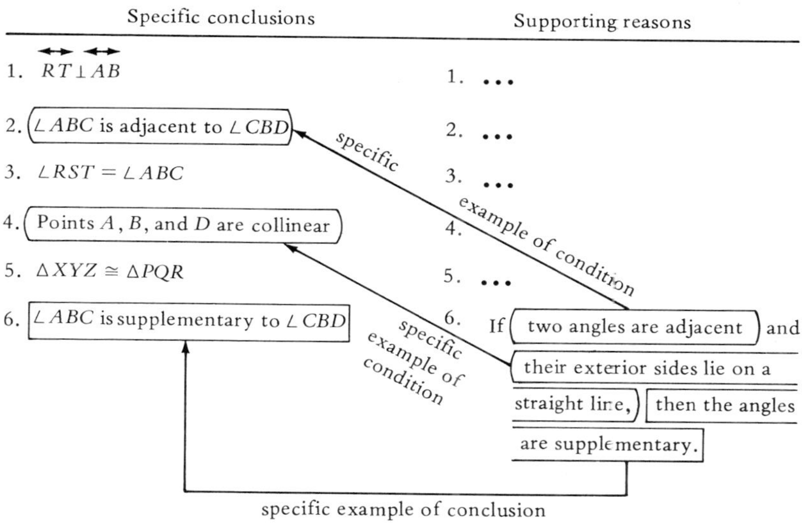

This reason was correctly used since both conditions of the hypothesis were satisfied.

Example 3

Is reason 4 a suitable supporting statement for conclusion 4?

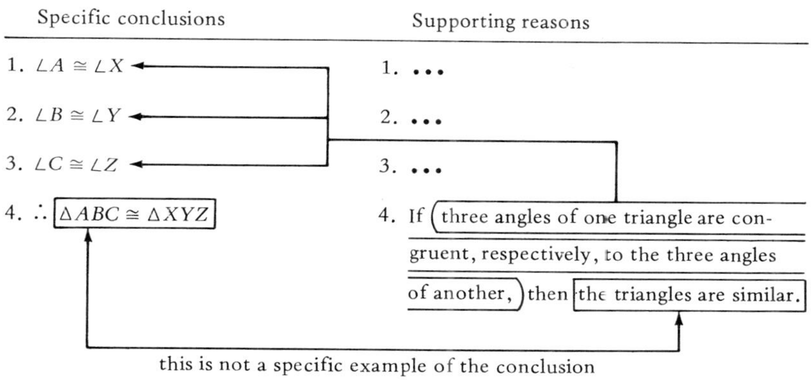

Here, although all the conditions of the hypothesis were established earlier in the proof, the reason does not support the specific conclusion since conclusion 4 is not a specific example of the conclusion of reason 4.

Example 4

Is reason 4 a suitable supporting statement for conclusion 4? (See Fig. 15.5.1.)

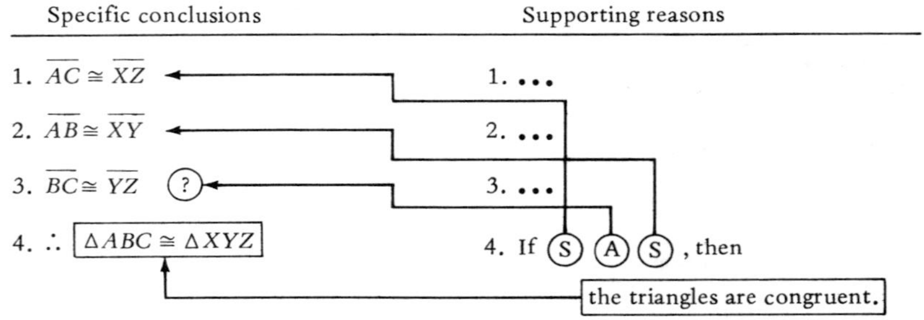

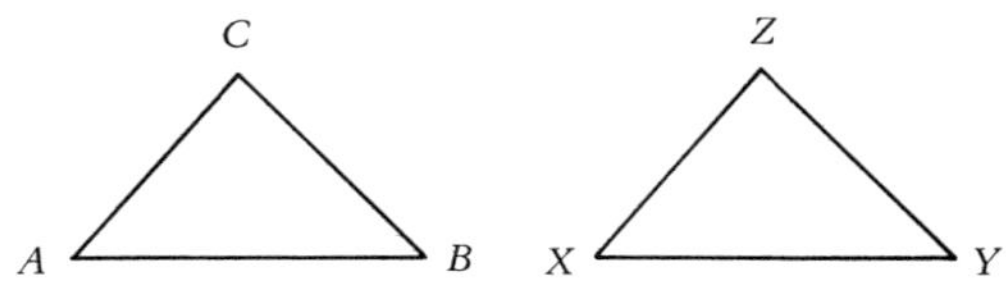

Figure 15.5.1

Here the reason does not support the conclusion for even though the conclusions correspond correctly, one of the conditions of the hypothesis has not been established. It is suggested that perhaps one reason why many teachers and authors prefer to offer proofs in the statement—reason format is because it provides one with an easier way of checking the validity of the supporting reasons.

Exercise 15.5.1

1. Examine the following excerpts of proofs. Decide which of the supporting reasons are valid and explain your decision.
   a. (See Fig. 15.5.2.)

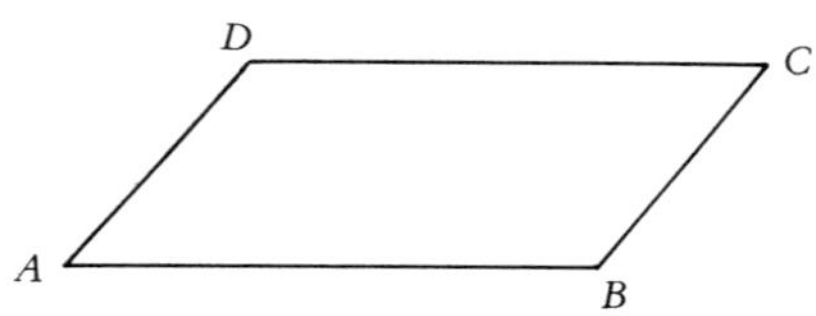

Figure 15.5.2

Specific conclusions	*Supporting reasons*
1. $\overleftrightarrow{AB} \parallel \overleftrightarrow{DC}$	1. . . .
2. $\overline{AD} = \overline{BC}$	2. . . .
3. ∴ *ABCD* is a parallelogram	3. If both pairs of opposite sides of a quadrilateral are parallel, then the figure is a parallelogram.

   b. (See Fig. 15.5.3.)

Figure 15.5.3

Specific conclusions	*Supporting reasons*
1. *ABCD* is a parallelogram	1. . . .
2. ∴ $\overleftrightarrow{AB} \parallel \overleftrightarrow{DC}$ and $\overleftrightarrow{AD} \parallel \overleftrightarrow{BC}$	2. If the opposite sides of a quadrilateral are parallel, then the figure is a parallelogram.

c. (See Fig. 15.5.4.)

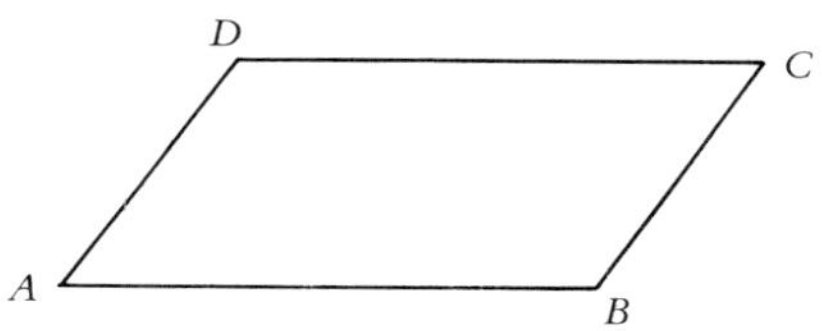

Figure 15.5.4

Specific conclusions	*Supporting reasons*
1. *ABCD* is a parallelogram	1. . . .
2. ∴ *ABCD* is a rhombus	2. If a quadrilateral is a rhombus, then it is a parallelogram.

d. (See Fig. 15.5.5.)

Figure 15.5.5

Specific conclusions	*Supporting reasons*
1. $\overleftrightarrow{AB} \parallel \overleftrightarrow{DC}$	1. . . .
2. $\overline{AB} = \overline{DC}$	2. . . .
3. ∴ *ABCD* is a parallelogram	3. If one pair of opposite sides of a quadrilateral is parallel and congruent, then the figure is a parallelogram.

2. Complete the following excerpts of proofs by supplying only the specific conclusions necessary for the given statement to be a valid reason. Each problem relates to the accompanying diagram. (You do not need to supply any supporting reasons and you may not need all the steps provided.)

   a. (See Fig. 15.5.6.)

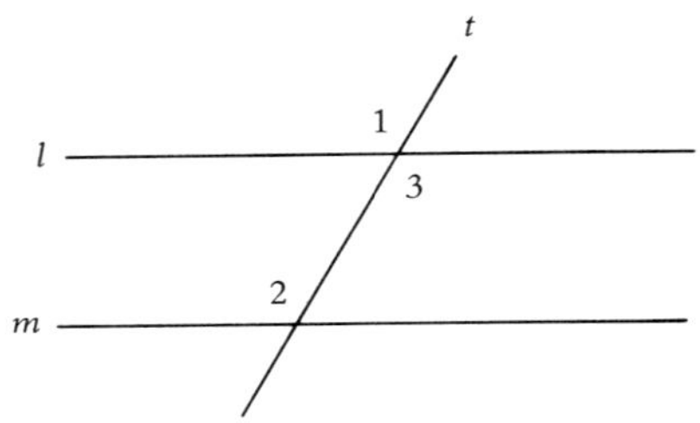

Figure 15.5.6

Specific conclusions	*Supporting reasons*
1. ________	1. . . .
2. ________	2. . . .
3. ________	3. If two parallel lines are cut by a transversal, then the corresponding angles are congruent.

   b. (See Fig. 15.5.7.)

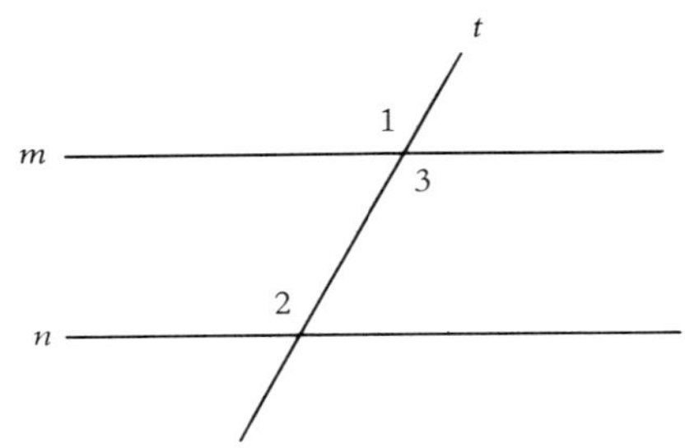

Figure 15.5.7

Specific conclusions	*Supporting reasons*
1. ________	1. . . .
2. ________	2. . . .
3. ________	3. If two lines are cut by a transversal so that a pair of alternate interior angles are congruent, then the lines are parallel.

c. (See Fig. 15.5.8.)

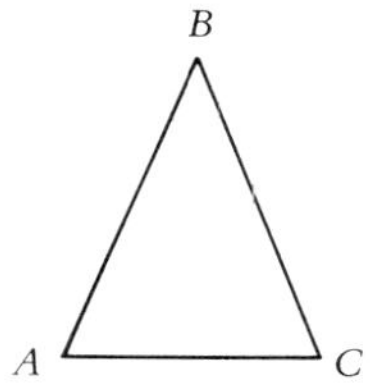

Figure 15.5.8

Specific conclusions	*Supporting reasons*
1. ________________	1. . . .
2. ________________	2. . . .
3. ________________	3. If two angles of a triangle are congruent, then the sides opposite these angles are congruent.

d. (See Fig. 15.5.9.)

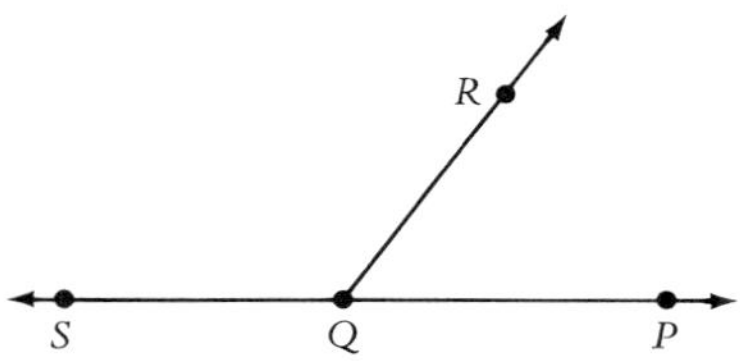

Figure 15.5.9

Specific conclusions	*Supporting reasons*
1. ________________	1. . . .
2. ________________	2. . . .
3. ________________	3. If two angles are adjacent and supplementary, then their exterior sides lie on a straight line.

e. (See Fig. 15.5.10.)

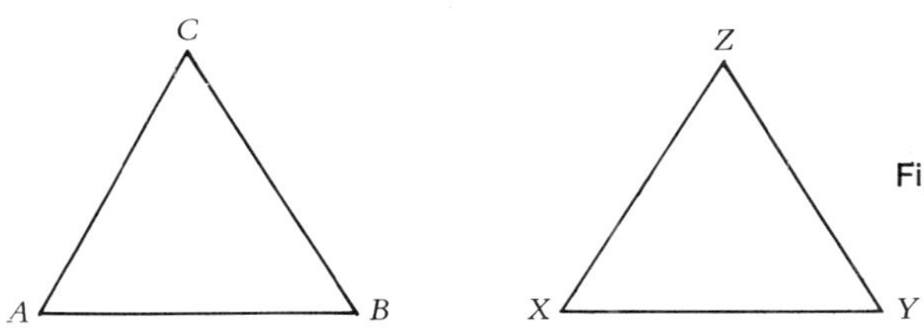

Figure 15.5.10

Specific conclusions	*Supporting reasons*
1. ____________	1. . . .
2. ____________	2. . . .
3. ____________	3. . . .
4. ____________	4. AAS

f. (See Fig. 15.5.11.)

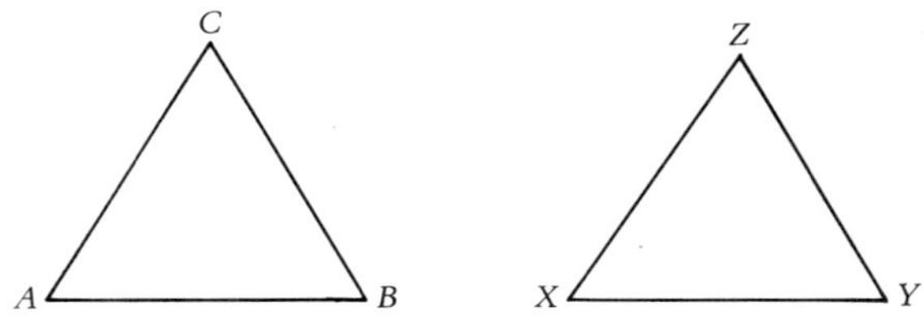

Figure 15.5.11

Specific conclusions	*Supporting reasons*
1. ____________	1. . . .
2. ____________	2. . . .
3. ____________	3. . . .
4. ____________	4. . . .
5. ____________	5. Corresponding angles of congruent triangles are congruent.

15.6 PROOF—PHASE FOUR: LOOKING BACK

Once you have finally progressed through the three earlier phases, you should take some time to contemplate what you have done. Once again, this phase can be performed by asking yourself a series of questions, some of which are listed below.

1. Look at your proof. Do you finally arrive at the conclusion of the statement you wanted to prove?
2. Look at each statement of your proof. Is it supported by a valid reason?
3. Would there have been another way you could have proved it in a shorter way?
4. Look at the final conclusion. Can you use it?
5. Look at your method. Can you use it in other proofs?

in our opinion . . .

It is our opinion that the key to success in proving statements is to ask yourself very intelligent questions about the problem you are confronted with. Indeed, a good problem solver is one who asks good questions. Frequently, a student will complain, "When I am in class, I can do the proof, but at home I can't." Often the reason for this is obvious. In class, there is the teacher who asks the right questions. In your attempt to answer these questions you solve the problem. When you are at home, there is no one to ask you these important questions. Thus, *you* must supply yourself with the proper questions. For any problem, whether you can solve it or not, you should be able to prepare a list of significant questions about the problem. If you can train yourself to do this, then you will have made significant progress in being a successful problem solver.

15.7 AN EXAMPLE OF A PROOF IN DETAIL

Prove the statement: Every point equidistant from the endpoints of a line segment is on the perpendicular bisector of the line segment.

Corresponding to each phase, the authors will include a few of the questions and some likely responses.

Phase One: Convincing Yourself

Questions	*Responses*
1. What is the hypothesis and conclusion?	1. That's difficult to say. I don't see any If—then clauses.
2. Can you write it in if—then form?	2. I think so. If a point is equidistant from the endpoints of a line segment, then it is on the perpendicular bisector of the line segment.
3. What is the hypothesis and conclusion?	3. The hypothesis is: A point is equidistant from the endpoints of a line segment. The conclusion is: The point is on the perpendicular bisector.
4. Can you draw a figure which satisfies the hypothesis and introduce suitable notation?	4. Let's see. We need a line segment $\overline{AB}$ and a point P which is equidistant from A and B (Fig. 15.7.1).

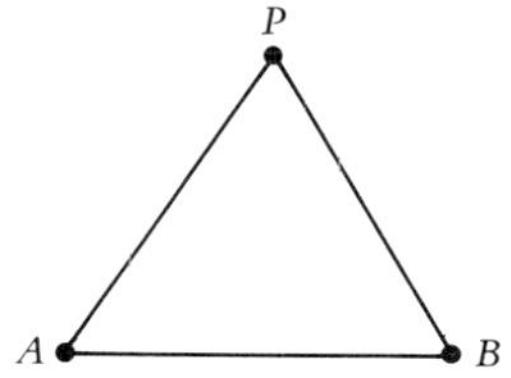

Figure 15.7.1

Questions	*Responses*
5. Separate the various parts of the hypothesis.	5. Given: Line segment $\overline{AB}$, point P, and $\overline{AP} \cong \overline{BP}$.
6. Separate the various parts of the conclusion.	6. To prove: P is on a line perpendicular to $\overline{AB}$. This line also bisects $\overline{AB}$.
7. Does the conclusion make sense?	7. Let me locate several points satisfying the hypothesis. . . . Yes, they all seem to lie in the perpendicular bisector (Fig. 15.7.2).

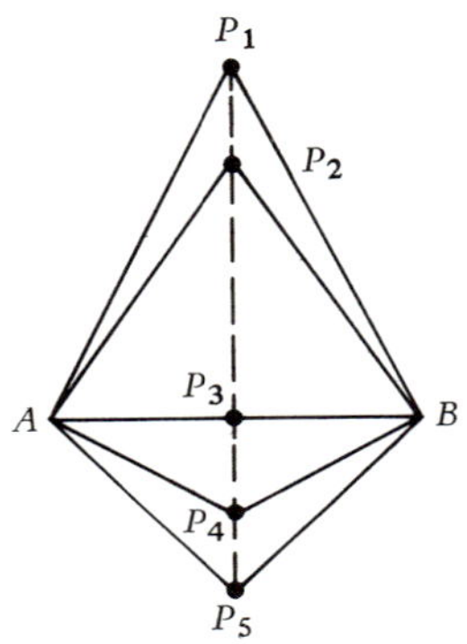

Figure 15.7.2

8. Can you contradict the conclusion?	8. Let me try to put a point off the perpendicular bisector. . . . But then P doesn't seem to be equidistant from A and B (Fig. 15.7.3).

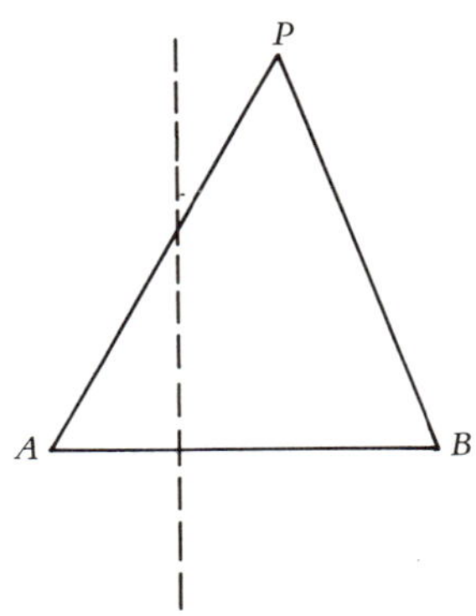

Figure 15.7.3

9. Does the conclusion make sense?	9. Yes, it seems very reasonable.

End of Phase One

Phase Two: Planning an Argument

Questions	*Responses*
1. Can you prove a special case of the statement?	1. Yes. If the point were on the segment, the conclusion is obvious (Fig. 15.7.4).

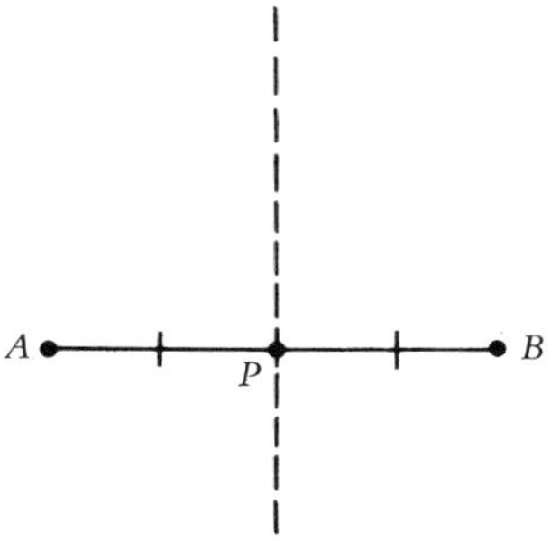

Figure 15.7.4

2. What remains to be done?	2. Prove the statement for P off the line.
3. Can you prove a part of the conclusion?	3. Yes. I can prove that P is on a bisector of $\overline{AB}$, but not necessarily perpendicular (Fig. 15.7.5).

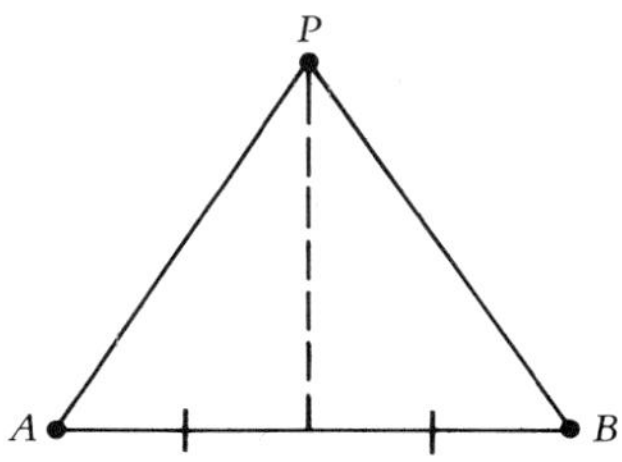

Figure 15.7.5

4. Can you prove a simpler but related problem?	4. Let me try to prove: Given: $\overline{AP} \cong \overline{BP}$ $\overline{AM} \cong \overline{MP}$ To prove: $\overleftrightarrow{PM} \perp \overleftrightarrow{AB}$ (Fig. 15.7.6).

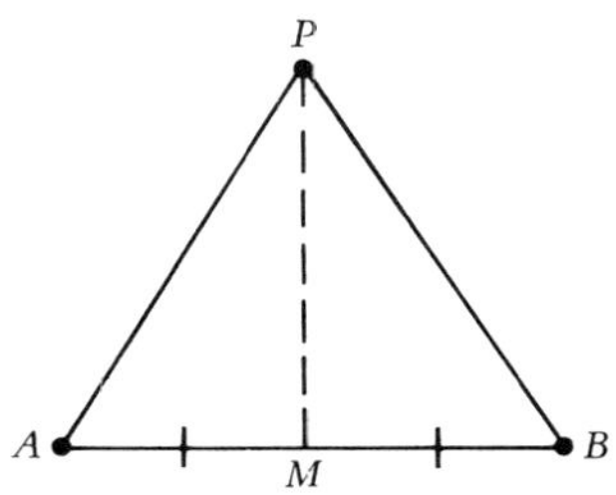

Figure 15.7.6

Questions	*Responses*
5. Look at this conclusion. Can you think of a statement having the same conclusion?	5. If two lines intersect to form congruent adjacent angles, then the lines are perpendicular.
6. Can you satisfy the hypothesis of this statement?	6. Obviously angles AMP and BMP are adjacent. Maybe I could prove them congruent by congruent triangles; by SSS, I think.
7. Can you put all these ideas together?	7. Let's see. If I select M, the midpoint of $\overline{AB}$, and draw $\overleftrightarrow{PM}$, I have two triangles which are congruent. Thus, we have two congruent adjacent angles, and, therefore, $\overleftrightarrow{PM} \perp \overleftrightarrow{AB}$. But $\overline{PM}$ also bisects $\overline{AB}$ so $\overleftrightarrow{PM}$ must be the perpendicular bisector of $\overline{AB}$.

End of Phase Two

Phase Three: Presenting Your Argument

Essentially, we will present the proof suggested on this page, but here we must accompany each step with a valid reason. So that the reader can more readily see the conditions for valid reasons are satisfied, we present it first in the statement—reason format.

15.7.1 Theorem

Every point which is equidistant from the endpoints of a line segment is on the perpendicular bisector of the line segment.

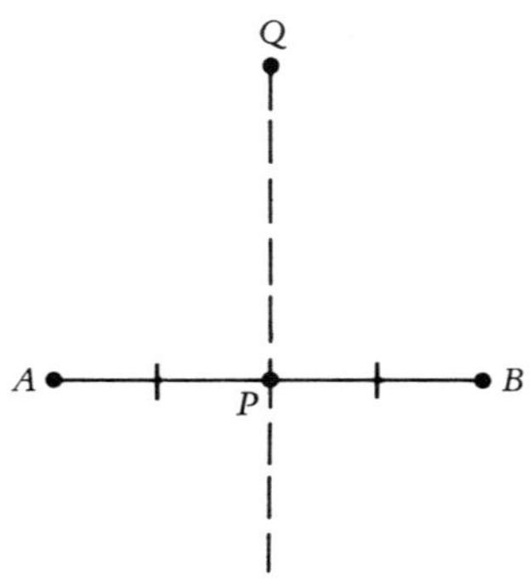

Figure 15.7.7

Part I: The Point Is on the Segment

Given

Line segment $\overline{AB}$. (See Fig. 15.7.7.)

$P \in \overline{AB}$
$\overline{AP} \cong \overline{BP}$

To prove

P is on the perpendicular bisector of $\overline{AB}$.

Specific conclusions	*Supporting reasons*
1. $P \in AB$	1. Given
2. From P construct a line $\overleftrightarrow{QP} \perp \overline{AB}$	2. At a point on a line one and only one line may be drawn perpendicular to the line.
3. $\overline{AP} \cong \overline{BP}$	3. Given
4. Then $\overleftrightarrow{QP}$ is the perpendicular bisector of $\overline{AB}$.	4. Definition of a perpendicular bisector of a segment.

Part II: The Point Is Not on the Line Segment

Given

Line segment $\overline{AB}$.

P not on $\overleftrightarrow{AB}$
$\overline{AP} \cong \overline{BP}$

To prove

P is on the perpendicular bisector of $\overline{AB}$ (Fig. 15.7.8).

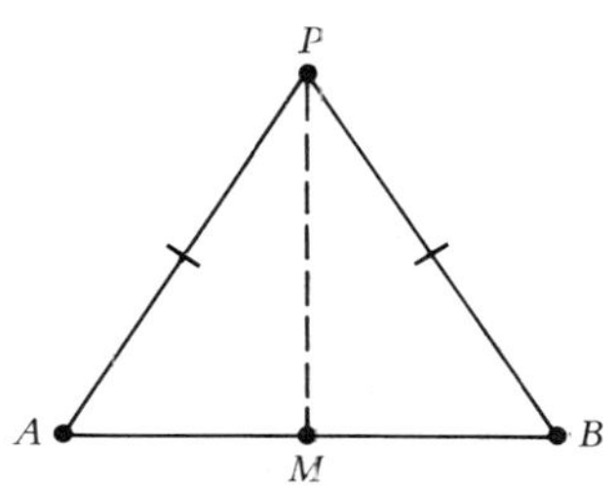

Figure 15.7.8

Specific conclusions	Supporting reasons
1. Let M be the midpoint of $\overline{AB}$	1. Every line segment has one and only one midpoint.
2. Draw $\overline{PM}$	2. Two points determine one and only one line.
3. $\overline{AP} \cong \overline{BP}$	3. Given.
4. $\overline{AM} \cong \overline{MB}$	4. Constructed in step 1, definition of midpoint.
5. $\overline{PM} \cong \overline{PM}$	5. Reflexive property of congruence
6. $\triangle AMP \cong \triangle BMP$	6. If (S.S.S.), then the △s are ≅.
7. $\angle AMP \cong \angle BMP$	7. If (two △s are ≅,) then the corresponding ∠s are ≅.
8. $\angle AMP$ is adjacent to $\angle BMP$	8. Definition of adjacent.
9. $\overline{PM} \perp \overline{AB}$	9. If two lines intersect to form (congruent)(adjacent angles,) then the lines are perpendicular.
10. ∴ $\overline{PM}$ is the perpendicular bisector of $\overline{AB}$.	10. If (a line is perpendicular to a line segment) and (goes through the midpoint of the segment,) then it is the perpendicular bisector of the line segment.

Now, it has been indicated several times that a proof need not be in the above form to be correct. Thus, there now will follow the same proof in paragraph form.

In considering the validity of the statement that every point which is equidistant from the endpoints of a line segment is on the perpendicular bisector of the line segment, it is helpful to consider two cases. Case I would be for the point on the segment and case II for the point off the segment.

The first case is relatively easy. If a point is on a segment and equidistant from the endpoints, then it is the midpoint of the segment, by definition of midpoint. Now, by a fundamental postulate it is possible to construct one and only one line perpendicular to this segment at this midpoint. Thus, this point is

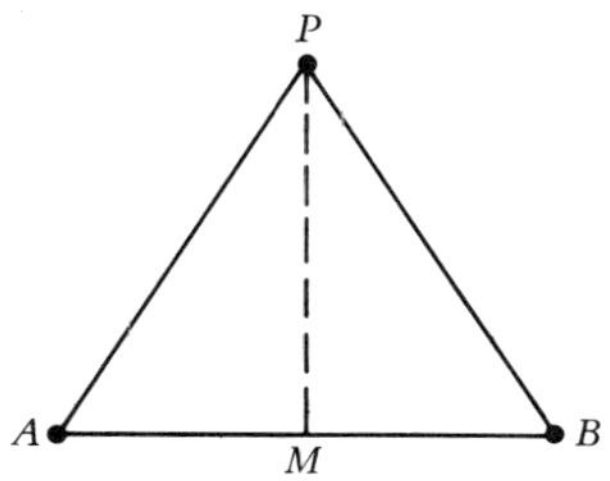

Figure 15.7.9

on the perpendicular bisector of the segment, by definition of perpendicular bisector.

For the second case, let us consider Fig. 15.7.9. Let $\overline{AB}$ be the line segment and P be the point not on $\overline{AB}$, such that $\overline{AP} \cong \overline{BP}$. By a postulate we know that $\overline{AB}$ has one and only one midpoint so let M be this point. Now construct $\overline{PM}$. This is possible since any two distinct points determine a unique line. Now in $\triangle$s AMP and BMP we have $\overline{AP} \cong \overline{BP}$ by hypothesis, $\overline{AM} \cong \overline{MP}$ by definition of midpoint, and $\overline{PM} \cong \overline{PM}$ by the reflexive property for congruence. Thus, we can say that $\triangle AMP \cong \triangle BMP$ by SSS. Now $\angle$s AMP and BMP are corresponding angles in the two congruent triangles and thus they are congruent. Also, these angles are adjacent by definition. Thus, $\overleftrightarrow{PM}$ and $\overleftrightarrow{AB}$ are two lines which intersect to form congruent adjacent angles and they are, therefore, perpendicular. Since $\overleftrightarrow{PM}$ is perpendicular to $\overleftrightarrow{AB}$ and since M is the midpoint of $\overline{AB}$, we see that, therefore, $\overleftrightarrow{PM}$ is the perpendicular bisector of $\overline{AB}$ by definition. Thus, our statement is proved.

End of Phase Three

Phase Four: Looking Back

If we reflect on our proof, we might ask ourselves if this proof could have been done differently. One possibility is that perhaps we could have first drawn the perpendicular through P to AB and then show that it intersects $\overline{AB}$ at its midpoint. This will be left as an exercise for the student.

Furthermore, we might wonder if this could not have been accomplished through a different approach. For example, we have learned much about perpendicular lines and lengths of segments when points are on a coordinate system. Perhaps our proof might have incorporated some of these ideas. Let us look at that. The first important point is to consider where we should locate the point on the Cartesian plane. The answer is to locate it so that your notation and coordinates are as simple as possible. We try to use the axes if possible. Thus, we might try to locate A and B on the x-axis. Furthermore, if we consider our earlier proof, we are reminded that our first step was to select the midpoint of AB. This suggests placing M at the origin and A and B on either side, an equal distance from the origin. Finally, we must locate point P. Now here we know it should be on the y-axis, but we will purposely keep it off and see if we can argue that it should be on the y-axis. Thus, our placement looks like Fig. 15.7.10. Since M is to be the midpoint of $\overline{AB}$, we can let the coordinates of A be $(-a, 0)$ and B be $(a, 0)$. Let the coordinates of P be (x, y). Now, if we can show that P is on the y-axis, then we will have shown that P is on the perpendicular bisector. This can be done by showing that $x = 0$. We can use our distance formula and the fact that $\overline{AP} \cong \overline{BP}$. The proof follows.

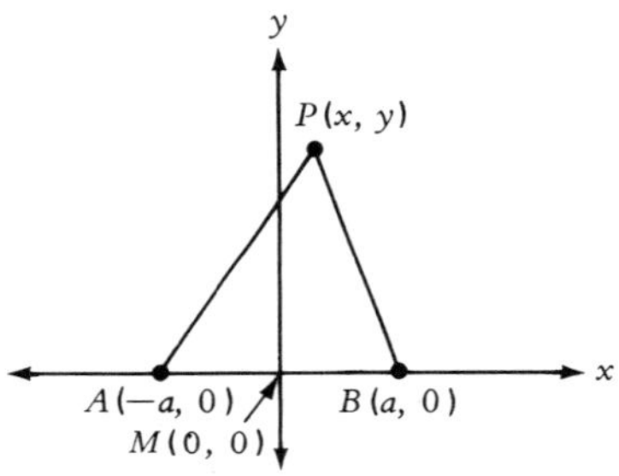

Figure 15.7.10

Let A and B be on the x-axis with coordinates $(-a, 0)$ and $(a, 0)$, respectively. Then $M(0, 0)$ is the midpoint of $\overline{AB}$. Also, let P have coordinates (x, y).

Now,

$$AP = \sqrt{(x - (-a))^2 + (y - 0)^2} = \sqrt{(x + a)^2 + y^2}$$

and

$$BP = \sqrt{(x - a)^2 + (y - 0)^2} = \sqrt{(x - a)^2 + y^2}$$

by the distance formula. By hypothesis $\overline{AP} \cong \overline{BP}$, we have

$$\sqrt{(x + a)^2 + y^2} = \sqrt{(x - a)^2 + y^2}$$

and then squaring both sides we get

$$(x + a)^2 + y^2 = (x - a)^2 + y^2$$

and simplifying

$$x^2 + 2ax + a^2 + y^2 = x^2 - 2ax + a^2 + y^2$$

or

$$4ax = 0$$

Since $4ax = 0$, either $a = 0$ or $x = 0$.

If $a = 0$, then $\overline{AB}$ is not a segment.

$\therefore x = 0$. But if $x = 0$, $(x, y) \varepsilon$ y-axis. Thus, P is on the y-axis.

$\therefore P$ is on the perpendicular bisector of $\overline{AB}$. Thus, the statement is proved.

15.8 ANALYTIC AND SYNTHETIC PROOF

There is one interesting point which should be made here. Did you notice that it was not necessary to consider the proof as two cases? The above argument is valid no matter if P is on the line segment or not!

What you have just seen is a proof technique which is called an *analytic proof*. The first two proofs you saw are called *synthetic proofs*. The difference between the two is quite obvious. The analytic proof utilizes techniques associated with coordinate geometry whereas synthetic geometry avoids such ideas. Both types of proof are correct since they are both convincing arguments. Sometimes you will find that one approach will have advantages over the other such as one may appear easier or shorter than the other. In the exercises that follow, you will be asked to do several proofs using both techniques in order to contrast them.

At this point, suppose we prove the converse of the last statement using both an analytic and a synthetic approach. We will omit Phases One and Two this time, but the reader is reminded that these activities did occur previous to what will follow. You are urged to proceed through Phases One and Two before reading the proofs that are here.

15.8.1 Theorem

If a point is on the perpendicular bisector of a line segment, then it is equidistant from the endpoints of the line segment.

First Proof: An Analytic Approach

Let A and B be on the x-axis with coordinates $(-a, 0)$ and $(a, 0)$, respectively. Let P be on the y-axis with coordinates $(0, p)$. Then P is on the perpendicular bisector of $\overline{AB}$ (Fig. 15.8.1). It now remains to be shown that $\overline{AP} \cong \overline{BP}$.

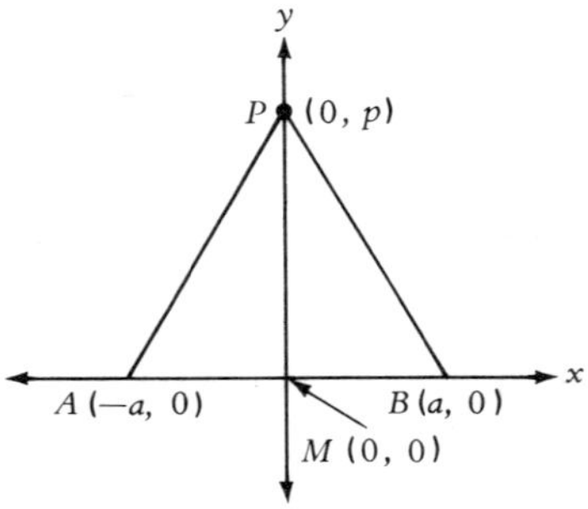

Figure 15.8.1

By the distance formula, we have

$$AP = \sqrt{(0 - (-a))^2 + (p - 0)^2} = \sqrt{a^2 + p^2}$$

and

$$BP = \sqrt{(0 - a)^2 + (p - 0)^2} = \sqrt{a^2 + p^2}$$

Thus, $AP = BP$ and $\therefore\ \overline{AP} \cong \overline{BP}$. Thus, the theorem is proved.

Second Proof: A Synthetic Approach

Case 1

P is on the line segment. (The proof is left to the reader.)

Case 2

P is off the line segment.

Given: $\overleftrightarrow{PM} \perp \overleftrightarrow{AB}$
$\overline{AM} \cong \overline{MB}$
To prove: $\overline{AP} \cong \overline{BP}$

Specific conclusions	*Supporting reasons*
1. $\overleftrightarrow{PM} \perp \overleftrightarrow{AB}$	1. Given
2. $\angle AMP \cong \angle BMP$	2. If two lines are perpendicular, they intersect to form congruent adjacent angles.
3. $\overline{AM} \cong \overline{MP}$	3. Given
4. $\overline{PM} \cong \overline{PM}$	4. Reflexive
5. $\triangle AMP \cong \triangle BMP$	5. If SAS, then congruency.
6. $\therefore\ \overline{AP} \cong \overline{BP}$	6. If two triangles are congruent, then the corresponding sides are congruent.

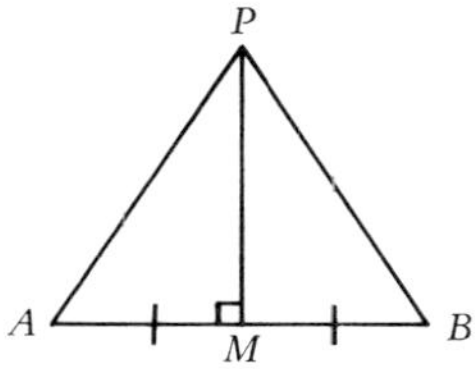

Figure 15.8.2

15.9 PROOF BY CONTRADICTION

Perhaps you recall that in Chapters 9 and 10 we made frequent use of a particular type of argument referred to as *indirect reasoning*. There is a rather intuitive discussion of this method in Chapter 9. We will now attempt to examine this method more closely. In order to do this, let us recall a proof which was presented in Chapter 10 of the theorem: If two lines are parallel, then the alternate interior angles are congruent. So that the argument will be fresh in our minds, let us reconsider the argument.

Let l and m be two parallel lines cut by a transversal t in points P and Q, respectively (Fig. 15.9.1). We will attempt to show that the angles denoted in the picture by 1 and 2 are congruent. It is certainly true that these two angles are congruent or they are not congruent. We will suppose that they are not congruent and show that this assumption leads to a contradiction.

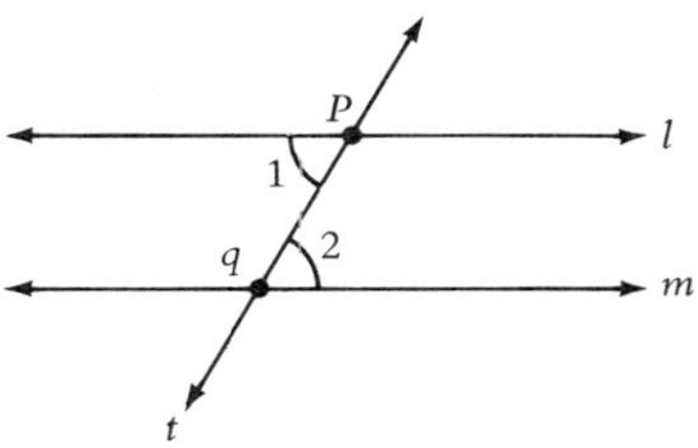

Figure 15.9.1

If $\angle 1$ and $\angle 2$ are not congruent, it is possible to construct a line through P, say n, for which alternate interior angles are congruent (Fig. 15.9.2). That is, $\angle 3 \cong \angle 2$. Note since $\angle 1$ and $\angle 3$ are not congruent, lines l and n are distinct. We know now that if two lines are cut by a transversal making a pair of alternate interior angles congruent, then the lines are parallel. Thus, $m \| n$. But also $m \| l$, and we have two lines through point P which are parallel to line m. This contradicts the Parallel Postulate. Thus, our assumption that $\angle 1$ and $\angle 2$ are not congruent has led to a contradiction. Hence, $\angle 1$ and $\angle 2$ are congruent.

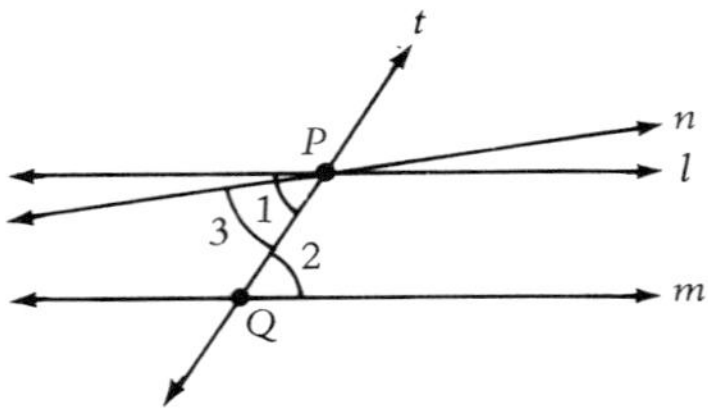

Figure 15.9.2

Perhaps the reader will recall that a statement of the If— then form can be expressed symbolically as $p \rightarrow q$ where p is the If statement (hypothesis) and q is the then statement (conclusion).

To every statement one of two truth values can be assigned. A statement is either true or false: never both. Thus, in examining the statement $p \rightarrow q$, Table 15.9.1, known as a *truth table,* exhibits all the possibilities of truth and falsity of the statements p, q, and $p \rightarrow q$. This table can be thought of as a definition.

Table 15.9.1

p	q	$p \rightarrow q$
T	T	T
T	F	F
F	T	T
F	F	T

Now our statement: If two parallel lines are cut by a transversal, then the alternate interior angles are congruent; they can be symbolized by $p \rightarrow q$, where p stands for: two parallel lines are cut by a transversal and q stands for: the alternate interior angles are congruent.

Note that statement p can be either true or false, statement q can be either true or false, and statement $p \rightarrow q$ can be either true or false subject to Table 15.9.1. In our argument, we used as the hypothesis that the alternate interior angles were not congruent and $l \| m$; then we concluded that through a point not on a given line there were two lines parallel to the given line. This relationship can also be expressed as $p \rightarrow q$, p standing for the hypothesis that the alternate interior angles were not congruent and $l \| m$, while q stands for the existence of two parallels.

Note that statement q is false; statement $p \rightarrow q$ is true (we were able to make a successful argument showing that q was a consequence of p). Thus, looking at our table, we find that the only case when both of these conditions hold is when statement p is false. That is, it is not true that lines l and m are parallel and the alternate interior angles are not congruent. It is true that

$l \| m$, so what must be false is that the alternate interior angles are not congruent. Thus, it follows that the alternate interior angles are congruent.

The above illustrates why an indirect argument is valid. In general, if we are given statement r, we know that either r or its negation $\sim r$ is true but not both. If we are inclined to believe that r is true, it may be possible to justify that by an indirect argument. We suppose that $\sim r$ is true, reach a consequence which is false by a correct argument, and must conclude that our assumption that $\sim r$ is true is incorrect. Thus, r is true.

Therefore, we see that we can also consider presenting our proof either in a direct argument or as an indirect one, depending on which form seems easier to use. It should also be emphasized that an indirect proof can either be done through synthetic or analytic means.

Exercise 15.9.1

1. Copy the synthetic proof of Theorem 15.8.1 and then indicate why reasons 2, 5, and 6 are valid, using the arrow diagrams as was done in the proof on page 364.
2. Complete the following synthetic proof. (See Fig. 15.9.3.)

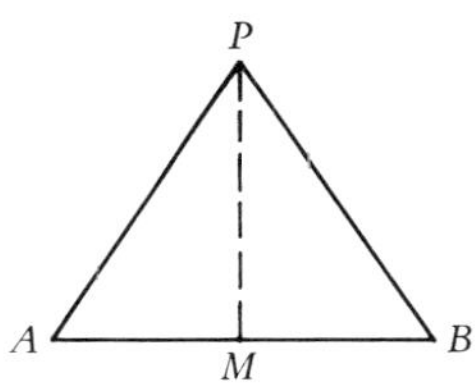

Figure 15.9.3

Given: Segment $\overline{AB}$

$\overline{AP} \cong \overline{BP}$

$\overleftrightarrow{PM} \perp \overleftrightarrow{AB}$

To prove: $\overline{AM} \cong \overline{MB}$.

Do you see that this suggests how you can prove Theorem 15.7.1 by first drawing the perpendicular through P?

3. The above problem can be proved analytically as well. See Fig. 15.9.4 and complete the proof.
Assume that A and B are on the x-axis, with coordinates $(-a, 0)$ and $(b, 0)$, respectively. P is on the y-axis with coordinates $(0, p)$. Also, $\overline{AP} \cong \overline{BP}$. Show that $|b| = |a|$, and that hence, M is the midpoint of $\overline{AB}$.
4. Supply the proof of case 1 for the synthetic approach to Theorem 15.8.1.

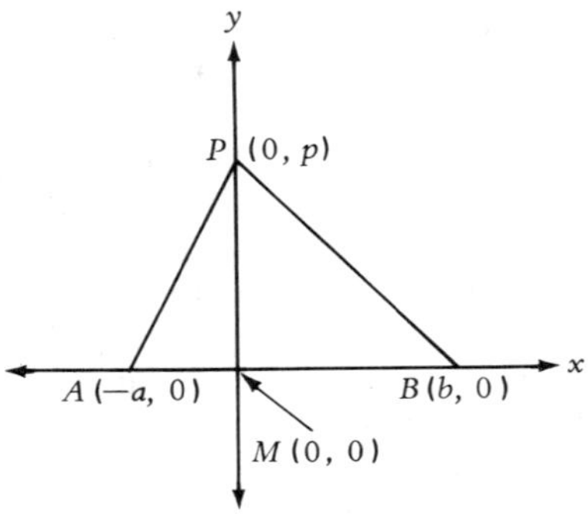

Figure 15.9.4

5. If you place a quadrilateral on a coordinate system as in Fig. 15.9.5:

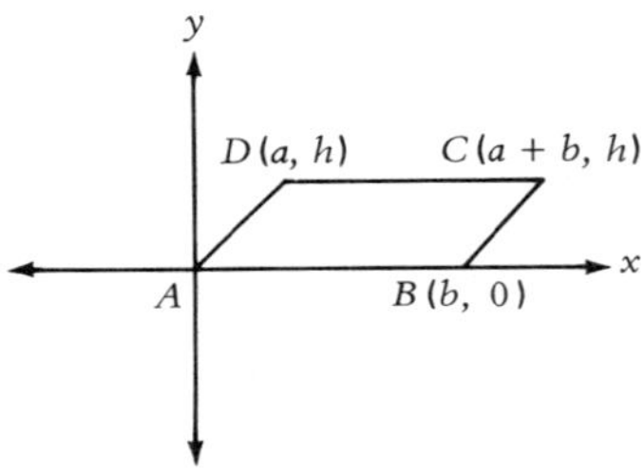

Figure 15.9.5

   a. Prove analytically that *ABCD* is a parallelogram. Thus, in future analytical proofs you may use the fact that any quadrilateral with the coordinates of (0, 0), (b, 0), ($a + b$, h) and (a, h) is a parallelogram.
   b. What special conditions would you impose on a, b, or h so that the figure will be a rectangle?
   c. What special conditions would you impose on a, b, or h so that the figure would be a rhombus?
   d. What special conditions would you impose on a, b, or h so that the figure would be a square?
6. Prove the following statements synthetically and analytically.
   a. The opposite sides of a parallelogram are congruent.
   b. If two opposite sides of a quadrilateral are congruent and parallel, the figure is a parallelogram.
   c. If two opposite sides of a quadrilateral are parallel and the remaining opposite sides are congruent, then the figure is a parallelogram.
   d. If a quadrilateral is not a rhombus, then the diagonals are not perpendicular.
   e. The diagonals of a parallelogram bisect each other.
   f. The diagonals of a rhombus are perpendicular.

g. The diagonals of a rectangle are congruent.
h. If a parallelogram is not a rectangle, then its diagonals are not congruent.

comment . . .

Do not forget Phase One. You do not want to prove something you do not believe, do you?

7. Prove: If the hypotenuse and leg of one right triangle are congruent, respectively, to the hypotenuse and leg of another, then the triangles are congruent. [*Hint:* Use a combined analytic and synthetic approach.] Figure 15.9.6 is suggested. Observe that △s ABC and QPR are right triangles and that $\overline{AB} \cong \overline{QP}$. Now assume that $\overline{AC} \cong \overline{QR}$ and show that $|h| = |c|$. Then complete the rest of the proof synthetically.

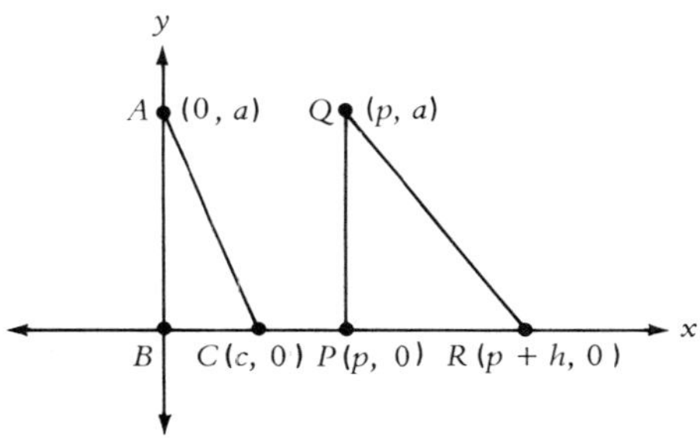

Figure 15.9.6

8. Prove the following by any means which seem most appropriate. For any statement you are unable to prove, prepare a list of intelligent questions about the problem.
   a. If a line segment is the diameter of a circle and P is any point of the circle except the endpoints of the line segments, then P is the vertex of a right triangle with the line segment as its hypotenuse.
   b. State and prove the converse of part a.
   c. If a point is on the bisector of an angle, then it is equidistant from the sides of the angle.
   d. State and prove the converse of part c.
   e. If the diagonals of a quadrilateral are perpendicular, then the figure is a rhombus.
   f. A line segment whose endpoints are midpoints of two sides of a triangle is parallel to the third side and equal in length to one-half the length of the third side.
   g. A line segment whose endpoints are the midpoints of the legs of a trapezoid is parallel to the bases and equal in length to one-half the sum of the lengths of the bases.

h. The midpoints of the sides of a quadrilateral are the vertices of a parallelogram.
i. If the opposite angles of a quadrilateral are congruent, the figure is a parallelogram.

15.10 SUMMARY

1. A proof is a convincing argument.
2. There are four essential steps to successful problem solving:
   a. convincing yourself
   b. planning an argument
   c. presenting your argument
   d. looking back
3. Successful problem solving requires one to engage in self-inquiry.
4. A proof may take several forms and may be either analytic or synthetic, direct or indirect, or any combination of these.

LOCUS AND CONIC SECTIONS

16

16.1 PROLOGUE

In preceding chapters, we have imposed various conditions on points. For example, we described a circle as the set of all points in a plane which are a fixed distance from a fixed point called the center. Perhaps you will recall that such a description implied the following two ideas relative to a given plane and a point in that plane.

1. If a point is a fixed distance from the point, then it is on the circle.
2. If a point is on the circle, then it is a fixed distance from the point.

From the last chapter we could thus reword the description as follows: A point is on a circle if and only if it is a fixed distance from a fixed point.

Another way to say this same thing is the following.

16.1.1 Locus

A circle is the locus *of all points in a plane which are a fixed distance from a fixed point of the plane.*

When we describe a set of points as a *locus* (plural is *loci*), we are implying that this set of points consists of all the points and only the points which satisfy certain conditions. In general, then, if we say S is the locus of points such that $C_1, C_2, \ldots, C_n$, we are saying that if a point $p \in S$, then p satisfies conditions $C_1, C_2, \ldots, C_n$ and if some point p satisfies $C_1, C_2, \ldots, C_n$, then $p \in S$ (that is, $p \in S$ if and only if p satisfies $C_1, C_2, C_3, \ldots, C_n$). In the remainder of this chapter some of the more well-known loci will be discussed. The chapter will be culminated by a series of experiments which are designed to lead the reader to a discovery of the conditions which describe the loci for the conic sections of the circle, ellipse, hyperbola, and parabola.

16.2 SOME SIMPLE LOCI

In the preceding chapter we proved the following two statements:

1. If a point is equidistant from the endpoints of a line segment, then it is on the perpendicular bisector of the line segment.
2. If a point is on the perpendicular bisector of a line segment, then it is equidistant from the endpoints of the line segment.

Expressed as a bi-conditional this becomes: A point is on the perpendicular bisector of a line segment if and only if it is equidistant from the endpoints of the line segment. We can use the word "locus" and say the following.

16.2.1 Locus

The locus of all points equidistant from the endpoints of a line segment is the perpendicular bisector of the line segment (Fig. 16.2.1). This statement was proved when Statements 1 and 2 above were proved in the last chapter.

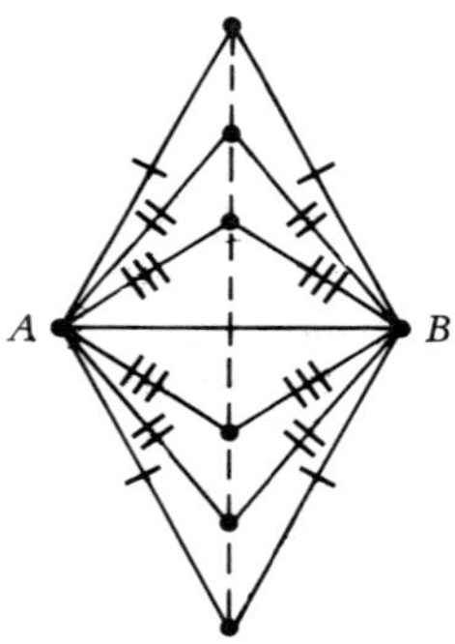

Figure 16.2.1

A second locus was suggested in Exercise 15.9.1 when the following statement and its converse were proved:

1. If a point is equidistant from the sides of an angle, then it is on the angle bisector.
2. If a point is on an angle bisector, then it is equidistant from the sides of the angle.

These two statements can combine to form the following locus description.

16.2.2 Locus

The locus of points equidistant from the sides of an angle is the angle bisector (Fig. 16.2.2).

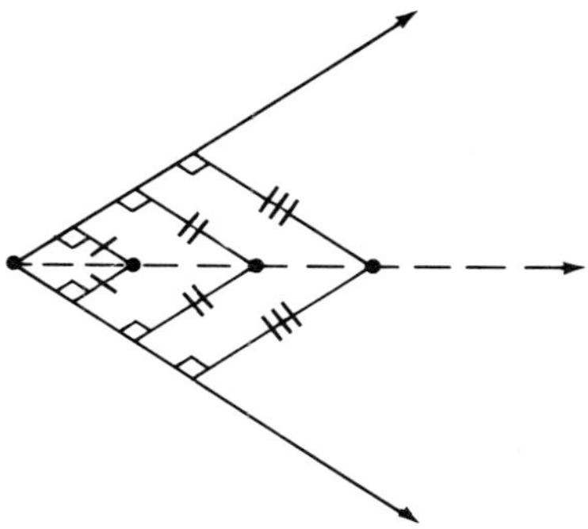

Figure 16.2.2

So far we have considered loci which were described synthetically. But we could also have analytical descriptions of loci. That is, we could consider the points corresponding to a coordinate system and describe them by describing the coordinates of the points. Thus, for analytical loci, the set-builder notation $\{(x, y)| \ldots\}$ becomes useful. Recall that in Chapter 5 we proved that $\{(x, y)|x^2 + y^2 = r^2\} =$ {the set of all points on a circle with radius r and center at (0, 0)}. A look at the proof will reveal that if A represents the first set and B the second set, we proved that $p \varepsilon A$ if and only if $p \varepsilon B$. In other words, we could say: A circle with radius r and center at (0, 0) is the locus of points corresponding to $\{(x, y)|x^2 + y^2 = r^2\}$.

Likewise, in Chapter 12 we essentially proved a locus theorem concerning a straight line. You will be asked to consider this one in an exercise.

Exercise 16.2.1

1. Prove that the locus of points in the plane of two parallel lines which are equidistant from the two lines is a line in the plane midway between the two lines and parallel to each of them.
2. Sketch the following loci in one plane.
   a. the locus of all points which are a fixed distance from a given line
   b. the locus of all points which are a fixed distance from a circle
   c. the locus of all points equidistant from two intersecting lines
   d. the locus of all points equidistant from the four vertices of a rectangle
   e. the locus of all points which are centers of a circle having a given line segment as a chord
3. Determine what the following locus is and prove it: the locus of all points which are the vertices of a right triangle with a given line segment as a hypotenuse.
4. Give a verbal description of the following loci in three dimensions.

a. the locus of all points which are a given distance from a fixed point
b. the locus of all points which are a given distance from a fixed line
c. the locus of all points which are equidistant from the endpoints of a line segment
d. the locus of all points which are equidistant from the sides of a given angle

16.3 CONIC SECTIONS

If we consider a pair of lines which intersect at the origin to be rotated about the y-axis, the two lines will sweep out a surface which looks like two cones joined at their apexes (Fig. 16.3.1). Such a surface is called a *cone of two nappes.* Now, if we intersect this surface with various planes, the different intersections will be a *circle, ellipse, parabola,* or *hyperbola,* depending on the angle of the intersecting plane (Fig. 16.3.2). For this reason, these four curves are often referred to as *conic sections.* Perhaps you have encountered some of these curves in graphing some of the relations in earlier chapters. One of these sections, the circle, has already been described as a locus of points which satisfy certain conditions. Many instruments which we use to make circles utilize these conditions. It so happens that the three remaining curves can be also described as loci. In the remainder of this chapter we shall suggest experiments which may be performed to discover the conditions which the points of these loci must satisfy.

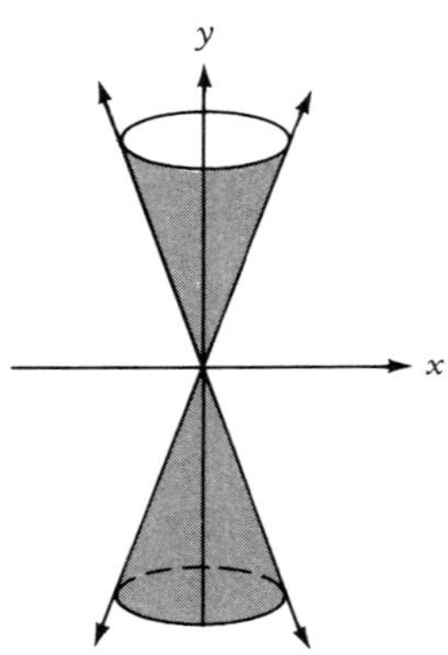

Figure 16.3.1

To provide you with a fuller understanding of what you are to gain from these experiments, an example now follows of a

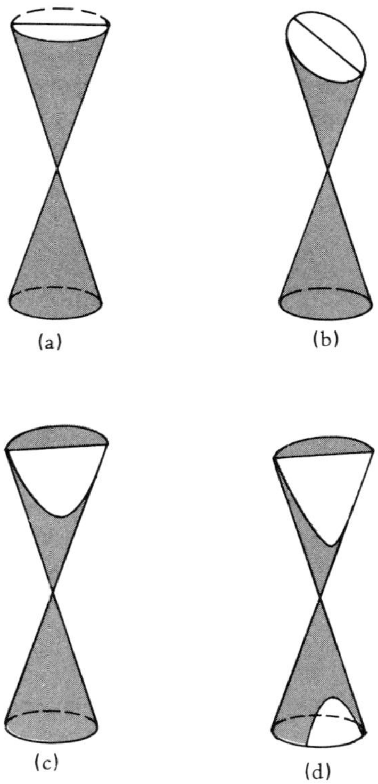

Figure 16.3.2
(a) Circle;
(b) Ellipse;
(c) Parabola;
(d) Hyperbola

similar experiment which would have been performed to discover the conditions which describe the points of a circle, assuming we did not already know them.

Example 1

An experiment to determine conditions which the points of a circle must satisfy.

Materials

1. peg board and paper
2. golf tees
3. string
4. pencil

Procedure

1. Place a large piece of paper on the peg board, fastening it with golf tees.
2. Pierce the paper somewhere near its center with another golf tee.
3. Cut a piece of string of any length which is less than half the width of the paper and tie a loop in each end.

4. Pass the golf tee through one loop of the string and pass a pencil through the other end.
5. Keeping your pencil perpendicular to the plane of the paper and always keeping the string taut, move the pencil so that it traces out a path (Fig. 16.3.3).

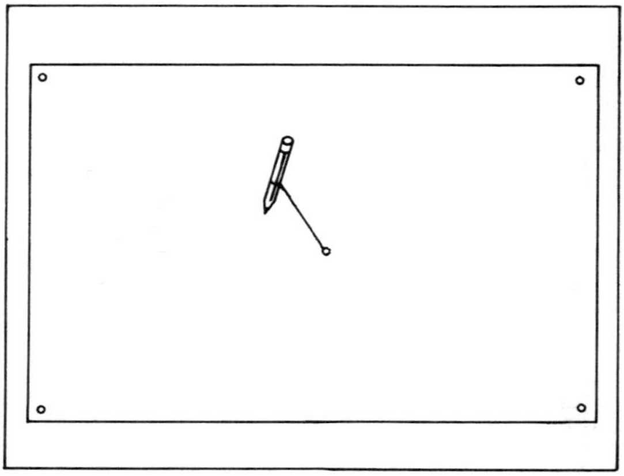

Figure 16.3.3

6. Now answer the following questions.
   a. What conic section is being traced out by this apparatus? *a circle*
   b. Indicate which geometric concepts are being represented by each of the parts of the apparatus listed below.
      i. golf tee: *fixed point*
      ii. string: *line segment or possibly a fixed distance*
      iii. pencil point: *the points of the locus*
      iv. paper: *a plane*
   c. What geometric conditions are being fulfilled by our insisting that the pencil is always perpendicular to the plane of the paper and the string is taut? *This locates all the points of the locus so that each point is the same distance from the fixed point.*
   d. From your observation of the apparatus and from your answers to the above questions, define this conic section as a locus: *A circle is the locus of all points in a plane whose distance from a fixed point of the plane is a constant.*

Now proceed to the next three experiments and see if you can discover the definitions of the remaining three conic sections as loci.

Experiment 1

To discover the definition of one of the conic sections.

Materials

1. peg board and paper
2. several golf tees
3. string
4. pencil
5. a right triangular piece of wood with a cup hook screwed into a leg

Procedure

1. Place a large sheet of paper on the peg board and fasten it with golf tees.
2. Arrange as many golf tees as possible in a straight line along the left-hand edge of the paper.
3. Place the triangular piece of wood so that the shorter leg is flush against the row of golf tees and the longer leg is about half-way down from the top of the paper and the vertex opposite the longer leg is toward the top of the paper.
4. Pierce the paper with a golf tee at some point which is flush with the longer leg of the right triangle between the row of golf tees and the cup hook.
5. Take a piece of string which is provided and pass one loop over the single golf tee and the other over the cup hook.
6. Place your pencil so that it is flush with the longer leg of the triangle and so that it holds a portion of the string flush against this leg.
7. Always keeping the string taut, move the triangle upward, being sure that the shorter leg is always against the row of golf tees. Your pencil should now be tracing out a portion of one of the conic sections.
8. When you have moved the triangle as high as you can, flip the triangle over and move it downward, tracing out the lower portion of the curve (Fig. 16.3.4).

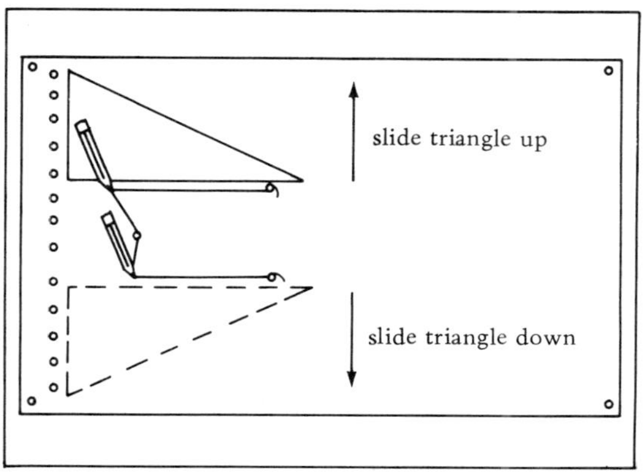

Figure 16.3.4

9. Now answer the following questions:
   a. What conic section is being traced out by this apparatus?

      ____

   b. Indicate which geometric concepts are being represented by each of the parts of the apparatus listed below.
      i. the row of golf tees: ________
      ii. the single golf tee: ________
      iii. the string: ________
      iv. the wood: ________
      v. the pencil point: ________
   c. Compare the length of the string with some part of the wooden triangle. What is the length of the string equal to? ________
   d. As you move the apparatus, the string's length does not change. What distances do change? ________
   e. What aspects or characteristics of the triangular piece of wood are essential for this apparatus to work?

      ____

   f. How is each point of the locus related to the single golf tee and the row of golf tees?

      ____

   g. From your observation of the apparatus and from your answers to the above questions, indicate what you think is the definition of this conic section.

      ____

   h. Change the position of the single golf tee and repeat the experiment. What happens to the locus as the tee moves farther away from the row of tees? ________________

      ____

Experiment 2

To discover the definition of one of the conic sections.

Materials

1. peg board and paper
2. golf tees
3. string
4. pencil

Procedure

1. Fasten a large sheet of paper to the peg board with some golf tees.
2. Pierce the paper with two golf tees on a line near the middle of the paper parallel to the bottom of the paper at some convenient distance apart.

3. Prepare a string with a loop at either end whose length is greater than the distance between the two golf tees.
4. Pass one loop around one tee and the other loop around the second tee.
5. Put the pencil against the string and move it so that the string is always taut. You should be tracing out one of the conic sections (Fig. 16.3.5).

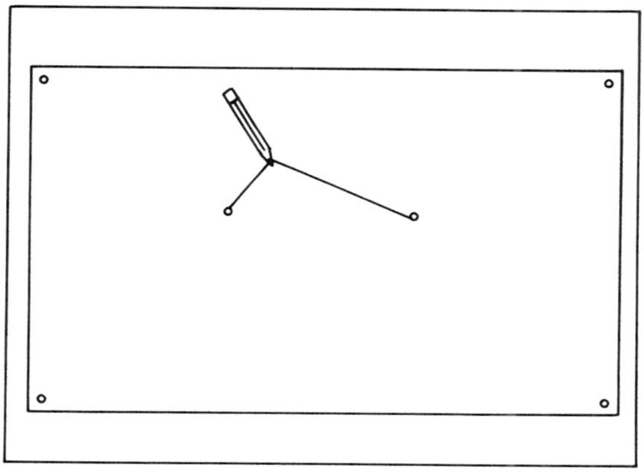

Figure 16.3.5

6. Now answer the following questions:
   a. What is the name of this conic section? ______
   b. What geometric concepts are represented by each of the following.
      i. the golf tees: ______
      ii. the string: ______
      iii. the pencil point: ______
   c. As you work the apparatus, what distances remain unchanged? ______
      What distances are changing? ______
   d. From your observations and your answers to the above questions, what do you think is the definition of this conic section? ______
   e. Repeat the experiment moving the golf tees further apart or closer together but keeping the length of the string the same. What do you observe? ______
   f. Repeat the experiment keeping the golf tees the same distance apart, but changing the length of the string. What do you observe? ______

Experiment 3

To discover the definition of a conic section.

Materials

1. peg board and paper
2. golf tees
3. string
4. pencil

Procedure

1. Fasten a large sheet of paper to the peg board with golf tees.
2. Pierce the paper with two golf tees on a line near the middle of the paper parallel to the bottom of the paper at some convenient distance apart.
3. Cut a piece of string whose length is at least four times as long as the distance between the two golf tees.
4. Join the ends of the string by knotting them together, thus forming one large loop.
5. Arrange the string around the golf tees according to Fig. 16.3.6.

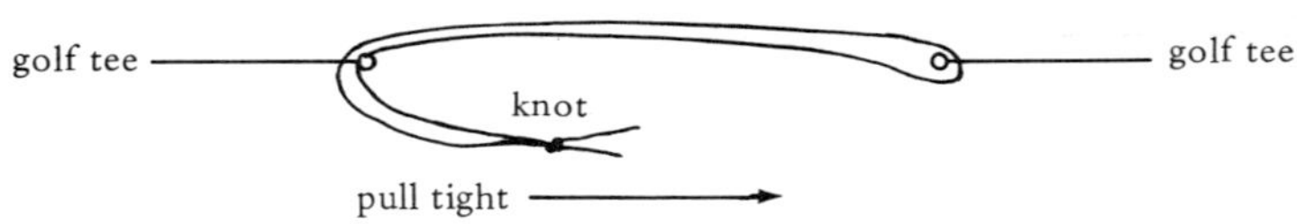

Figure 16.3.6

6. Now grasp the knotted ends with one hand and pull the string tight in a direction parallel to the bottom edge of the paper.
7. With your other hand, place your pencil at some point between the golf tees so that it is closer to the left tee than it is to the right one.
8. Hold *one* of the two strings tight against the pencil with your thumb so that the string will not slide along the pencil.
9. Still holding the string tight with the one hand and the pencil and string with the other, slowly move the hand holding the knotted ends to the left, keeping the string tight with the hand holding the pencil. Your pencil point should be tracing out a portion of one of the conic sections. To trace out another portion, move your pencil so that it is closer to the right golf tee and so that its distance from the right tee is the same as it was from the left tee originally (Fig. 16.3.7).
10. Answer the following questions:
    a. What conic section is being traced by this apparatus? ____________
    b. Carefully observe how the apparatus works and then define the conic section as a locus. ____________
    ____

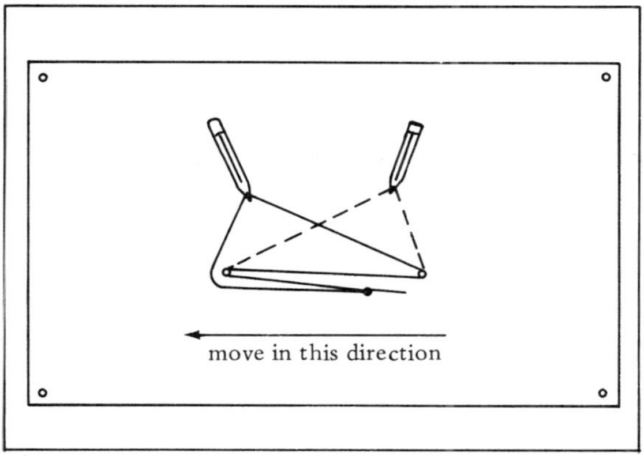

Figure 16.3.7

c. What locus is described if you place your pencil at the midpoint between the two golf tees? ____________________

d. Move the golf tees further apart. What do you observe?

____

16.4 SUMMARY

1. A locus is the set of all points and only those points that satisfy a certain set of conditions.
2. A circle is the locus of points in a plane which are a fixed distance from a fixed point of the plane.
3. A perpendicular bisector of a given line segment is the locus of points which are equidistant from the endpoints of the line segment.
4. An angle bisector is the locus of points which are equidistant from the sides of the angle.
5. A parabola is the locus of points which
6. An ellipse is the locus of points which
7. A hyperbola is the locus of points which

comment . . .

We leave it to you to complete the last three statements.

INDEX

74 75 76 9 8 7 6 5 4 3 2 1